Microgrids Including Hybrid Resources

Protection and Stability

Microgrids Including Hybrid Resources

Protection and Stability

Majid Sanaye-Pasand
University of Tehran
Tehran, Iran

Reza Bekhradian
University of Tehran
Tehran, Iran

IEEE Press
WILEY

Published by John Wiley & Sons, Inc., Hoboken, New Jersey.

Library of Congress Cataloging-in-Publication Data applied for:

Hardback ISBN: 9781394278589

Cover Design: Wiley
Cover Image: © noppadon/stock.adobe.com

Set in 9.5/12.5pt STIXTwoText by Straive, Chennai, India

To our Families

Contents

About the Authors

Majid Sanaye-Pasand is a Professor at the School of Electrical and Computer Engineering, University of Tehran, Iran. His research interests include power system protection and control, smart grid, digital protective relays, power system transients, and power system automation. He has been the Member of the Editorial Board of various academic journals and has served as Associate Editor for different prestigious journals, including *IEEE Transactions, TPWRD* for more than 10 years, and *TSG* since 2021. He has served as the Keynote Speaker, Invited Speaker, and Member of the Technical Program Committee for numerous national and international conferences. He has supervised more than 110 graduate students in the power system area. Many of his former students have high-ranking jobs at well-known universities and in the power industry worldwide.

His contributions to research have been appreciated by the academic authorities, and he has been rewarded several significant promotions and prizes in his carrier. He not only has been a quite successful professor in the academia by publishing 350 papers and some books and book chapters, but also his proficiency has been quite beneficial for the power system industry and society. He has managed about 20 funded research projects with excellent outcomes and has been awarded various prestigious and national prizes from the industry for his practical research innovations. In recognition of the novelty of his power system expertise, he has been appointed an adviser and member of research and development board in several power utilities and companies where he contributes significantly to the system protection and stability studies. More importantly, he has been selected as the member of the board and consultant to CEO, Tavanir Company, the highest-ranking power utility in Iran, responsible for the management of national grid generation, transmission, and distribution.

Reza Bekhradian received the MSc and PhD degrees in Electrical Engineering from the University of Tehran, Tehran, Iran, in 2015 and 2021, respectively. He is currently an Assistant Professor with the School of Electrical and Computer Engineering at the University of Tehran.

His research interests include the protection, stability, and control of modern microgrids and power systems, with particular emphasis on hybrid power systems comprising synchronous generator- and inverter-based resources. As part of his teaching activities, he has developed a graduate-level course titled *Fundamentals of Renewable Power Plants*, which covers the principles of design and operation of renewable power plants, including systems dominated by inverter-based generation. He has also presented different technical lectures for industry engineers in the areas of protection and automation.

In addition to his academic activities, he has been involved in the construction of high-voltage substations and renewable power plants, including photovoltaic and hydropower facilities. He has also served as a consultant for Tavanir, Iran's highest-ranking power utility responsible for generation, transmission, and distribution, where he contributed to the development of grid codes and technical requirements for the integration of hybrid inverter-based resources into the power system.

Preface

The electric power system is undergoing a major transformation driven by the increasing penetration of distributed energy resources, renewable energy technologies, energy storage systems, and advanced power electronic converters. As power systems evolve from centralized hierarchical structures toward more decentralized architectures, microgrids have emerged as one of the most promising solutions for improving reliability, resiliency, efficiency, and sustainability.

The growing demand for electrical energy, together with the need to reduce operating costs, minimize losses, improve efficiency, and enhance service reliability, has accelerated the deployment of distributed generation resources throughout modern distribution networks. Today, distributed generators are widely used in residential, commercial, and industrial facilities. Among the most common technologies are synchronous-generator-based resources, such as diesel generators and gas engines, and inverter-based distributed generations, such as photovoltaic and battery energy storage systems. Historically, distributed generators were mainly operated in grid-connected mode, where the utility network maintained system voltage and frequency while local generators supplied part of the demand. In recent years, however, their role has expanded considerably. Modern microgrids are increasingly expected to continue supplying critical and essential loads during utility outages through islanded operation. Consequently, reliable performance under both grid-connected and islanded conditions has become one of the primary requirements of modern microgrid design and operation. Moreover, in many practical applications, different distributed generation technologies operate together within the same microgrid. Such hybrid structures allow operators to exploit the advantages of multiple energy sources while improving flexibility and reliability. At the same time, they introduce new technical challenges associated with system dynamics, stability, control, and protection.

Among these challenges, stability is one of the most critical issues. Modern microgrids often contain a combination of synchronous generators and inverter-based resources whose dynamic characteristics differ significantly. While synchronous generators inherently contribute electromechanical inertia and support system stability, inverter-based resources depend largely on control algorithms and power electronic converters to regulate their behavior. Consequently, many conventional methods of stability assessment and control design must be revised to accommodate the characteristics of converter-dominated systems. Moreover, protection represent another essential aspect of microgrid operation. A properly designed protection system must rapidly detect faults and abnormal operating conditions while maintaining adequate security and dependability against non-faulty conditions. The integration of distributed generators changes fault-current signatures including magnitude and phase, alters power-flow patterns, and introduces bidirectional current paths that may affect the performance of conventional protection schemes. These challenges have motivated the development of new protection philosophies, especially for setting adjustments and adaptive approaches capable of maintaining reliable operation under changing system conditions.

Modern distributed generation technologies are increasingly capable of supporting utility systems through voltage regulation, frequency support, disturbance ride-through capability, and enhanced system resiliency. Therefore, properly designed microgrids can serve not only as beneficial local energy systems but also as valuable assets for the broader power network.

The primary motivation behind writing this book is to provide a comprehensive and practically oriented treatment of protection and stability of microgrids that integrates both synchronous-generator-based and inverter-based technologies within a unified framework. Although numerous publications discuss distributed generations, renewable energy systems, and microgrid control, relatively few references address protection-system design, stability assessment, islanding detection, and stability enhancement in an integrated manner. This book seeks to bridge that gap by combining fundamental concepts, analytical methods, and practical engineering considerations within a single reference. In addition, the book has been developed with a strong educational perspective. It is intended to serve as a valuable resource for both undergraduate and graduate students, researchers, and practicing engineers who are involved in the analysis, planning, design, operation, and protection and stability studies of modern microgrids.

By providing detailed models, engineering insights, solved examples and practical methodologies, the book aims to support both advanced academic studies and professional engineering practice. To help the readers better comprehend the provided material, different problems are provided at the end of each chapter. In addition, simulation files for some sample distribution systems are provided in the book companion website. Moreover, the supplied presentation slides in the website would help the instructors to teach the book material more efficiently.

It is our hope that the material presented herein will assist readers in developing a deeper understanding of microgrid behavior and enable them to design more reliable, secure, and resilient microgrid systems in an increasingly decentralized power industry. To achieve this objective, the book is organized into six chapters.

Chapter 1 introduces the major fundamentals and basic structure of distributed generation technologies employed in modern microgrids, including photovoltaic systems, wind generation technologies, small-scale synchronous generators, and inverter-based distributed generators. The chapter also presents dynamic models, control structures, and protection architectures that form the foundation for subsequent discussions.

Chapter 2 focuses on the stability of synchronous-generator-based microgrids. Various forms of stability are defined and analyzed, together with the associated control systems and assessment methodologies. Control system and power supply stability, voltage and frequency stability, as well as transient stability are studied carefully and induction motor instability detection is discussed. The chapter also investigates dynamic phenomena that may threaten stable operation under different operating conditions.

Chapter 3 examines the stability of inverter-based microgrids. Particular attention is devoted to inverter control systems, grid-following and grid-forming strategies, synchronization mechanisms, transient stability, and small-signal stability. The implications of the control systems behavior on modern protection schemes are also discussed. In addition, various power and current control loops as well as different phenomena including voltage disturbance ride-through and fault-induced delayed voltage recovery are investigated.

Chapter 4 addresses the challenges associated with microgrid protection systems. Widely used protection schemes are reviewed, and practical issues such as relay miscoordination, sympathetic tripping, reduced protection sensitivity, unsuccessful reclosing operations, and the limitations of conventional load-shedding methods are investigated. Moreover, various concepts such as backup and loss of field protection, as well as synchronous-generator-based and inverter-based distributed generations protection challenges are discussed.

Chapter 5 investigates islanding detection in both synchronous-generator-based and inverter-based microgrids. Various local/remote and active/passive detection techniques are evaluated using appropriate performance criteria, and their applicability to different microgrid configurations is discussed. Different concepts such as nondetection zone, islanding detection time and unsymmetrical islanding are studied. Several approaches for improving islanding detection performance are also presented.

Chapter 6 presents practical methods for enhancing both microgrid protection and stability. Relay-setting methodologies, adaptive protection schemes, solutions for back-feed, sympathetic tripping and fault-induced delayed voltage recovery are discussed. Advanced approaches for improving system security, dependability, and resiliency are reviewed together with practical solutions for improving microgrid stability during severe disturbances.

May 2026

Majid Sanaye-Pasand
Reza Bekhradian
University of Tehran

Acknowledgments

This book represents the culmination of years spent by the authors teaching and engaging in research works in academia and industry in the related areas. Therefore, we would like to extend our gratitude to all people who have provided valuable discussions throughout these years and have direct or indirect participation in preparing this content. We acknowledge Mr. Armin Nuri Imani and Mr. Mehrab Mojeni for their helpful assistance in preparing some figures, tables and slides of this book. Furthermore, we are appreciative to our colleagues at the ECE School, University of Tehran, especially Dr. Moein Abedini and Dr. Mahdi Davarpanah, for their beneficial consultations. Special thank is devoted to Dr. Ali Hooshyar for his valued support.

List of Abbreviations

51VC	Voltage-controlled overcurrent
51VR	Voltage-restrained overcurrent
ACB	Air circuit breaker
ADMS	Advanced distribution management system
AI	Acceleration index
ANFIS	Adaptive network-based fuzzy inference system
ANN	Artificial neural network
AREP	Auxiliary winding regulation excitation principle
AVR	Automatic voltage regulator
BESS	Battery energy storage system
CAPEX	Capital expenditure
CB	Circuit breaker
CCT	Critical clearing time
CHP	Combined heat and power
CLPU	Cold load pickup
CPS	Centralized processing server
CT	Current transformer
CTI	Coordination time interval
DC	Direct current
DG	Distributed generation
DGR	DG relay
DT	Decision tree
EAF	Electrical arc furnace
EI	Extremely inverse
EkNN	Evidence-theoretic k-nearest neighbor
ELM	Extreme learning machine
EMD	Empirical mode decomposition
EMF	Electromotive force
ESS	Energy storage system

ETU	Electronic trip unit
FCT	Fault clearing time
FDR	False detection ratio
FDS	Fault detection signal
FESS	Flywheel ESS
FI	Fault indicator
FIDVR	Fault-induced delayed voltage recovery
FL	Fuzzy logic
FOS	FIDVR occurrence signal
FPV	Floating photovoltaic
FRT	Fault ride-through
FT	Fourier transform
FTRT	Faster than real-time
GFL	Grid-following
GFM	Grid-forming
GOOSE	Generic object oriented substation event
GPI	Generator performance index
GPRS	General packet radio service
GPS	Global positioning system
HHT	Hilbert–Huang transform
HIL	Hardware-in-the-loop
HMGs	Hybrid microgrids
HST	Hyperbolic S-transform
HV	High voltage
IBDG	Inverter-based distributed generation
IBMG	Inverter-based microgrid
IDS	Islanding detection scheme
IDT	Islanding detection time
IE	Internally excited
IED	Intelligent electrical device
IG	Induction generator
IM	Induction motor
IMF	Intrinsic mode function
LBS	Load break switch
LOF	Loss of field
LPF	Low-pass filter
LR	Load relay
LSIG	Long-time, short-time, instantaneous, and ground
LSIP	La salle's invariance principle
LTI	Long-time inverse
LV	Low voltage

LVRT	Low-voltage ride-through
MCB	Miniature circuit breaker
MCCB	Molded case circuit breaker
MCT	Maximum clearing time
MG	Microgrid
MM	Mathematical morphology
MMT	Minimum melting time
MV	Medium voltage
NC	Normally closed
NO	Normally open
NCT	Neutral current transformer
NDI	Nondetection index
NDZ	Nondetection zone
OCR	Overcurrent relay
OF/UF	Over/under frequency
OFR	Overcurrent feeder relay
OV/UV	Over/under voltage
PCC	Point of common coupling
PERC	Passive emitter rear cell
PF	Power factor
PI	Proportional-integral
PIAPC	Proportional-integral active power controller
PIRPC	Proportional-integral reactive power controller
PLL	Phase-locked loop
PMG	Permanent magnet generator
POC	Point of connection
POI	Point of interconnection
POM	Point of measurement
PQ	Active/reactive power control mode
PS	Pickup setting
PT	Potential transformer
PV	Photovoltaic
PWM	Pulse-width modulation
RBF	Radial basis function
RCS	Remote-control(led) switch
RDI	Reacceleration detection index
RF	Random forest
RLC	Resistor-inductor-capacitor
RLOF	Recoverable LOF
ROA	Computed region of attraction
ROCOF	Rate of change of frequency

RTDS	Real-time digital simulator
SCADA	Supervisory control and data acquisition
SE	Self-excited
SEF	Sensitive earth fault
SEP	Stable equilibrium point
SG	Synchronous generator
SGBMG	Synchronous generator-based microgrid
SI	Standard inverse
SPM	Single-phase motor
SRIG	Squirrel-cage rotor induction generator/Squirrel-ring induction generator
SSSG	Small-scale synchronous generator
ST	S-transform
STFT	Short-time Fourier transform
SVM	Support-vector machine
TMF	Transient monitoring function
TMS	Time multiplier setting
TR	Transformer
TTT	Time-time transform
TVF-EMD	Time-varying filter-based empirical mode decomposition
UEP	Unstable equilibrium point
UFLS	Underfrequency load-shedding
UI	Unrecoverable index
ULOF	Unrecoverable LOF
USS	Upstream sub-transmission substation/Upstream system
UVLS	Undervoltage load-shedding
VI	Very inverse
VS	Vector shift
VS-OC	Voltage-supervised overcurrent
VSM	Virtual synchronous machine
WRIG	Wound rotor induction generator
WSE	Wavelet singular entropy
WT	Wavelet transform
WTG	Wind turbine generator
μPMU	Micro phasor measurement unit

Nomenclatures

Chapter 1

τ_1	Actuator time constant
τ_2	Engine dead time (time delay)
E_{fd}	Exciter field voltage
E_c	Regulator output
K_I	Integrator gain of PI controller
K_P	Proportional gain of PI controller

Chapter 2

P_m	Mechanical power of SSSG
V_t	Voltage at PoC before SSSG connection
ΔV_{pu}	Per-unit change of voltage magnitude at PoC
P_{SSSG}, Q_{SSSG}	Output active and reactive powers of SSSG
R, X	Real and imaginary parts of equivalent impedance seen from SSSG terminal
P_{ref}, Q_{ref}	Reference setpoints of output active and reactive powers
V_{ref}	Reference of voltage magnitude
S_i, S_j	Rated apparent power of the *i*th and *j*th SSSGs
R_i, R_j	Droop coefficient of the *i*th and *j*th SSSGs
f	SGBMG frequency
f_{ref}	Reference setpoint of SGBMG frequency
P_{pre}	Output active power of SSSG before the disturbance occurrence
t_{RT}	Ride-through duration
V_{L1}, V_{L2}	Voltage thresholds for ride-through regions
P_{min}	Minimum amount of SSSG output active power during low-frequency events

S_{nom}	SSSG nominal power
ΔP	Active power change of SSSG due to frequency change
H	SSSG inertia constant
ω_0	SSSG rated electrical speed
K_S	Synchronizing torque coefficient
K_D	Damping torque coefficient
δ	SSSG rotor angle
u	PI controller output of the SSSG governor system
K_u	Constant coefficient to determine contribution of controller efforts on GPI
K_I, K_P	Integral and proportional gains of PI controller
P_e	Electrical power of SSSG
ω	Electrical angular speed
$\tau_{P,Q}$	Constant paramaeters for representing the dynamic behavior of load restoration
E_{total}	SSSG rotor total energy
δ_{max}	Maximum amount of SSSG rotor angle for stable operation
P_{max}	Maximum active power of synchronous generator
γ	Angle of equivalent impedance
φ	Load angle
P_0, Q_0	Load active and reactive power consumptions at nominal voltage V_0
P_s, Q_s	Static active and reactive power consumptions at voltage V
P_t, Q_t	Transient active and reactive power consumptions
x_P, x_Q	Generic dimensionless state variables
α_P, α_Q	Constant exponents representing static load dependency on voltage
α, β	Percentage of aggregated static and dynamic loads
P_{IM}, Q_{IM}	Active and reactive powers of aggregated induction motor load
P_{ZIP}, Q_{ZIP}	Active and reactive powers of aggregated static load
Y_{load}	Equivalent static load admittance
X', X	Transient and open-circuit reactances
R_s	Stator resistance
E_d', E_q'	d-axis and q-axis transient EMFs
U_d, U_q	d-axis and q-axis terminal voltages
I_d, I_q	d-axis and q-axis stator currents
T_L	Load torque
T_p, T_n	Positive and negative electromagnetic torques

ω_s	Synchronous speed
I_p, I_n	Motor positive- and negative-sequence currents
R_r', X_r'	Rotor resistance and reactance
s	Rotor slip
T_e	Total electromagnetic torque
ω_r	Rotor speed of the aggregated induction motor
J	Total inertia of the motor and its corresponding mechanical load
s_{Tmax}	Critical slip for maximum electromagnetic torque
V_p, V_n	Positive and negative sequence voltages after fault clearance
s_{ef}	IM slip at the fault clearance instant
t_R	Reacceleration time of aggregated IM
Y_{IM}	Admittance of dynamic aggregated induction motor
X	State vector of the nonlinear system
X^*	Equilibrium point of the system
U	Lyapunov function
$K(\omega), E(\delta)$	Kinetic and potential energy components of Lyapunov function
X_{ls}, X_{lkq}, X_{lfd}	Leakage reactances of stator, damper, and field windings
X_{md}, X_{mq}	Magnetizing reactance of *d*-axis and *q*-axis
ω_b	Base rotor speed
T_m	Mechanical torque
ω_e	Angular frequency of upstream system
ω^*_{DG}	Reference angular frequency of DG
ω_{nl}	Angular frequency under no-load condition
m_p	Active power droop gain
V^*_{tr}	Reference value of the terminal voltage magnitude
n_q	Reactive power droop gain
P^*_{Gov}	Reference of the SSSG governor active power
k_{pGov}, k_{iGov}	Proportional and integral gains of governor PI controller
V^*_{AVR}	Reference voltage of excitation system and AVR
k_{pAVR}, k_{iAVR}	Proportional and integral gains of AVR PI controller
E_{Th}, X_{Th}	Upstream system Thevenin voltage and reactance
θ_{Th}	Angle of Thevenin voltage
u_{FTRT}	Faster than real-time (FTRT) input vector
V_d, V_q	SSSG stator voltages in *d-q* rotating frame
I_d, I_q	SSSG stator currents in *d-q* rotating frame

Chapter 3

V^d, V^q	d-axis and q-axis voltages at PCC
δ_{PCC}	Actual amount of voltage angle at PCC
δ_{PLL}	Angle estimated by PLL
G_{PLL}	Transfer function of PLL
G_{LPF}	Transfer function of low-pass filter
P, Q	Active and reactive powers
P_i, Q_i	Active and reactive powers of the ith DG
S_i	Rated power of the ith DG
V_{PCC}	Voltage at PCC
i^*_d, i^*_q	d-axis and q-axis current references
P_{IBDG}, Q_{IBDG}	Output active and reactive powers of IBDG
P_{SSSG}, Q_{SSSG}	Output active and reactive powers of SSSG
P^*_{IBDG}, Q^*_{IBDG}	IBDG active and reactive power setpoints
S^*_{IBDG}	Reference apparent power of IBDG
φ	Power factor angle
$I^*_{PCC}\ (\lvert I_{PCC}\rvert \angle\theta_I)$	Reference current phasor generated by the power control loop
ω_{PLL}	Frequency estimated by PLL
ω_G	Grid angular frequency
ω_{nom}	Nominal value of angular frequency
P_i^{set}, Q_i^{set}	Nominal steady-state amounts of output active and reactive powers of IBDG
m_{pi}	Droop coefficient of frequency/Watt control loop
m_{qi}	Droop coefficient of Volt/VAr control loop
P_{max}, P_{min}	Maximum and minimum allowable amounts of ouput active power
q_{max}, q_{min}	Maximum and minimum allowable amounts of ouput reactive power
$F[x]$	Saturation function
λ_p, λ_q	Power-sharing parameters for the active and reactive powers
P_{iE}, Q_{iE}	Active and reactive powers for equal power sharing
P_{DC}	Injected power to the DC-link from PV or wind resource
P_{Cap}	Power absorbed by the DC-link capacitor
P_{Loss}	Inverter and its associated filter losses
C_{DC}, V_{DC}	DC-link capacitance and voltage
V^*_{DC}	Reference of DC-link voltage
$PI_{APC}(s)$	Transfer function of PI active power controller

$K_P{}^{APC}$, $K_I{}^{APC}$	Proportional and integral gains of active power controller
X	State vector
U	Input vector
A	State matrix
B	Input matrix
$CE(s)$	Characteristic equation
ω_{d_IBDG}	Oscillation frequency of IBDG
α_{IBDG}	Damping coefficient of IBDG
ζ_{IBDG}	Damping ratio of IBDG
v_d, v_q	*d*-axis and *q*-axis voltages of the grid
R_{filter}, L_{filter}	Filter resistance and inductance
e_d, e_q	*d*-axis and *q*-axis inverter voltages
i_d, i_q	*d*-axis and *q*-axis currents
$K_p{}^{CCd}$, $K_I{}^{CCd}$	Proportional and integral gains of *d*-axis current controller
$K_p{}^{CCq}$, $K_I{}^{CCq}$	Proportional and integral gains of *q*-axis current controller
ω_s	Angular frequency of upstream grid
τ	Time constant
J	Virtual inertia
P_{ref}, Q_{ref}	Active and reactive power setpoints
D_p	Virtual damping coefficient
ω_g	Angular frequency setpoint
θ_{ref}	Reference electrical angle
V_g, δ_g	Grid voltage magnitude and angle
θ_I	Angle of current phasor
φ_Z	Angle of Thevenin equivalent impedance
K_p, K_i	Proportional and integral coefficients of controller
Z_{Th}, X_{Th}	Thevenin impedance and reactance of upstream system
L_g	Equivalent grid inductance
t_{FCT}	Fault clearing instant
K_q	Constant representing voltage–reactive power droop
Q_0	Reference of output reactive power
E	Thevenin voltage of upstream system

Chapter 4

E_{SSSG}	Internal voltage of SSSG
E_{SGBMG}	Thevenin voltage of E_{SGBMG}
θ, θ_k	Transformer and ambient temperatures

I_0	Zero-sequence current
I_1, I_2	Positive- and negative-sequence currents
I_{1s}	Maximum permissible transformer current for 1 s
I_F	Fault current
I_{grid}	Fault contribution of the upstream grid
$I_{loadmax}$	Maximum load current
I_{max}, I_{min}	Maximum and minimum fault currents
I_n	Nominal current
I_{pickup}	Pickup current
$I_{primary}$	Pickup current at the primary side of CT
I_{SGBMG}	SGBMG back-feed fault current
I_{MG}	MG current contribution
I_{grid}	Grid current
TMS	Time multiplier setting of overcurrent relay
CTI	Coordination time interval
E_{DG}	Internal voltage of DG
E_{Th}	Thevenin voltage of the upstream system
I_A, I_B	Feeder currents
I_{DG}	DG current
K	MCCB time–current characteristic constant parameter
T_f	Fault occurrence time
f_{min}	Minimum frequency threshold
V_{min}	Minimum voltage threshold
V	Voltage magnitude
t	Operation time
I_a, I_b, I_c	Phase currents
V_a, V_b, V_c	Phase voltages
k	Parameter used to specify the curve type
τ	Transformer thermal time constant
t_{1R}	Operation time of the first stage of recloser (recloser fast curve)
t_{2R}	Operation time of the second stage of recloser (recloser slow curve)
t_{MCT}	Maximum clearing time of the fuse
t_{MMT}	Minimum melting time of the fuse
MCT	Maximum clearing time
MMT	Minimum melting time
t_{MMT}	MMT of the fuse at I_{min}
t_{MCT}	MCT of the fuse at I_{max}
t_{op}	Operation time
t_{reset}	Reset time

t_S	Maximum permissible voltage recovery time
V_0	Displacement voltage
V_s	Nominal voltage of the system
Z_{Line}	Line impedance
Z_{reach}	Relay reach
Z_{SSSG}	SSSG impedance
Z_1, Z_2	Impedance between bus B1/B2 and MG connection point
Z_{DG}	Thevenin impedance of DG
Z_{Th}	Upstream system Thevenin impedance
Z_f	Fault impedance
Z_{reach}	Maximum reach of relay
Z_{SGBMG}	Thevenin impedance of SGBMG

Chapter 5

$\Delta P, \Delta Q$	Active and reactive power mismatches
$\Delta P_{1,2}$	Minimum detectable active power mismatch
$\Delta Q_{1,2}$	Minimum detectable reactive power mismatch
β	Relay threshold
t_d	Required islanding detection time
H	SSSG inertia
f_0	Nominal frequency
Δt	Intentional delay
E_{Th}, Z_{Th}	Equivalent upstream grid voltage and impedance
E_{SSSG}	Internal voltage of SSSG
P_{SSSG}, Q_{SSSG}	Active and reactive power outputs of SSSG
P_{Load}, Q_{Load}	Load active and reactive powers
P_{sys}	Power exchange with upstream grid
ω_0	Synchronous angular frequency
ω	Angular frequency
δ	Rotor angle
$\Delta\omega$	Angular frequency deviation
Δf	Frequency deviation
$K(t), K(s)$	Filtered ROCOF signal in time and Laplace domains
V_0	Steady-state voltage of SSSG
V_t	Terminal voltage of SSSG
T'_d, T''_d	Transient and subtransient short-circuit time constants of SSSG
i_{SSSG}	Armature current of SSSG
λ	Initial phase angle of i_{SSSG}

D	Damping coefficient
P_M	Mechanical power of SSSG
P^*	Active power setpoint of SSSG
R	Droop gain
P_{output}	Electrical power output of SSSG
$T(s)$	Turbine transfer function
$C(s)$	Power controller transfer function
K_i, K_p	Integral and proportional gains
$s_{1,2}$	Eigenvalues
Q_f	Load quality factor
P_G, Q_G	IBDG active and reactive powers
P_L, Q_L	Load active and reactive powers
V	Voltage
f	Frequency
$\hat{V}$	Postislanding voltage
$\hat{f}$	Postislanding frequency
$\Delta P_G, \Delta Q_G$	Change in generated active and reactive powers
$\Delta P_L, \Delta Q_L$	Change in load active and reactive powers
L	Inductance
C	Capacitance
R_f	Active power–frequency droop coefficient
R_V	Reactive power–voltage droop coefficient
P_{nom}	Nominal Power
V_{min}, V_{max}	Minimum permissible voltage, maximum permissible voltage
f_{min}, f_{max}	Minimum permissible frequency, maximum permissible frequency
T_f	Time constant of the preprocessing low-pass filter
N_T	Total number of analyzed islanding or nonislanding case studies
N_F	Number of false decisions
T	Observation time interval
FDR	False detection ratio
NDI	Nondetection index
I_1, I_2	Positive- and negative-sequence currents
dp/dt	Rate of change of active power
df/dt	Rate of change of frequency
dq/dt	Rate of change of reactive power
dv/dt	Rate of change of voltage
dq/dp	Rate of change of reactive power over active power
dq/dv	Rate of change of reactive power over voltage

df/dq	Rate of change of frequency over reactive power
dv/dp	Rate of change of voltage over active power
dp/dq	Rate of change of active power over reactive power
dv/dq	Rate of change of voltage over reactive power
dq/df	Rate of change of reactive power over frequency
dp/dv	Rate of change of active power over voltage
dp/df	Rate of change of active power over frequency
dv/df	Rate of change of voltage over frequency
df/dp	Rate of change of frequency over active power
df/dv	Rate of change of frequency over voltage
P	Active power
X_d	Direct axis synchronous reactance
X'_d, X''_d	Direct axis transient and sub-transient reactances
X_q, X''_q	Quadrature axis synchronous and sub-transient reactances
U_k	Transformer short-circuit voltage
S	Transformer apparent power
X/R	Upstream grid reactance to resistance ratio
c_f	Chopping factor in AFD method
t_z	Dead-time in AFD method
T_{grid}	Period of the grid voltage signal in AFD method
i_{inv}	Inverter's ouput current
θ_{SMS}	Intentional perturbation injected into the i_{inv} used in SMS method
I_p	Perturbation current signal used in SVS method
V_e	Voltage error
k_e	Constant gain used in SVS method
θ_m	Maximum phase shift
f_m	Corresponding frequency at θ_m
f_{PCC}	Frequency of the PCC voltage

Chapter 6

I_{fmax}	Maximum three-phase short-circuit current
I_{nTr}	Nominal current of transformer
S_{nTr}	Nominal capacity of transformer
U_k	Per-unit impedance of transformer
$I^{50}_{pickup-R1}$	Pickup value of the instantaneous stage of relay R1
$I_{Saturation}$	Maximum AC current for which the CT does not saturate

I_{MR}	Feeder maximum current under the motor reacceleration condition
I_{feeder}	Steady-state current of feeder, all static and dynamic loads are connected
k	Induction motor penetration factor
n	Locked rotor current of induction motors
K_{VD}	Distribution substation bus voltage during motor reacceleration
$I^{51}_{pickup-R1}$	Pickup value of the inverse stage of relay R1
I_{nCT}	Nominal current of the CT
$K_{OverLoad}$	Coefficient indicating the presence of an overload function
t_{op}	Operation time of the inverse stage of overcurrent relay
$t_{op-51(R1)}$	Operation time of the inverse stage of relay R1
I	Measured current
k, α, a, b, c	Constant parameters in time-current characteristic of overcurrent relay
CTI	Coordination time interval
TMS	Time multiplier setting of overcurrent relay
V_{nom}	Nominal voltage
$I_{fmax-R1}$	Maximum short-circuit fault at the location of overcurrent relay R1
K_{SF}	Safety margin to cover the impact of the DC component and CT error
$I_{nFeeder}$	Rated current of the feeder
$I^{50N}_{pickup-R1}$	Pickup value of the instantaneous stage of earth fault protection of relay R1
$I^{51N}_{pickup-R1}$	Pickup value of the inverse stage of earth fault protection of relay R1
I_{nZCT}	Nominal current of core-balance/neutral CT
I_0	Zero-sequence current
K_{CLPU}	Safety factor to consider the impact of cold load pickup
$\|Z_{AF1}\| \angle\theta_1$	Impedance between bus A and fault location F1
$\|V_A\| \angle\delta_A$	Voltage phasor at bus A
I_{F1}, I_{F2}	Fault current at locations F1and F2
I_{Ref}	Reference current for determining fault direction
Δ	Index for discriminating forward faults from the reverse ones
$\angle I_{Fault}$	Phase angle of the fault current phasor
N_{CP}	Number of main-backup coordination pairs
N_V	Number of miscoordinations or violating constraints

t_{pa}	Operation time of the main relay *a* for the maximum near-end three-phase fault current
t_{bb}	Operation time of the backup relay *b* for the remote-end phase-phase fault
h_1, h_2	Constant weighting factors
$I_{sc}^{2\varphi}$	Phase-phase fault current occurring at the remote-end
$\text{TMS}_{\text{min}}, \text{TMS}_{\text{max}}$	Minimum and maximum adjustable settings of TMS
I_{pickup}	Pickup value of overcurrent relay
$I_{pickupmax}, I_{pickupmin}$	Maximum and minimum values of I_{pickup}
k	Temporary overload factor of the backup relay
$I_{loadmax}$	Maximum load current passing through the backup relay
x_i	Data point
c_j	Center point of data
D_{ij}	Distance between the center point c_j and the data point x_i
d	Space dimension
V_t	Pickup value of voltage
V	Positive-sequence value of the measured voltage
λ	Slope of the VS-OC function
PS	Pickup setting of VS-OC function
PSmax	Maximum pickup setting of VS-OC function
ξ	Bias of the VS-OC function
t_i	Time delay of the main relay for the near-end three-phase fault
t_j	Operation time of the backup relay for the fault occurring at the remote-end of the main relay
S	Set of various topologies of the system
$\lambda_{min}, \lambda_{max}, \xi_{min}, \xi_{max}$	Positive constant parameters
I_{th}^{50}	Overcurrent element threshold for FIDVR detection
I_{th}^{37}	Undercurrent element threshold for FIDVR detection
T_1	Predefined duration time for detecting prolonged fault event and FIDVR
T_2	Time utilized to reset FDS signal
I_1	Positive-sequence current
T_3	Time interval for disabling set signal
C_{TH}	Reset counter threshold
f_c	Sampling frequency
$t_R\|_{@fi}$	Operation time of the main relay for close-in faults
$t_R\|_{@fj}$	Operation time of the main relay for remote-end faults
CCT_{DGn}	Critical clearing time of the *n*th DG

t_{margin}	Time margin to cover unintentional delays
Y	System admittance seen by the load-shedding relay
Y_{th}	Admittance constant threshold
φ_c	DG load angle at the fault clearing instant
φ_{Th}	DG load angle threshold
t_R	Reacceleration time of the aggregated IM
Y_{ss}	Steady-state admittance of the feeder
E_c	Output signal of the AVR
V_{max}	Maximum value of AVR output
E_{fd}	Excitation system voltage
LOFDI	LOF detection index
e_q'	q-axis component of generator internal voltage
E_i	Generator internal voltage
T_{th}	Adaptive time delay for LOF detection
θ	Rotor position
ω_r	Rotor mechanical speed in pu
δ	Rotor angle
ω_s	Synchronous speed
P_t	SSSG output active power
P_m	Mechanical input power
H	Inertia constant
P_e	Electrical power
D	Damping coefficient
Q_t	SSSG reactive power
X'_d, X'_q	Sub-transient d-axis and q-axis reactances
δ_{LOF}	SSSG rotor angle during the LOF event
δ_{max}	Maximum critical value of rotor angle
T_{ps}	Time until loss of synchronism
$s_{nom}, s_{critical}$	Nominal and critical slips
T_{Stall}	Deceleration time from s_{nom} to $s_{critical}$
J	Inertia of aggregated motor
Th_1	Predefined threshold for de'_q/dt
Th_2	Predefined threshold for $d(E_i/V_t)/dt$

About the Companion Website

This book is accompanied by a companion website:

www.wiley.com/go/sanaye-pasand/microgrids

To access the website materials, click on the link under Related Resources on the book's page on wiley.com. You can also find this by searching www.wiley.com for the ISBN 9781394278589.

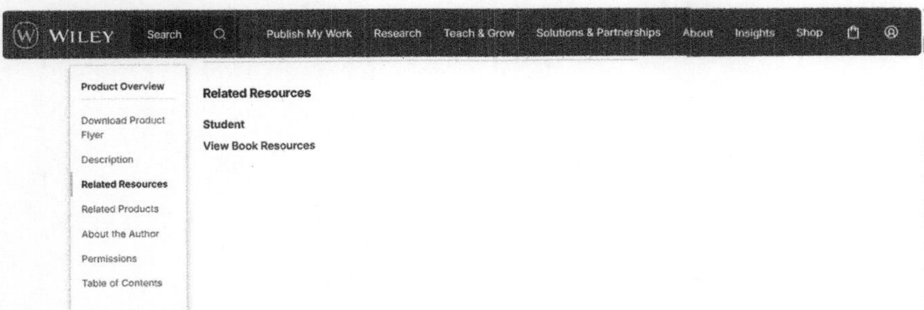

The website includes:

- Case studies simulation files
- Tutorial presentation slides

1 Introduction

1.1 Introduction

Around the world, the penetration of microgrids (MGs) has been significantly increased recently by both the government and private sectors. This is due to various factors and requirements, including enhancement of power supply reliability and considerations regarding environmental issues and different concerns such as a reduction in greenhouse gas emission, peak shaving, power quality improvement, loss reduction, an uninterruptible power supply for essential loads, and financial benefits [1–3]. MGs contain self-controlled distributed generations (DGs) and local loads, which may operate in either the grid-connected or islanded, i.e., standalone mode. Different DG types and technologies may be adopted for integrating MGs with renewable and nonrenewable energies, including small-scale synchronous generators (SSSGs), induction generators (IGs), and inverter-based DGs (IBDGs) [4, 5].

The choice of DG type and its technology depends on various factors, such as governmental policies, energy cost, climate conditions, region topology, and geographical features. The application of these technologies, along with different types of energy sources, including solar irradiation, wind, hydropower, geothermal, biomass, and fossil fuels, is briefly introduced in this chapter, considering the technical and financial issues, and then their benefits and shortcomings are outlined. Investigation of various types of DG and MG technologies properly reveals that SSSGs have an essential role in generating highly reliable and smooth electricity power from both renewable energy sources, such as hydropower, and nonrenewable energy sources, such as natural gas. On the other hand, to take the benefits of the renewable resources, particularly wind and solar irradiation, and also to reduce the dependence on nonrenewable energies, the use of IBDGs is crucial. Therefore, in modern MGs, hybrid energy resources and technologies, including both SSSGs and IBDGs, are employed to enhance system reliability and stabilize power generation, while also minimizing operational costs, reducing reliance on nonrenewable energy sources, and mitigating environmental impacts.

To meet the main goals of modern MGs, especially diminution of greenhouse gas emission, high-efficiency technologies using low-carbon generation are commonly utilized. In this regard, combined heat and power (CHP) systems are widely used in modern MGs [6], which usually contain SSSG and its associated prime mover, equipped with heat recovery systems to utilize the emitted heat. A gas turbine or reciprocating engine is usually used as the SSSG prime mover. Meanwhile, various technologies such as absorption cooling and heat storage may be exploited for further improvement of the CHP system performance [7]. In this regard, the CHP systems and their fundamentals are briefly explained in this chapter, too.

In this chapter, the structure of various electricity generation technologies from different resources is presented. First, the structure of photovoltaic (PV) power plants, developed to generate electricity from solar irradiation, is explained, along with a brief description of their equivalent model. Next, various types of wind generation technologies are introduced, including a brief explanation of the equivalent model of IBDGs. Meanwhile, detailed discussions on the modeling and control of IBDGs are provided in Chapter 3 of this book. Afterward, the structure of SSSGs, including their excitation systems and prime movers, is discussed. Different types of modern excitation systems, such as the shunt technique, auxiliary winding excitation, and permanent magnet generators (PMGs), are described. A complete model for SSSG packages is then presented, which can be effectively used for dynamic and transient stability assessments. Finally, the structure of MGs, including their configuration, control systems, and protection schemes, is explained.

1.2 Distributed Generations Technologies

1.2.1 Solar Radiation Energy

Solar radiation energy is a well-known type of renewable energy that has been widely utilized in the past decade to generate electricity on both small and large scales. Various types of technologies have been recently developed to generate electricity from solar radiation energy [8, 9]. The most widely used commercial technologies are solar thermal and PVs, which are briefly explained in the following paragraphs.

Solar thermal power plants employ concentrators such as mirrors, reflectors, and lenses to produce heat from the solar radiation energy. The heat produced will be utilized in a thermodynamic cycle to generate electricity. The schematic of a typical solar thermal power plant is depicted in Fig. 1.1 [9]. As shown in this figure, the working principle of such a solar power plant is very similar to that of

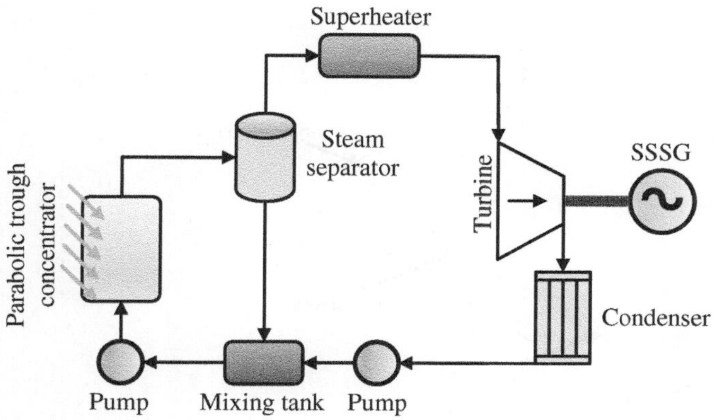

Fig. 1.1 Typical schematic of a solar thermal power plant. Adapted from [9].

the conventional combined cycle power plants. In fact, the main equipment of the thermodynamic cycle of a thermal solar power plant including turbines and synchronous generators (SGs) are similar to those of conventional power plants. In other words, the solar thermal power plant can be considered a conventional thermal power plant in which the input fossil fuel cell is replaced by solar radiation energy. The efficiency of the solar thermal power plant is normally between 15% and 30%, which can also be improved using thermal energy storage systems (ESSs) [8, 9]. Saving solar thermal energy by employing these storage systems helps these power plants to generate electricity, even when solar radiation is not available (e.g., during the night). Moreover, for further improvements of the capacity factor and efficiency, a combination of solar thermal with other renewable/nonrenewable energies can be employed, which is called hybridization. However, the solar thermal power plant may not be suitable for small-scale applications due to the required high amount of capital cost. Furthermore, this type of solar power plant may not be efficient in low-radiation regions.

The second category for generating electricity from solar radiation energy is PV technology in which solar PV cells are employed. Unlike solar power plants, PV cells are able to directly convert solar radiation to electricity, which is suitable for both low- and high-radiation regions. PV cells generally exhibit a basic structure analogous to that of semiconductor diodes and can, in most cases, be modeled as photodiodes. The core component of a PV cell is the p–n junction, as shown in Fig. 1.2. The output power of PV cells significantly depends on the PV cell surface area, solar irradiance, and other environmental factors [9]. Since the output voltage and power of each PV cell are very small and are not suitable for practical applications, a series combination of PV cells, called PV modules, is employed to be assembled into the PV modules.

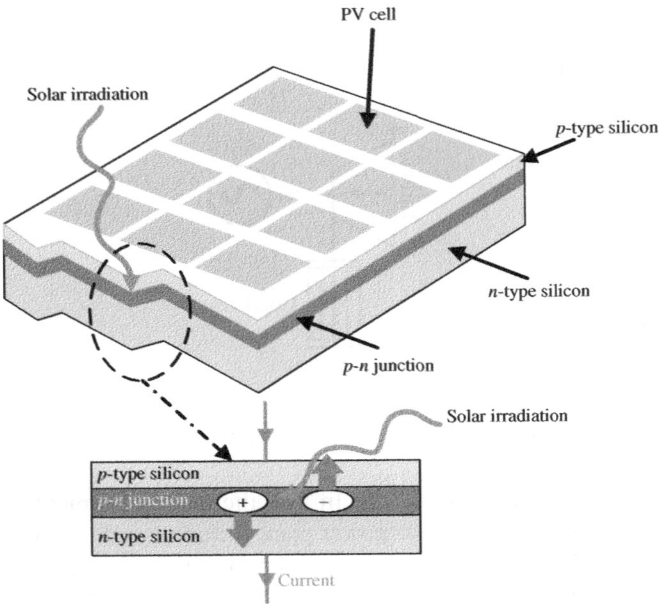

Fig. 1.2 Structure of a PV module.

Figure 1.2 illustrates a PV module in which multiple PV cells are connected in series. When photons strike the p–n junction, they generate electron–hole pairs. The built-in electric field of the p–n junction exerts a force on these charge carriers, causing electrons and holes to move in opposite directions. When the PV cell is connected to a closed circuit, this movement of charge carriers produces an electric current. It is important to note that this current generation occurs only under solar irradiation and while the PV cell remains part of a closed circuit.

Various technologies have been developed in recent years for PV modules. The commercial widely used technologies of PV modules can be mainly categorized into two groups: monocrystalline silicon and polycrystalline silicon. The efficiency and price of the PV panels based on polycrystalline silicon are lower than those of monocrystalline silicon. The average efficiency of PV panels is between about 13% and 27%. However, in recent years, some new technologies have been utilized in PV panels, which considerably enhance the PV panel performance and reliability. These new techniques include half-cut cells, bifacial PV panels, multi-busbar cells, and passive emitter rear cells [10].

As discussed, a PV cell can be modeled as an active, nonpassive diode. Consequently, an ideal PV cell can be represented by a current source in parallel with a diode, as illustrated in Fig. 1.3(a). In the absence of irradiation, the current

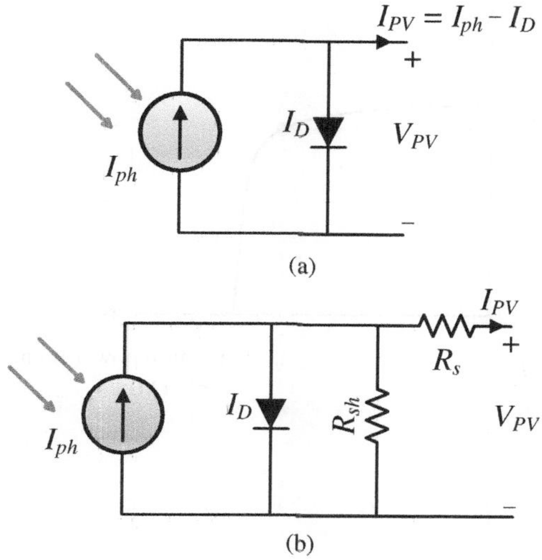

Fig. 1.3 Equivalent model of PV cell: (a) ideal model and (b) nonideal model.

source is open-circuited, and the PV cell behaves like a diode. As solar irradiation increases, the current generated by the PV cell rises proportionally. A more accurate model of the PV cell that accounts for its nonidealities is shown in Fig. 1.3(b). In this model, the added resistances represent effects such as contact resistance, leakage paths across the p–n junction, and resistive losses associated with busbars and other internal structural elements.

The current–voltage (I_{PV}–V_{PV}) characteristic and the output power–voltage (P_{PV}–V_{PV}) curve of a PV module are shown in Fig. 1.4. As illustrated, the knee point of the I–V curve corresponds to the maximum power point (MPP). To extract the maximum power from the PV module, its operating point should be maintained at the MPP. This requires regulating the module voltage to the corresponding V_{mpp}. Adaptation of the PV module voltage at V_{mpp} is performed by the inverter, the operation of which is described in the following paragraphs. This regulation process is known as maximum power point tracking (MPPT). The value of V_{mpp} varies with changes in solar irradiation and environmental conditions such as ambient temperature and wind speed. Therefore, an appropriate algorithm should be employed in the inverter to track the MPP and determine V_{mpp} in real time.

For constructing a PV power plant, a series/parallel combination of PV modules should be utilized to generate the expected energy. For example, PV modules connected in series form a PV string, and multiple strings connected in parallel form a PV array. Since the PV panels generate direct current (DC), proper inverters should be employed to connect them to an alternating current (AC) power system. To take the maximum efficiency of PV modules, their optimum tilt angle should

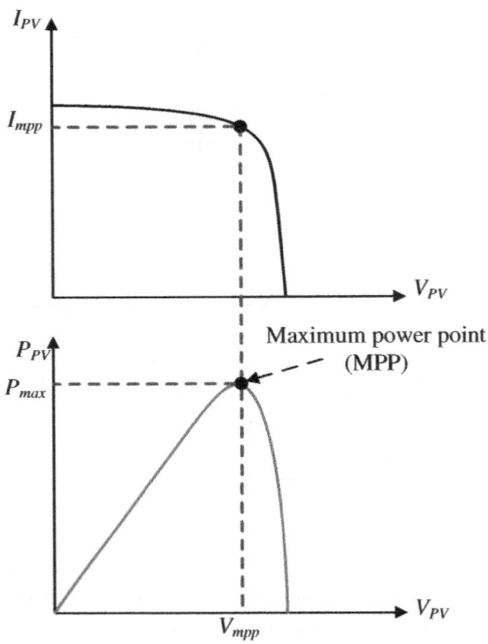

Fig. 1.4 The current–voltage (I_{PV}–V_{PV}) characteristic and the output power–voltage (P_{PV}–V_{PV}) curve of a PV cell.

be adjusted based on the latitude and other environmental factors. Meanwhile, this optimum angle varies in various seasons of the year. With the recent advancements in technology and to increase the PV power plant efficacy, tracking structures can be employed instead of fixed ones. However, these structures have a higher cost and also require more installation area. The total installation cost of PV power plants highly depends on the PV module technical specifications, such as efficiency, output power, and bifacial capability.

The inverter regulates the DC-side voltage of the string at V_{mpp} to track the MPP and extract the maximum available power from the PV string. In practice, multiple PV strings are often connected in parallel to the inverter. Because the MPP and corresponding V_{mpp} of each string may differ, each string is equipped with a dedicated DC/DC converter to interface it with the inverter's DC bus, as illustrated in Fig. 1.5. The control systems of the inverter are thoroughly discussed in Chapter 3.

The PV solar power plant can operate in both the grid-connected and standalone modes. However, for the standalone operation, ESSs such as batteries are required, which significantly increases the required capital expenditure of the PV power plant. Therefore, the standalone operation of a PV power plant may not be economically feasible, especially for large-scale applications, i.e., greater than 1 MW. On the other hand, the PV power plants, which are not equipped with ESS, cannot

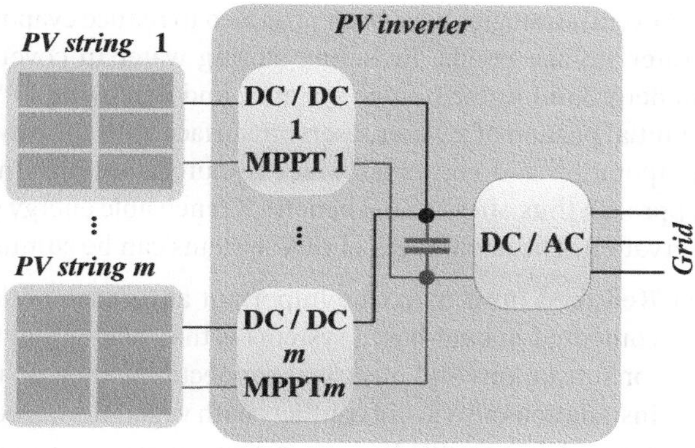

Fig. 1.5 PV inverter schematic.

provide smooth constant power output; thus, they may not be appropriate for supplying the essential loads. To cope with these issues, other types of technologies that can generate constant smooth output power including SSSGs, such as geothermal and fossil fuels, can be employed along with PV power plants.

PV solar power plants can be classified based on the type of mounting into two categories: (i) ground-mounted PV and (ii) floating PV (FPV). Ground-mounted PV power plants, which constitute the majority of solar power installations, can be deployed on rooftops, coastal areas, deserts, plains, and other suitable locations. However, the deployment of such power plants faces several challenges, including [11]:

- Limited availability of land
- Visual impacts and resident discomfort due to reflected sunlight
- Significant increase in ambient temperature at installation sites
- Alterations to the local vegetation cover
- Potential damage during storms, including the displacement of panels toward nearby communities.

One approach to mitigating these challenges is the adoption of floating solar power plants on the water surface. The use of FPV systems has gained significant attention in many countries, particularly in Asia, including India, China, Singapore, and South Korea [11].

In recent years, global warming has led to increased water evaporation, resulting in severe water scarcity in several regions. To address water shortage, the construction of dams has been a key strategy. However, surface water evaporation from reservoirs remains a major source of water loss. Solar radiation is a primary factor driving evaporation. Evaporation is increasingly recognized as one of the most significant factors aggravating drought in arid and semi-arid regions worldwide.

Several strategies have been proposed to reduce evaporation from reservoirs and water storage ponds, including storing water in covered reservoirs, developing underground water storage systems, and deploying FPV panels. Covering a substantial portion of a water reservoir surface with PV panels not only reduces water evaporation and conserves water resources but also inhibits algae growth. This approach thus offers a dual benefit of renewable energy generation and water conservation. The advantages of FPV systems can be summarized as follows [11]:

1) **Reduced land use:** One important advantage of floating solar power plants compared to land-based systems is that, aside from the limited areas required for substations and electrical connections, they do not occupy any land. FPV installations also avoid conflicts with vegetation, green spaces, and agricultural areas. Additionally, floating systems have a significantly lower visual impact than that of ground-mounted plants.
2) **Installation and decommissioning:** FPV systems are generally more compact, denser, and easier to manage, install, and decommission than land-based systems. This is due to the absence of fixed mounting structures. The modular floating platforms are connected through flexible anchoring systems, which can be easily dismantled or reconfigured when relocation or design modifications are needed.
3) **Water conservation and water quality effects:** Covering reservoir surfaces with FPV modules provides an important benefit in reducing water evaporation. The magnitude of this reduction depends on climatic conditions and the percentage of surface coverage. In arid regions, evaporation reduction can exceed 80% of the vaporization of the covered area, which is highly valuable for reservoirs used for irrigation and agriculture. The impact on water quality depends on parameters such as reservoir size, surface coverage, type of floating structure, water-level variations, circulation patterns, and other environmental factors.
4) **Passive cooling and solar tracking:** The floating nature of FPV systems enables simple and cost-effective cooling. Due to the cooling effect of water and increased air flow above the surface, PV module temperatures remain lower than those in land-based systems, resulting in higher electrical efficiency. In some installations, additional cooling is provided through water spraying. Large floating platforms facilitate single-axis solar tracking, and small systems can also achieve tracking via mobile mounting structures. Energy gains from tracking generally range from 15% to 25%, sometimes exceeding those of land-based installations.
5) **Hybrid systems and synergy with hydropower:** Floating solar power plants integrate effectively with other renewable technologies. Solar and wind power generation profiles often complement each other; therefore, hybrid FPV–wind installations, especially in coastal regions with variable weather,

can reduce generation uncertainty from the perspective of the grid. FPV systems can also be combined with hydropower plants. Deploying FPV arrays on hydropower reservoirs reduces evaporation and enhances pumped-storage operations, improving overall efficiency and reliability.

6) **Environmental benefits:** A significant environmental advantage of FPV deployment is the reduction of algal growth. By partially shading the reservoir and limiting sunlight penetration, FPV systems help mitigate algae formation and improve water quality.
7) **Greenhouse effect reduction:** Ground-mounted PV plants reduce surface albedo and create localized negative radiative forcing, which leads to increased surface heating and contributes to global warming. In contrast, FPV systems do not significantly alter the radiative balance of water surfaces and, therefore, do not introduce the same localized warming effects. This is an important environmental benefit, particularly for large-scale FPV installations.
8) **Cost reduction:** Capital cost is the most critical factor in renewable energy deployment. In regions with high land prices, FPV systems can be more cost-effective than land-based alternatives. Recent technological advancements, especially in floating platform design, have significantly reduced FPV installation costs.
9) **Dust reduction:** FPV installations experience lower dust accumulation compared to land-based systems, particularly in dusty regions. This reduces cleaning requirements and contributes to higher long-term energy yield.

Traditionally, solar power was considered "renewable," "zero-carbon," and "free of fossil fuel resources," which has no adverse environmental impact (e.g., carbon emission). However, recent field investigations and measurements reveal that these traditional ideas are not correct. These power plants have various adverse environmental impacts, especially "carbon emissions" in the manufacturing, construction, operation, and disposal stages. Based on the field investigations reported by [5], the amount of renewability factor and carbon emissions for a pilot solar power plant during the life cycle are about 45% and 36% of those of the coal-fired power plants, respectively. Besides, the following environmental, health, and safety issues have also been reported [12]:

- Various toxic chemical materials, e.g., Cd, Pb, and Hg, are exploited in the manufacturing stages of PV panels, which may cause cancer and other health issues.
- Solar power plants, especially large-scale ones, have direct adverse effects on the ecology such as wildlife, native vegetation, and loss of habitat.
- Decomposition and disposal of solar modules at their end-of-life cycle leads to the release of toxic materials, which may pollute soil and groundwater.
- For constructing PV power plants, a larger area is required compared to those of the other technologies.

1.2.2 Wind Energy

Wind energy is categorized as a renewable energy type that can be utilized to generate electricity in regions with an average wind speed of about 5–7 m/s [13]. Considering environmental conditions, various wind turbine–generator (WTG) technologies have been developed and commercially utilized. The simple schematic of a typical WTG is illustrated in Fig. 1.6. Various technologies have been introduced for each part of the WTG, including its turbine, tower, gearbox, and generator, and thus, WTG can be classified from various points of view. Commonly, based on the orientation of the mechanical parts (e.g., rotor shaft and generator) viewpoint, the WTG can be vertical or horizontal. Generally, the widely utilized configurations of WTG are classified into four major groups: types I, II, II, and IV [14]. In the following paragraphs, these various WTG configurations are briefly described.

Figure 1.6(a) represents the topology of type I as the oldest version of WTGs. This topology contains a squirrel-cage rotor induction generator (SRIG), which is directly connected to the upstream system. The SRIG absorbs reactive power from the main grid and, thus, a parallel capacitor is employed to compensate the reactive power. Since an IG cannot inject reactive power into the system, the fault ride-through (FRT) and low-voltage ride-through (LVRT) capabilities of this type are weak. Moreover, the turbine speed should be constant in this configuration, and it cannot operate at variable turbine speeds. Due to these serious shortcomings, this configuration is not widely utilized [15].

The general configuration of WTG type II, shown in Fig. 1.6(b), is very similar to that of WTG type I, and only the SRIG is replaced by a wound rotor IG (WRIG). The turbine of WTG type II, unlike the preceding case, can operate at various speeds due to the use of the wound rotor. However, similar to the WTG type I, the FRT and LVRT capabilities of WTG type II may not be suitable for practical applications. Meanwhile, the control mechanism of the blade pitch angle is commonly employed for type II and the newer versions of type I to prevent speed increment during the grid voltage dips.

To resolve the shortcomings of types I and II and to enhance the FRT and LVRT capabilities, WTG types III and IV have been developed using AC/DC/AC converters. Figure 1.6(c) depicts the configuration of WTG type III, which includes a WRIG. This generator is directly connected to the upstream network, and its rotor is supplied by an AC/DC/AC converter. In such a topology, unlike the previous configurations, both the injected real and reactive powers can be controlled without requiring a parallel capacitor bank. Furthermore, the WRIG is able to inject reactive power during voltage dips. For further improvement of speed control and FRT capability, WTG type IV is developed in which the

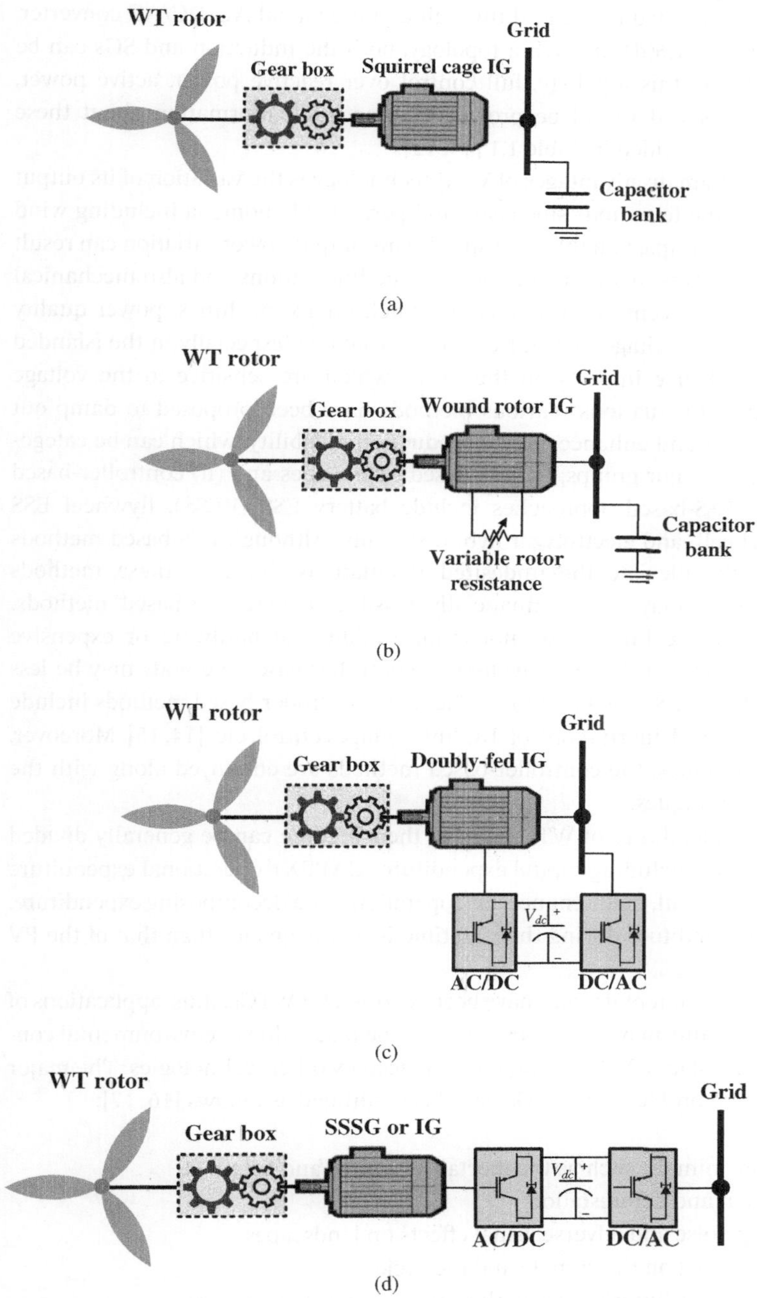

Fig. 1.6 Various types of wind turbines, (a) Type I, (b) Type II, (c) Type III, and (d) Type IV [14, 15].

generator is connected to the grid through a bidirectional AC/DC/AC converter, as shown in Fig. 1.6(d). In such a topology, both the induction and SGs can be employed. Using this topology, full control over reactive power, active power, and operating speed would be provided. Some brief information about these technologies is provided in Table 1.1 [14, 15].

One of the main disadvantages of WTG technology is the variation of its output active power due to various stochastic and periodic phenomena including wind shear and tower impact on the wind speed. This output power variation can result in various problems including torsional torque fluctuations and also mechanical fatigue in the adjacent synchronous or asynchronous machines, power quality problems such as voltage flicker, frequency variations (especially in the islanded mode), and adverse impacts on the loads, which are sensitive to the voltage and frequency fluctuations. Various methods have been proposed to damp out these variations and enhance energy production reliability, which can be categorized into two major groups: (i) ESS-based approaches and (ii) controller-based techniques. ESS-based approaches include battery ESS (BESS), flywheel ESS (FESS), fuel cell, and electrolyzer hybrid systems. Although ESS-based methods can efficiently alleviate the undesired fluctuations, some of these methods (such as BESS) may not be financially feasible. Unlike ESS-based methods, controller-based techniques do not require additional hardware or expensive equipment. However, the effectiveness of controller-based methods may be less than that of the ESS-based methods. The main controller-based methods include pitch angle control, inertia control, DC link voltage control, etc. [14, 15]. Moreover, in some techniques, the controller-based methods are employed along with the ESS-based techniques.

The total expenditure of WTGs during their lifetime can be generally divided into three parts, including capital expenditure (CAPEX), operational expenditure (including overhaul, maintenance, and operation), and decomposing expenditure. The WTG expenditure during their lifetime is highly greater than that of the PV power plants.

Serious environmental issues have been reported for WTGs; thus, applications of these power plants may not be possible in some places due to environmental concerns. In such places, WTGs should be replaced by other technologies. The major environmental problems of WTGs have been outlined as follows [16, 17]:

- Effects on animals, such as the mortality of birds and bats
- Soil erosion and deforestation
- Annoying noise and adverse visual effects on landscapes
- Adverse impact on the climate and life cycles
- Electromagnetic interference with waves (such as radio waves).

Table 1.1 Summary of various wind turbine characteristics [14, 15].

Parameter	Type I		Type II	Type III		Type IV	
	IA	IB		IIIA	IIIB	IVA	IVB
Generator type	Induction, squirrel cage		Induction, wound rotor	Induction, doubly fed		Either induction or synchronous	
Speed	Fixed (1–2% slip)		Fixed (up to 10% slip)	Variable (−30% to +40% slip)		Fully variable	
Fault ride-through capability	No	Yes (pitch control)	No	Yes (crowbar)	No (without crowbar)	Yes (chopper)	No

1.2.3 Geothermal Energy

Geothermal energy is a renewable energy type, which is defined as "the heat energy of Earth's interior" [18]. This energy may be utilized for various heating applications. Producing electricity from the heat energy is one of the common applications of geothermal resources. Various types of geothermal resources based on their temperatures are tabulated in Table 1.2. Commonly, three types of technologies have been employed to generate electricity from geothermal resources, including dry steam, flash steam, and binary organic Rankine cycle [18]. The proper technology for electricity generation should be selected based on the type of geothermal resources, as summarized in Table 1.2. The information in this table and the working principles of each technology are briefly described in [18].

Dry steam technology is the simplest technology requiring lower capital expenditure, which has higher efficiency compared to that of the other types of technology. The schematic of this technology is depicted in Fig. 1.7. In this technology, geothermal resources with high-pressure superheated steam are required, which are few in number around the world, and this challenge can be considered as the main limitation of dry steam technology.

Table 1.2 Various types of geothermal resources [18].

Type of geothermal resource	Temperature (°C)
Hot water	20–350
Vapor	240
Sedimentary basin	20–150
Radiogenic	30–350
Geopressured	90–200
Solidified (hot dry rock)	90–650
Parts still molten	>600

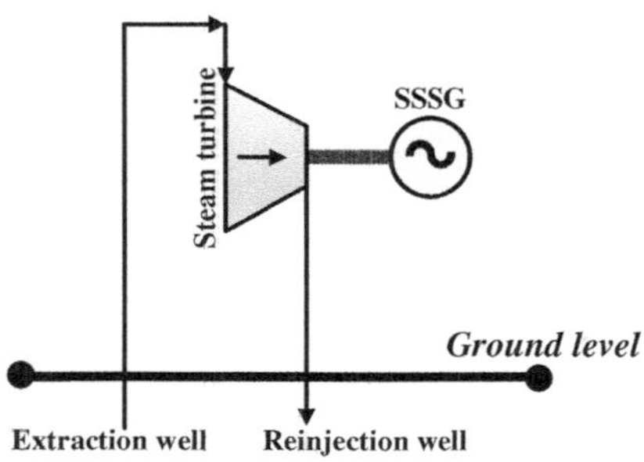

Fig. 1.7 Dry steam technology. Adapted from [18].

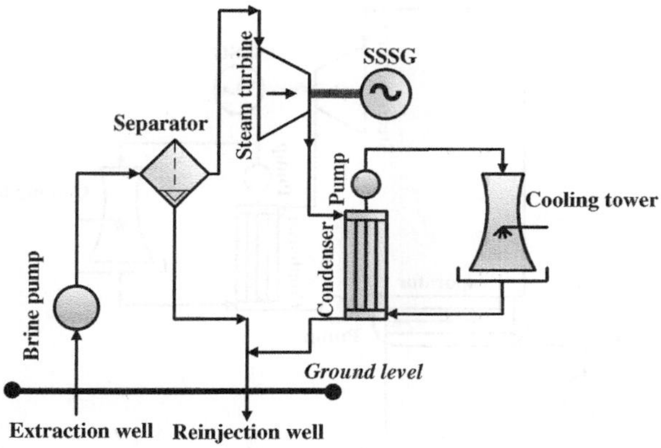

Fig. 1.8 Flash steam cycle. Adapted from [18].

Another type of technology that is widely utilized for producing electricity from geothermal resources is the flash steam schematic, as shown in Fig. 1.8 [18]. A high-pressure and high-temperature mixture of steam and water, which is referred to as "two-phase mixture," is employed by the flash steam system to generate electricity. The temperature of this mixture is commonly greater than 220 °C. After transferring the mentioned mixture to the surface, the fluids will be expanded through a separator, which leads to evaporation of the fluids. Then, the resultant steam will be exploited within a turbine to generate electrical power. To generate electricity from single-phase geo-fluid with medium/low temperature of geothermal resources (i.e., 100–220 °C), a binary organic Rankine cycle is utilized. In this technology, the Rankine cycle is employed to transfer heat from the geo-fluid into the organic fluid using a vaporizer. In fact, the main principle of this technology is utilizing organic fluids with low amounts of critical temperatures and pressures to employ the thermal energy of medium/low temperatures of geothermal resources [18]. The schematic of this technology is illustrated in Fig. 1.9. For all of the geothermal technologies, the employed steam turbine is coupled with a SG to produce electricity.

1.2.4 Biomass Gasification

Biomass gasification is another type of renewable energy that can be utilized for providing suitable gas fuels to be employed by internal combustion engines. The internal combustion engine should be coupled with an SSSG to generate electricity. Biomass gasification can be employed at large scales with the output electrical power over 2 MW. Moreover, it can be utilized in small-scale applications whose output power is less than 200 kW [19]. Small-scale utilization may be more effective for small communities such as homes due to low capital expenditures. The availability of biomass is more than that of the other types of

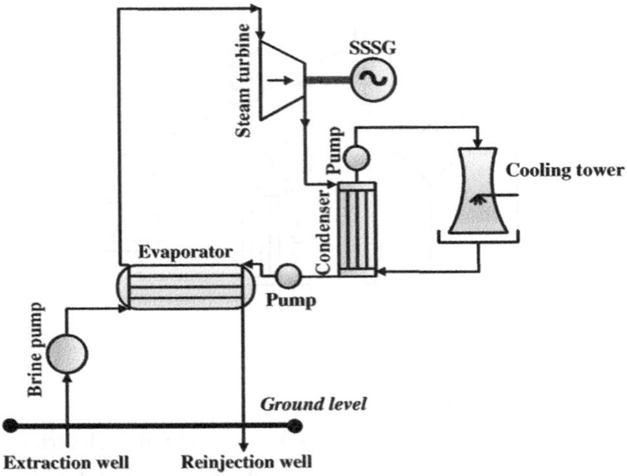

Fig. 1.9 Organic Rankine cycle technology. Adapted from [18].

renewable energies, and its share is about 50% over the other forms of renewable energies. In fact, biomass can be considered a "biodegradable organic material," which is produced from animals, plants, etc. Biomass has been widely used in various countries, such as Asian countries, to diminish environmental issues of fossil fuels. Although environmental issues of biomass are considerably lower than those of fossil fuels, it may still have some adverse impacts on the soil, water, and land [19].

1.2.5 Hydropower

Hydropower is a widely used renewable energy type, which is also known as the cleanest energy, with minimum greenhouse gas emissions. Hydropower has the largest share among the other types of renewable energies due to its reliability and cost [19]. In fact, hydropower is the most reliable and cost-effective type of energy, which is utilized along with SGs to generate electricity with a maximum capacity factor. Hydropower plants can be classified into various groups depending on their nominal capacity. The output power of small-scale hydropower plants is commonly limited to 30 MW, while large-scale power plants can generate above 100 MW. For implementing large hydropower plants, appropriate dams should be constructed. On the other hand, for the small-scale ones, direct river flows are commonly utilized through diversion channels or penstocks. The components of a sample hydropower plant are illustrated in Fig. 1.10.

Various types of hydropower turbines have been developed to take the maximum efficiency of the hydropower, which are generally classified into two

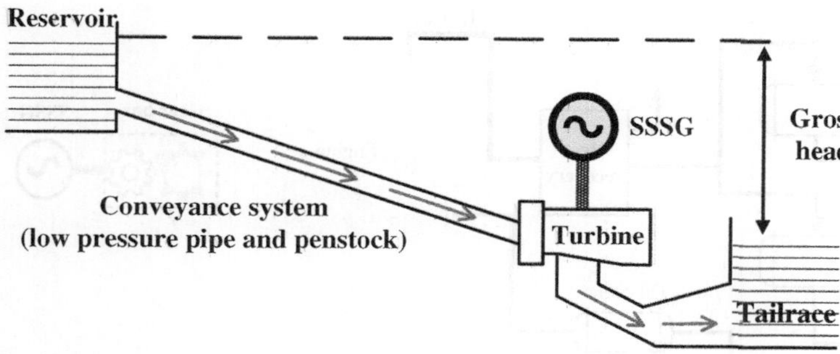

Fig. 1.10 Sample hydropower plant components. Adapted from [20].

major groups, including impulse and reaction turbines [20]. Different factors should be considered to select the most suitable hydropower turbine for a sample project, namely, head of water, volume and flow rate of water, installation depth, required efficiency, and cost. It is worth mentioning that small-scale hydropower plants can be considered an environmentally friendly solution for reliable power generation in any region, even with low water heads. Small-scale hydropower turbines are commonly coupled with a salient pole SG to integrate synchronous generator-based microgrids (SGBMGs) and generate electricity in both off-grid and on-grid modes.

1.2.6 Reciprocating Engines

Reciprocating internal combustion engine is a widely used technology for generating electricity, and its capacity commonly varies from 10 kW to over 18 MW. Reciprocating engines can be exploited with both SSSGs and IGs to produce reliable electricity for various applications including industrial, commercial, and residential. Various types of reciprocating engines have been developed, most of which operate based on four strokes of the piston into the cylinder [21]. Generally, reciprocating engines are divided into two major groups, i.e., "natural gas engines" and "diesel engines." The first group exploits spark plugs for ignition of the compressed mixture of natural gas and air within the cylinder. The capacity of natural gas engines is normally less than 5 MW, and based on the engine size, two major techniques are used for ignition, including "open chamber" and "precombustion chamber." In addition, depending on the gas engine technology, a compression range of 9:1–12:1 is required for the ignition procedure. Furthermore, the efficiency of small gas engines (i.e., less than 50 kW) is about 28%, while for the larger gas engines, it is about 42% [21].

The second group, i.e., diesel engines, is a high-efficiency simple-cycle power generation technology in which the mixture of injected diesel fuel and air

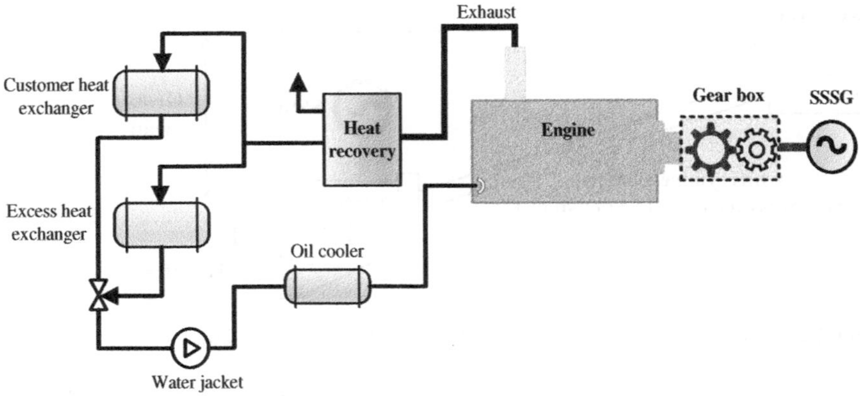

Fig. 1.11 Sample CHP system using combustion engines. Adapted from [21].

is ignited in the cylinder through a compression range of 12:1–17:1. Diesel engines are normally classified into high-speed (about 1200 rpm) and low-speed (60–275 rpm) categories. The capacity of high-speed engines is typically up to 4 MW, while low-speed engines can be designed for up to 65 MW. It is worth mentioning that the energy conversion efficiency of small high-speed diesel engines is about 30%, while for large diesel engines, it is about 42–48%, which highly depends on the engine technology, fuel quality, and environmental factors [20]. To increase the reliability of internal combustion engines, dual-fuel combustion engines have been developed. These engines are normally fueled by natural gas as the main fuel, and diesel oil is used as the backup fuel.

Reciprocating engines can also be classified based on their operating speed. High-speed engines have lower capacity, and their speed is about 1000–3600 rpm. In addition, the operating speed range of medium- and low-speed engines is about 275–1000 and 58–275 rpm, respectively. Furthermore, these engines have higher capacities compared to the high-speed engines [21]. To take maximum efficiency from reciprocating internal combustion engines, the thermal energy of the exhaust gas can be utilized in a separate loop for various applications, such as heating purposes, absorption refrigerators, and steam production. This cycle is commonly known as the simultaneous cogeneration of heat and power. A sample CHP heat recovery system is depicted in Fig. 1.11.

1.2.7 Gas Turbine

As shown in Fig. 1.12, gas turbines consist of three major parts, including an air compressor, combustion chamber, and turbine, which work based on a thermodynamic cycle known as the "Brayton cycle" [22]. Various types of fuel

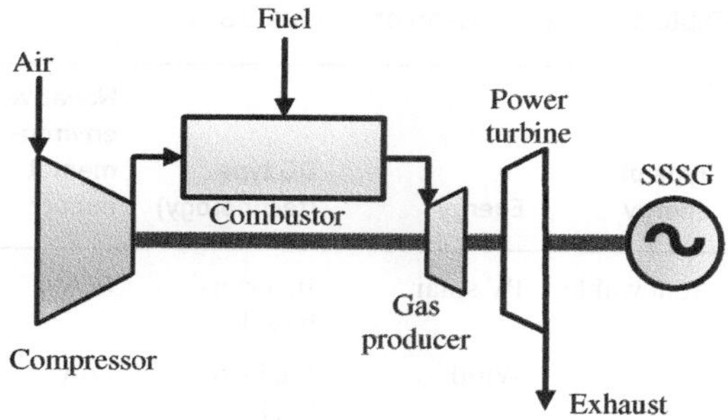

Fig. 1.12 Sample gas turbine. Adapted from [22].

can be exploited for gas turbines such as natural gas, synthetic gas, fuel oils, and landfill gas. Gas turbines are commonly used with SGs in various capacities ranging from 500 kW to 350 MW where their efficiency in a simple cycle is about 35–40%. The gas turbine produces high-temperature exhaust whose heat can be exploited for power generation through steam turbines. Using gas turbines in a combined cycle with steam turbines can enhance the energy conversion efficiency up to 60%. Generally, gas turbines are suitable for CHP applications due to their high-temperature exhaust. Using gas turbines in a separate CHP loop can improve the overall efficiency of the system to 70%.

As reported by [21], gas turbines can be exploited to produce process steam, which may be used directly in industrial processes for heating or drying. A widely used CHP application of gas turbines is producing enough steam for campus heating in winter. Moreover, absorption chillers based on the CHP loop can provide suitable cooling in summer.

1.2.8 Comparing Various Technologies

Various factors should be considered to select the most suitable technology and energy resources, which are exploited to form MGs. These factors include (i) quality of supply, (ii) outage duration, (iii) capital expenditure, (iv) O&M cost, and (v) environmental factors (especially carbon footprint) [12]. For various DG types, these factors are compared in Table 1.3. Based on this comparison, a suitable framework can be arranged to achieve optimum design, sizing, and long-term operation (at least 20 years) of the MGs.

The selection of the DG type and its technology may be different for various applications, such as commercial, residential, military, and industrial plants. In this regard, proper approaches should be utilized to minimize the costs, considering the reliability and environmental constraints. Since the MG is a local energy system and its size is relatively smaller compared to the upstream grid, the power fluctuations and uncertainty of the renewable resources have a

Table 1.3 Comparison of various DG types.

Type of energy	Energy	DG type (technology)	Negative environ-mental impact	CAPEX	O&M cost	Reliability
Renewable	PV solar	Inverter-based	Low	Medium	Low	Low
	Wind	Back-to-back converter	Low	High	Medium	Low
		Induction generator	Low	High	Medium	Low
	Hydropower	SSSG	Zero	Medium	Medium	Medium
	Geothermal	SSSG	Zero	Very high	High	High
	Biomass	SSSG	Low	High	Medium	High
Non-renewable	Gas and oil	SSSG	High	Low	Medium	High

considerable impact on MG operation. Furthermore, the reliability of MG, which is defined as the MG's ability to continuously supply the local loads, may be endangered. It is worth mentioning that the unavailability of renewable resources is one of the main reasons for the power outage of MGs (especially standalone ones), which may finally lead to MG instability and load rejections. In other words, the stable operation of MG and continuity of its supply, without load shedding, relies on the availability of its resources. This motivates MG operators to use fossil-based generation, i.e., oil and gas, along with renewable resources to enhance MG flexibility and resiliency. To generate electricity from gas and oil fuels, reciprocating engines, gas turbines, and CHPs are commonly exploited with SSSGs. In addition, using SSSGs increases the MG's mechanical inertia and enhances its stability during perturbations. Furthermore, since renewable generations, especially PV and wind turbines, may cause serious power variations in the MG, SSSGs can be effectively exploited to alleviate these oscillations.

In addition to the widespread use of SSSGs with fossil fuel-based prime movers, they are also deployed with various renewable energy sources. Notably, a considerable global installed capacity of SSSGs is associated with renewable resources, such as hydropower [23]. Despite their high reliability, the applicability of SSSGs may be limited in some cases due to their dependence on fossil fuels, especially when environmental concerns, fuel costs, and long-term sustainability are taken into account. Furthermore, the use of SSSGs with certain renewable resources (e.g., hydropower) is site-dependent and may not be feasible in all locations. Therefore, the hybrid integration of IBDGs using renewable energy sources alongside SSSGs presents an effective and practical solution, combining the stability benefits of SSSGs with the flexibility and sustainability of renewable-based generation.

Consequently, in modern MGs, the hybrid utilization of both resource types is considered an appropriate approach to enhance supply reliability, reduce fossil fuel consumption, and mitigate environmental impacts. This hybrid structure leverages the inherent stability and inertia characteristics of SSSGs while benefiting from the efficiency and sustainability of renewable energy sources, ultimately leading to improved system flexibility, operational robustness, and overall MG performance under various operating conditions.

1.3 Small-Scale SGs

1.3.1 Fundamentals

The fundamentals and basic structure of SSSGs are mostly similar to those of the large-scale SGs, which contain a stationary stator and a nonstationary wound rotor as shown in Fig. 1.13. The SSSG stator includes the armature core, which is mounted into a cast-iron frame. Furthermore, the armature conductors are placed in the slots on the inner periphery of the stator as illustrated in Fig. 1.13. The rotor part operates similarly to a large magnet where its *N* and *S* poles are located in the outer edges of the rotor. The magnet poles can be excited employing various techniques and sources, which may be completely different from those of the large-scale SGs.

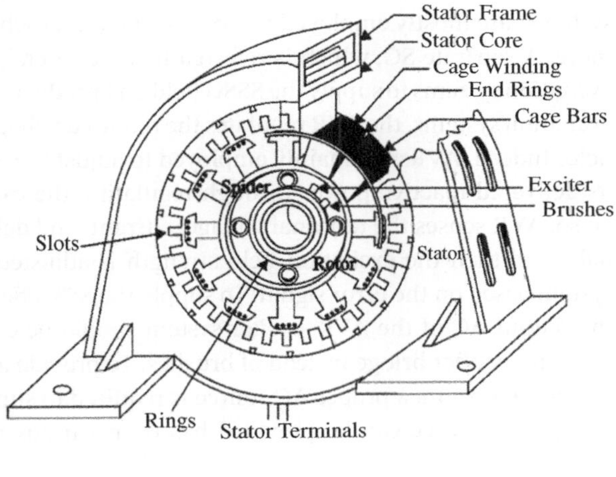

Fig. 1.13 Structure of a typical SSSG. [26] / NIT.

Commonly, SSSGs can be categorized into three major groups based on their excitation source: (i) self-excited (SE), (ii) internally excited (IE), and (iii) permanent magnet excited [25, 26]. These technologies are described in the following sections. On the other hand, based on the electromagnetic laws, a rotating magnetic field is required for producing a three-phase voltage in the armature conductors. Therefore, the SSSG rotor shaft and the magnet poles should be rotated by a prime mover. To do so, various renewable and nonrenewable energies may be employed, including hydropower, geothermal, solar thermal irradiation, biomass, gas, oil, etc. Therefore, the type and technology of the prime mover and turbine may be different for each type of the mentioned energies. The excitation system and prime mover are also briefly described in the following sections.

1.3.2 Excitation System

To generate a rotating field flux, the excitation system injects DC into the SSSG rotor windings. To do so, various methods have been employed for different SSSGs, which can be categorized into two major groups referred to as "direct excitation" and "indirect excitation." In the first technique, the field is directly powered by the SSSG stator through brushes and slip rings, which are mainly employed to convert the stator AC voltage to the DC voltage. Large DC field currents can be properly generated by the direct excitation configuration [27]. However, this technique commonly requires significant periodic maintenance due to its applications of the brushes, and this can be considered the main challenge of this excitation system. It should be noted that in this method, the automatic voltage regulator (AVR) capacity should be determined based on the nominal current of the SSSG field [25, 26].

With the recent development in the excitation system, indirect excitation systems in which the brushes have been removed (which is also known as brushless systems) are mostly employed for SSSGs, and direct schemes are not utilized anymore. A separate SG, which is referred to as "exciter," is required in the indirect excitation systems to supply the SSSG field and produce a suitable flux. For indirect excitation systems, the AVR regulates the required voltage for supplying the exciter field. Indeed, AVRs are mainly employed to adjust the SSSG terminal voltage and its delivered reactive power through regulating the excitation field strength. To do so, AVR senses the terminal voltage, current, and delivered reactive power signals, and then the excitation field strength is adjusted by a closed-loop control system based on the error signal. To supply the SSSG field using indirect schemes, the output AC of the AVR–exciter system should be converted to DC through a rotating rectifier bridge instead of brushes. To provide a suitable DC input for the excitation system, a proper AC source is required to supply the excitation system. The power source can be provided based on various methods, including PMG,

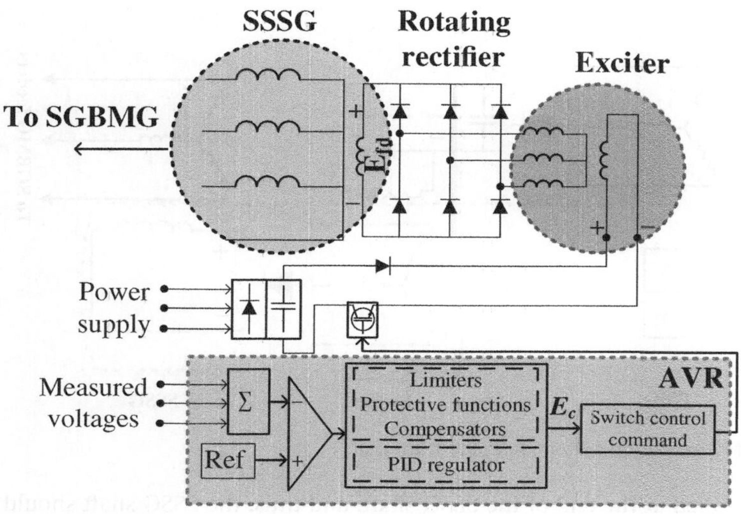

Fig. 1.14 Equivalent model of SSSG indirect excitation system [25, 26].

shunt excitation, and auxiliary winding regulation excitation principle (AREP) techniques [25, 26].

The complete model of various subsystems and components of the SSSG indirect excitation system is demonstrated in Fig. 1.14. As shown, the digital AVR contains various functional blocks, including regulators, limiters, compensators, etc. [27, 28]. However, some of these components may be overlooked for SSSGs.

The regulator is the main subsystem of the excitation system adopted to regulate the SSSG terminal voltage and reactive power depending upon the SSSG control mode. An error signal obtained based on comparing the reference value *Ref* with the measured signal is provided to the regulator. Afterward, the exciter field voltage E_{fd} is adjusted by the regulator output E_c. It should be noted that the grid-connected SSSGs commonly operate under the constant reactive power mode. Moreover, various limiters, including underexcitation limiter and overexcitation limiter, are used in the regulator to restrict its output during abnormal conditions of the generator and upstream system. If these limiters cannot preserve generator variables within the permissible ranges, the related protection functions should operate to prevent any generator damage. These functions shall be coordinated with the control limiters to enhance the security and dependability of the protection system.

1.3.2.1 Permanent Magnet Generator

PMG is the most widely used system among the externally supplied excitation techniques. In the PMG excitation method, the exciter is powered by a PMG,

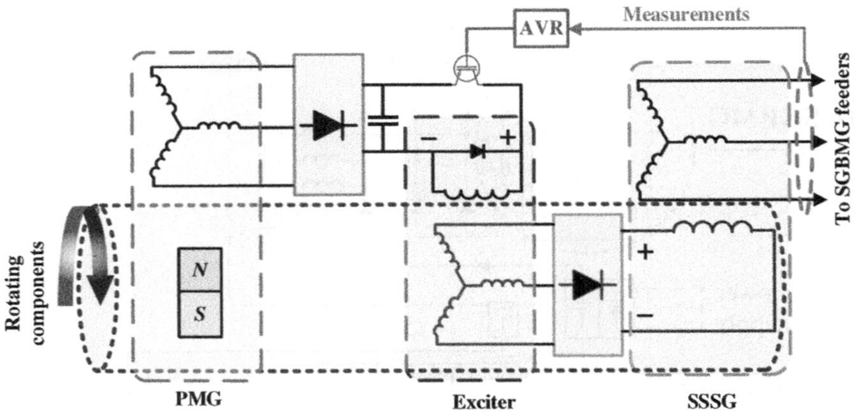

Fig. 1.15 PMG excitation system. Adapted from [25].

which is installed at the end of the SSSG shaft, and thus, the SSSG shaft should be extended for mounting the PMG. The schematic diagram of a PMG excitation system is depicted in Fig. 1.15 [26]. The PMG generates a proper voltage value for supplying the exciter, regardless of the SSSG main winding, and thus, the exciter supply would not be affected by the SSSG terminal voltage drop or system disturbances and faults. Moreover, it can provide a suitable low-voltage ride-through capability for the SSSG, which leads to injecting the required reactive power during the startup of large induction motors, fault-induced voltage recovery conditions, system faults, and other abnormal conditions. Since the PMG contains a permanent magnetic field, it has the startup capability after a long-duration shutdown without requiring any external sources, which is called the black-start capability. The main disadvantage of the PMG excitation system is the SSSG shaft extension for installing the PMG and additional parts. This leads to a length increment of the SSSG package, which decreases the mechanical parts reliability. Furthermore, the PMG system can increase the capital expenditure and maintenance costs of the SSSG package.

1.3.2.2 Shunt Excitation

To reduce the capital expenditure and maintenance costs, self-excitation techniques such as the shunt-excitation method may be preferred in some systems. The shunt method is the simplest technique for supplying the excitation system in which the exciter is supplied by the SSSG terminals. In other words, the shunt system provides both AC supply power and the measured signals required by the AVR controller to adjust the SSSG field flux, as illustrated in Fig. 1.16 [26].

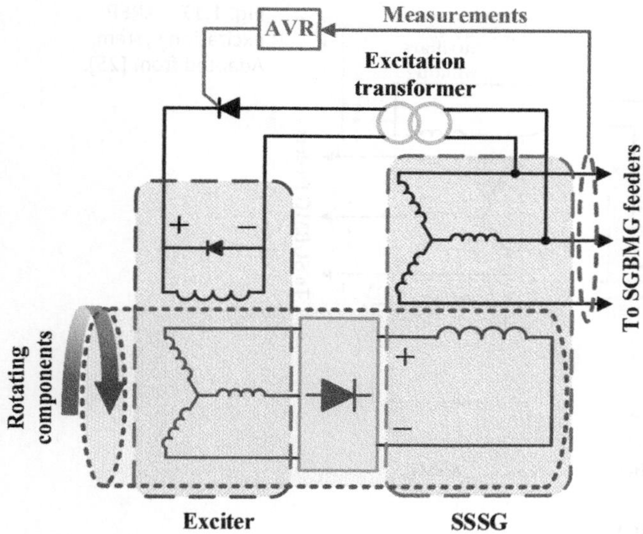

Fig. 1.16 Shunt-excitation system. Adapted from [25].

Therefore, the power of the excitation system is highly affected by the system events, disturbances, and even load dynamics such as induction motor starting. Furthermore, after severe voltage drops, the recovery time of the shunt system is much slower than that of the other methods. Furthermore, the highly nonlinear dynamic loads, which produce considerable harmonics, may have adverse impacts on the operation of the shunt method.

A proper residual magnetic flux should exist for starting the SSSGs equipped with the common types of shunt systems. This residual flux is normally maintained in both the exciter and SSSG fields from the previous operation, and its amount decays over time. Consequently, after a long period of time, the SSSG using a shunt system may not be able to start and build a suitable voltage at its terminal. To improve the performance of SSSG during the black-starting procedure, periodic maintenance and operation should be performed. After long shutdown periods and when the residual flux decays to the point at which the SSSG cannot properly start and deliver output power, the field should be "flashed" through an external source. Meanwhile, the time period in which this residual flux can be properly employed for starting the SSSG mainly depends on the construction materials. Although this method is cheap and simple, its improper performance during voltage drops and short-circuit faults is arguable. Therefore, this technology is only employed in SSSG rated up to 700 kVA [24–28].

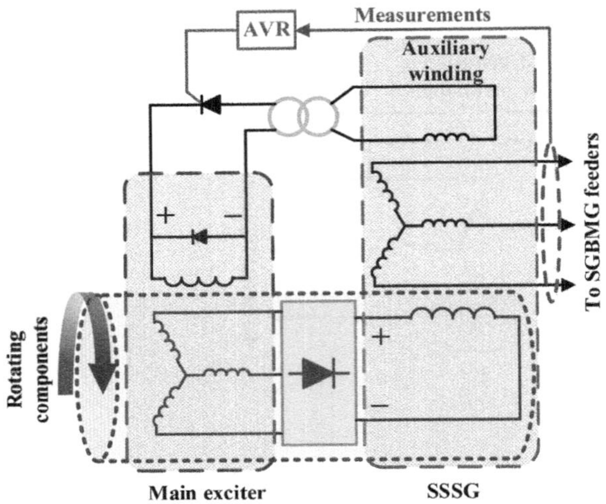

Fig. 1.17 AREP excitation system. Adapted from [25].

1.3.2.3 Internally Excited

To resolve the shortcomings of the shunt-excitation method, especially its challenging performance during voltage drop, a separate single-phase auxiliary winding can be utilized in the stator slots of the SSSG, as shown in Fig. 1.17 [25]. This method is referred to as IE or AREP [25, 26]. In this method, the output voltage of the auxiliary winding is utilized for powering the exciter, and thus, it is not affected by system disturbances in this condition. In fact, similar to PMG, this technique can provide a suitable performance during the voltage-degraded conditions.

To enhance the capability and reliability of the AREP method, some manufacturers use the combination of the main and auxiliary windings of the SSSG stator or a pair of auxiliary windings to supply the exciter. The auxiliary additional windings, which are located within the stator slots, are electrically isolated from the main windings through an insulation process. Unlike the shunt-excitation method, AREP is widely employed for the SSSGs rated larger than 1 MVA. Since additional space is required for locating the auxiliary winding, using the AREP system may result in increasing the dimensions of the SSSG. Although the AREP is cost-effective compared to the PMG excitation method, it does not support black-start capability due to the lack of remnant flux [25, 26].

1.3.2.4 Dynamic Model

As suggested by the standard IEEE Standard 421.5, the AVR excitation system can be dynamically modeled using the IEEE-AC5A whose block diagram is shown in Fig. 1.18. Definition of each parameter and its typical value is presented by this standard [28].

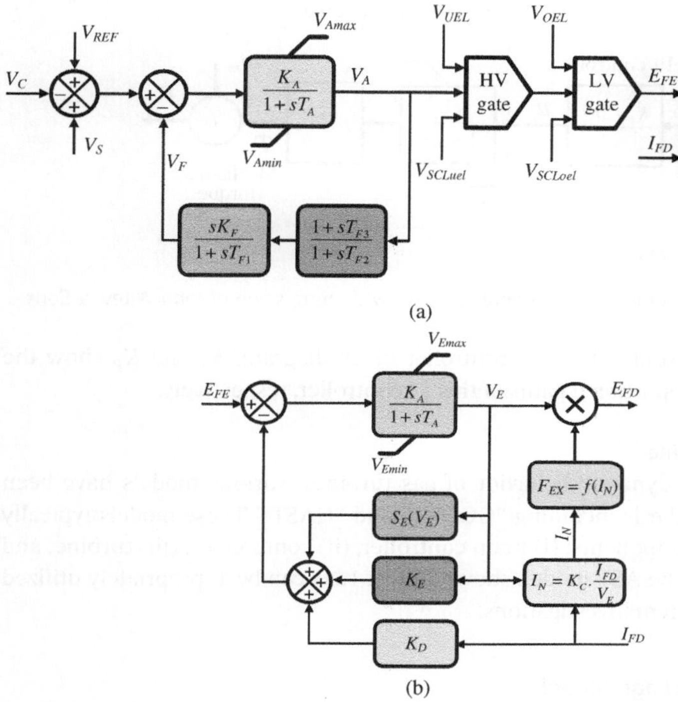

Fig. 1.18 Block diagram of excitation system: (a) AVR and (b) exciter. Adapted from [28].

1.3.3 Prime Mover

Gas turbines and reciprocating internal combustion engines are the widely used prime movers of SSSGs whose dynamic models are discussed in the following sections.

1.3.3.1 Reciprocating Internal Combustion Engines

The dynamic model of a combustion engine mainly includes three main parts: (i) the regulator (proportional-integral [PI] controller), (ii) the actuator, and (iii) the engine's combustion [29]. The PI controller is adopted in the governor system to regulate the amount of fuel injection. The main purpose of this controller is to adjust the SGBMG frequency under the islanded mode operation and SSSG output power under the grid-connected mode operation. The simplified block diagram of the governor–prime mover system is depicted in Fig. 1.19 [29]. As shown, the actuator of the fuel injection system is represented by a first-order transfer function with the time constant τ_1. In addition, the engine dead time is modeled by a time delay block with the time constant τ_2. Furthermore, the

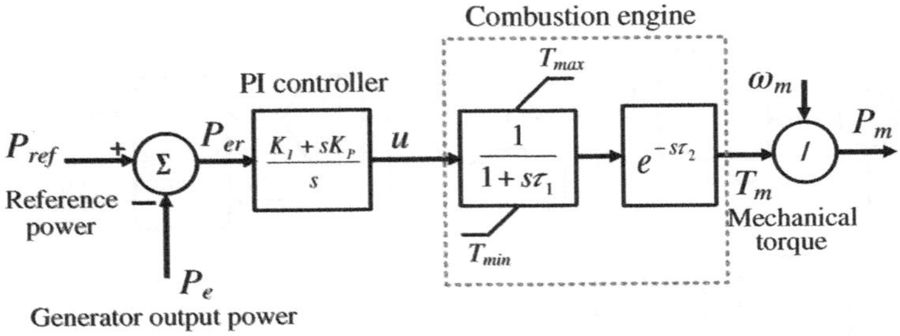

Fig. 1.19 Block diagram of diesel engine. [29] / with permission of John Wiley & Sons.

governor is represented by a PI controller block diagram. K_I and K_P show the integrator and proportional gains of this PI controller, respectively.

1.3.3.2 Gas Turbine

To represent the dynamic behavior of gas turbines, various models have been proposed in standards including "GGOV1" and "GAST." These models typically consist of four components: (i) main controller, (ii) compressor, (iii) turbine, and (iv) sensor [30]. The AST model, shown in Fig. 1.20, can be appropriately utilized for protection system investigations.

1.3.4 SSSG Package Model

To investigate and simulate the SGBMG behavior, a suitable dynamic model for the SSSG package should be employed, including the model of SSSG, excitation system, and prime mover. The relation between these parts is shown in Fig. 1.21. The dynamic model of SSSG includes two major parts: (i) electrical system and

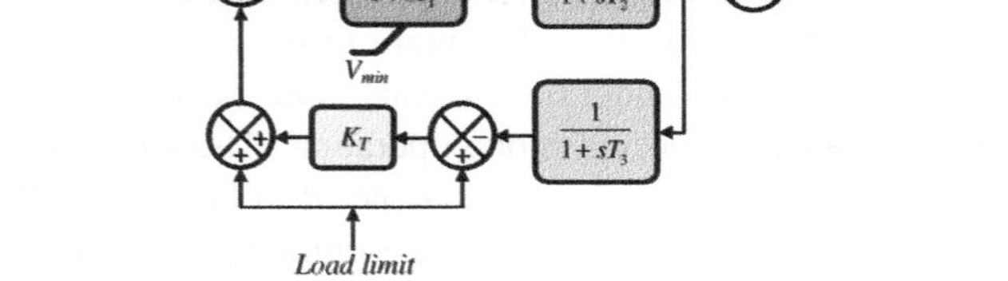

Fig. 1.20 Block diagram of a gas turbine. [30] / with permission of Elsevier.

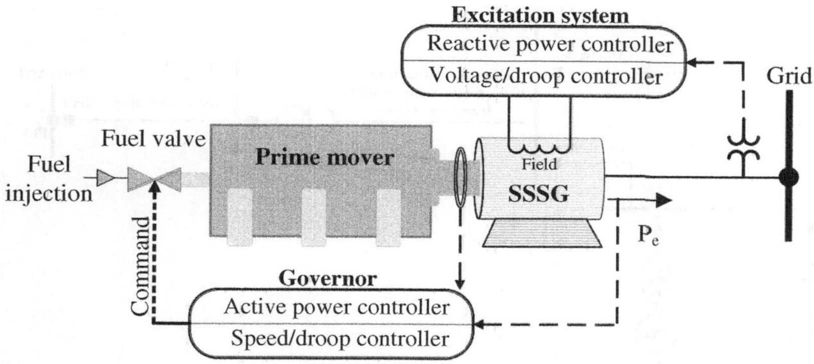

Fig. 1.21 SSSG package model. Adapted from [31].

(ii) mechanical system. The complete model of an SG is presented in [31], which can be properly employed for SSSGs. Similar to the other SGs, electrical and mechanical systems can be appropriately represented by sixth- and second-order models, respectively. It should be noted that the dynamic model of excitation system and prime mover has been presented in the previous sections.

1.4 Microgrid Structure

A typical configuration of an MG is illustrated in Fig. 1.22. As shown, MG contains DGs, local static and dynamic loads, and other power equipment, such as transformers, lines, and capacitor banks. Furthermore, for proper operation of the MG, some other equipment, including controllers, protection systems, communication systems, micro phasor measurement units (μPMUs), and monitoring systems are required, as shown in Fig. 1.22. The MG structure and the associated subsystems are briefly explained as follows.

The power equipment, which is employed to collect the produced electrical active/reactive power from DGs and supply the local loads or to exchange active/reactive power with the upstream system, is known as the "collector system." The main equipment of the collector system includes power transformers, switchgears, feeders, cables, and reactive compensation devices. The measurement devices for measuring the injected/consumed energy, active power, and reactive power are located at the point of measurement (POM). Moreover, an individual DG or groups of DGs are connected to the MG at the point of connection (POC), as shown in Fig. 1.22. In other words, POC is the electrically closest point in the MG at which an individual DG or groups of DGs are connected to the MG, where the operational requirements and technical specifications shall be

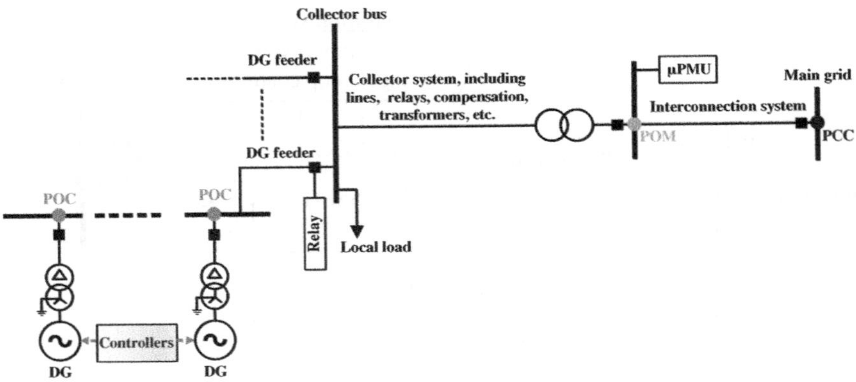

Fig. 1.22 Typical configuration of an MG. Adapted from [32].

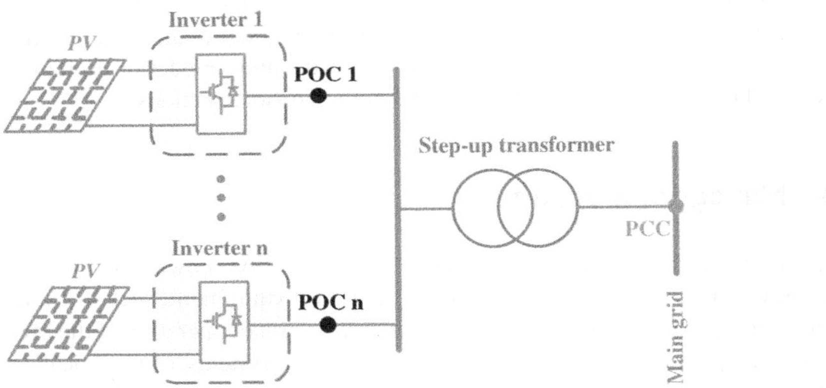

Fig. 1.23 Typical structure of a grid-connected PV plant.

satisfied including fault-ride-through, and control, and protection requirements. Moreover, this point may be equipped with appropriate switching equipment for connecting and disconnecting the DG unit. Under conditions where multiple DG units exist in a local network, several connection points for distributed generation units may exist. Examples of MGs with multiple POCs are shown in Fig. 1.22 and Fig. 1.23. Moreover, the POC could be located on either side of the unit transformer considering its definition. The whole MG, including the collector system, DGs, and loads, is connected to the upstream system at the point of interconnection and through an interconnection system such as transmission lines. It is worth mentioning that according to the IEEE Standard 1547 [33], this interconnection point is known as the "point of common coupling" (PCC). In fact, PCC is the point at which the MG is connected to the main grid, where other consumers or DG units may also connect to the same point of the main network. This point could be considered the electrical boundary (not the ownership boundary) between the local network, i.e., MG, and the upstream network. By opening the switch at the PCC, the MG becomes electrically separated from the main grid, and the grid configuration returns to a condition similar to the state before MG connection.

The local loads of MG can be divided into essential and nonessential loads. The priority of the control and protection systems is to supply essential loads in both the islanded and grid-connected operation of the MG. Moreover, the connection point of the DGs may be specified considering various factors such as (i) maximizing the reliability of the essential loads, (ii) minimizing the voltage drop, (iii) minimizing the loss, and (iv) maximizing MG stability.

1.4.1 Control System

MG may operate under the grid-connected or standalone (islanded) modes. In the grid-connected mode, MG is coupled with the distribution system through the PCC breaker where the energy can be exchanged between the MG and the upstream system. After disconnection from the upstream system by opening of the PCC breaker, the MG load demand should be supplied by the local resources, i.e., DGs. Moreover, the MG voltage and frequency are regulated by the DGs. If the DG generation is not sufficient to supply all of the loads, proper control/protection actions should be performed to retain the MG stability.

Two types of control systems are commonly utilized to provide a suitable operation of the MG during both the grid-connected and islanded mode operation, including the self-controller of DG and central controller of MG. Self-controllers of each DG independently regulate frequency/active power and voltage/reactive power based on the local measurement data and without any communication with the other controllers. For instance, during the grid-connected mode of the SSSG, its governor system commonly operates under the constant active power control mode or active power/frequency droop mode. Moreover, in this mode, its AVR–excitation system operates under the constant reactive power, constant voltage, or reactive power/voltage droop mode. After an islanding occurrence, the voltage and frequency of the MG should be directly regulated by one of the DGs, which is referred to as an "ISO synchronous controller" operation. Moreover, the other DGs can continue their operation under the constant active/reactive power or droop control modes. Detailed descriptions of the control systems and operating modes of SGBMGs and IBMGs are provided in Chapters 2 and 3 of this book.

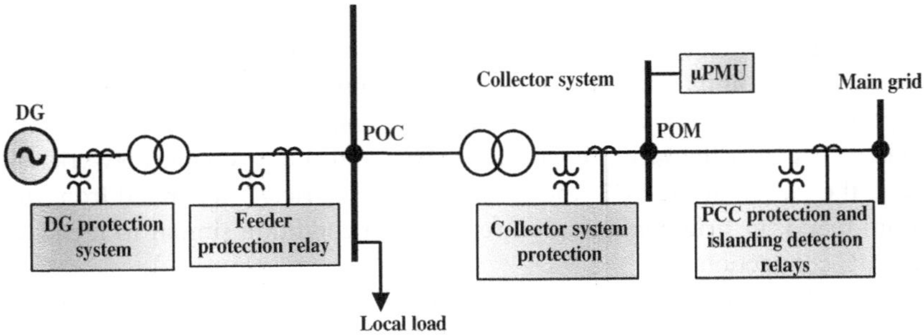

Fig. 1.24 MG protection system.

The central controller of MG is connected to the self-controllers via suitable communication links. The main function of central control is to coordinate the self-controllers to (i) enhance the MG stability, (ii) improve the reliable supply of essential loads, and (iii) provide energy optimization and loss minimization. To do so and provide proper coordination between the self-controllers, the central controller will determine the operation mode and reference values of the self-controllers. Moreover, the power sharing between DGs, voltage profile, and MG frequency will be regulated by the central controller by sending suitable commands to the self-controllers. In addition, the required input signals of the central controller will be transmitted from various points of the MG through μPMUs and suitable communication links.

1.4.2 Protection System

The MG should be equipped with a proper protection system to retain MG stability and prevent mechanical and thermal damage to the equipment. The common configuration of the MG protection system is depicted in Fig. 1.24. Generally, the MG protection system can be divided into three main parts: (i) DG protection relays, (ii) collector system protection relays, and (iii) PCC protection system relays. The DG protection relay should properly cover the following fault conditions [33–36]:

- Internal faults of the DG
- Abnormal operation conditions such as over-frequency, overvoltage, and loss of synchronism
- Backup protection of the collector system faults.

Moreover, the DG protection relays should remain stable during the nonfault disturbances, such as power swings and load switching.

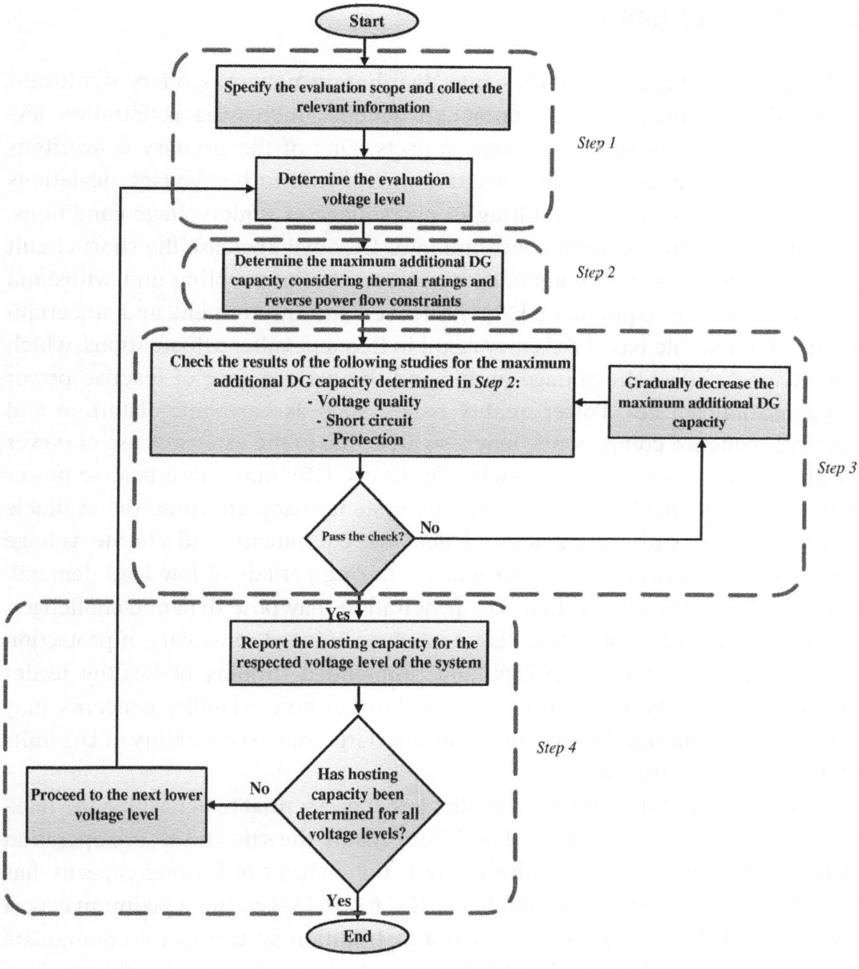

Fig. 1.25 Flowchart for determining the hosting capacity of a distribution network.

The main function of the collector system protection is to clear the short-circuit faults between the POC and the PCC. In addition, it provides a suitable backup protection for short-circuit faults, which occur in power equipment such as transformers. The protection relay of the PCC provides a suitable backup for the faults occurring in both the upstream system and the MG feeders. Moreover, this relay should detect the unintentional islanding condition of the MG. More detailed information about MG and DG protection relays is provided in the next chapters of the book.

1.5 Hosting Capacity

Although the integration of DGs into distribution networks offers significant technical, economic, and environmental benefits, increasing penetration levels may give rise to several adverse impacts. One of the primary concerns is voltage regulation, as high DG penetration can lead to bus voltage deviations beyond permissible limits, resulting in overvoltage or undervoltage conditions. In addition, the fault current contribution of DGs may increase the short-circuit capacity of the network, potentially exceeding the interrupting and withstand ratings of existing equipment [37]. Moreover, the intermittent and uncertain nature of renewable-based DGs may result in frequent voltage fluctuations, which can adversely affect the voltage profile and the performance of reactive power compensation devices. Power quality issues, such as harmonic distortion and negative sequence components, may also arise due to the extensive use of power electronic interfaces. At high penetration levels, DGs may cause reverse power flow toward upstream networks, increase system losses, consume the available thermal and operational capacity of network equipment, and elevate voltage magnitudes at network buses, particularly during periods of low-load demand. Beyond these static effects, high DG penetration may pose dynamic challenges. These include protection-related issues such as reduced sensitivity of protection schemes, coordination difficulties, and unintended tripping of healthy feeder relays due to faults on adjacent feeders. Furthermore, stability concerns may emerge, including small-signal and transient (large-signal) instability of DG units under system disturbances.

To address the aforementioned challenges, it is essential to determine the maximum allowable penetration level of DGs to ensure the safe and secure operation of the distribution system. In this context, the concept of hosting capacity has been introduced and formally defined in IEC 63276 [38] as "the maximum export capacity of distributed generation that a distribution system can accommodate while operating safely and reliably." Based on the issues discussed in the previous section, the determination of DG hosting capacity requires consideration of both the static and dynamic characteristics of the power system. The primary objective of hosting capacity assessment is to support the planning and integration of distributed energy resources in a manner that is consistent with actual system operating conditions and technical constraints. Notably, the hosting capacity of a power network is not a fixed value; rather, it is a time-varying quantity influenced by factors such as time of day, seasonal variations, and consumer behavior. Consequently, hosting capacity evaluation should be conducted within a time-dependent framework to accurately capture the impacts of load variability and generation uncertainty.

Depending on the performance indices considered in hosting capacity analysis, different evaluation approaches may be required. For certain indices, a worst-case scenario-based assessment may be appropriate, whereas for others, probabilistic metrics, such as the severity and frequency of constraint violations, should be employed. Accordingly, numerous standards, textbooks, and research studies have proposed both deterministic and probabilistic methodologies for assessing the hosting capacity of power systems.

The framework proposed in IEC 63276 follows a straightforward deterministic approach, the flowchart of which is illustrated in Fig. 1.25 [38]. This methodology can be easily applied to power networks for hosting capacity assessment. The evaluation is conducted in a hierarchical manner, starting from higher voltage levels and proceeding toward lower voltage levels. After defining the evaluation scope and the voltage level under study (*step* 1), the maximum thermal capacity of system equipment is assessed. Subsequently, under minimum load conditions, the maximum allowable reverse power flow from DGs is evaluated to ensure compliance with thermal capacity constraints at each voltage level (*step* 2). Once the maximum DG capacity is determined from a thermal standpoint, additional constraints, including voltage quality, short-circuit levels, and protection system performance, are examined in *step* 3. Voltage quality assessments are performed in accordance with IEC 62786 [39] and IEC 62749 [40] and generally include evaluation of (i) voltage magnitude deviations from nominal values, (ii) voltage harmonic distortion, and (iii) negative sequence voltage components. It should be noted that negative sequence voltages are primarily introduced by single-phase DGs connected to low-voltage (secondary) distribution networks. Short-circuit studies involve determining the maximum breaking, thermal, and mechanical short-circuit currents, as specified in IEC 60909 [41]. In addition, the performance of protection systems should be verified to ensure reliable fault detection without loss of sensitivity, while maintaining security under nonfault conditions.

If any of the constraints related to voltage quality, short-circuit levels, or protection coordination are violated, the DG capacity should be gradually reduced until all requirements are satisfied. After finalizing the maximum permissible DG capacity at a given voltage level, the algorithm proceeds to the next lower voltage level. For a specified system zone, the highest voltage level is selected as the starting point. Furthermore, the hosting capacity at a lower voltage level must not exceed that of the corresponding higher voltage level; otherwise, it should be reduced accordingly to maintain hierarchical consistency.

It should be noted that the IEC 63276 framework does not explicitly consider stability constraints, including small- and large-signal stability. However, in practical applications, stability issues may limit the safe and reliable operation of DGs in synchronism with the distribution system and, therefore, represent a critical factor in hosting capacity determination.

Based on the foregoing discussion, accurate determination of DG hosting capacity is essential for effective distribution system planning and operation. Hosting capacity should be evaluated and periodically updated while considering a comprehensive set of system aspects, encompassing both steady-state and dynamic phenomena. Moreover, such analyses require appropriate and detailed models of system components, particularly DG units and their associated control systems. Accordingly, subsequent chapters of this book focus on the protection and stability aspects, which play a key role in hosting capacity assessment, and provide detailed modeling approaches for system equipment, especially SSSGs and IBDGs, including their control structures.

1.6 Summary

In this chapter, a brief overview of small-scale power generation technologies, including solar irradiation, wind, hydropower, geothermal, biomass, and gas, has been presented. Various types of DGs can be employed with these resources, including SSSGs, IBDGs, and IGs. As discussed, modern MGs incorporate hybrid energy resources, combining both SSSGs and IBDGs, to ensure reliable power supply, smooth power generation, reduce operational costs, minimize the dependence on nonrenewable sources, and mitigate environmental impacts. Because both technologies play essential roles in contemporary MGs, this book addresses the protection and stability challenges associated with SSSGs and IBDGs, along with their corresponding solutions.

Problems

1.1 A sample SGBMG is illustrated in Fig. 1.26, which contains an SSSG, some static loads, and an induction motor. It is assumed that the SGBMG operates under the grid-connected mode. The parameters of this SGBMG are given in Table 1.4.

a) Simulate this system through an appropriate software, such as PSCAD/EMTDC, DIgSILENT, and ETAP, considering the following guidelines:
   - Use the library components of the software to model the SG, induction motor, and static load.
   - The block diagram shown in Fig. 1.18 can be exploited to model the AVR–excitation system of the SSSG. It is assumed the PMG-type excitation system is utilized and the field voltage of the SSSG is not affected by system disturbances.

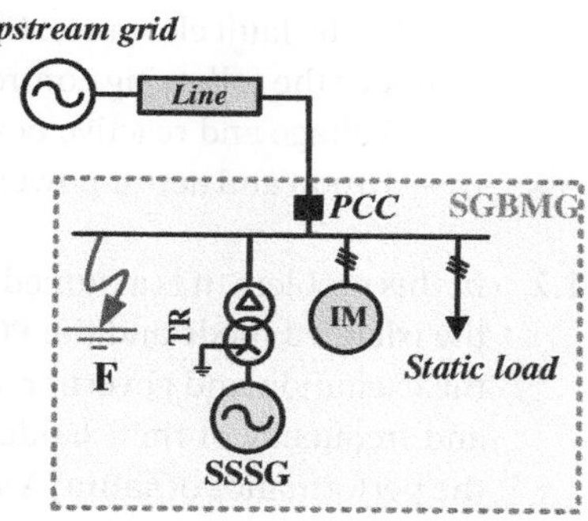

Fig. 1.26 Sample SGBMG and distribution system.

Table 1.4 System parameters.

Equipment name	Parameters values
Line	Length = 5 km, impedance = $0.35 + j0.4$ (Ω/km)
Static loads	$P = 4$ MW, PF = 0.95
Induction motor (IM)	Nominal power = 5 MW, inertia = 0.4 s
SSSG	Voltage = 11 kV, $P = 10$ MW, inertia = 0.57 s $X_d = 1.9$ pu, $X'_d = 0.28$ pu, $X''_d = 0.18$ pu, $X_q = 0.99$ pu, $X''_q = 0.18$ pu
Transformer (TR)	Ratio = 20/11 kV, $S = 15$ MVA, $U_k = 10\%$
Upstream grid	Voltage = 20 kV, short-circuit capacity = 170 MVA, $X/R = 5$

- The block diagram of the diesel engine and its PI controller shown in Fig. 1.19 can be utilized to model the turbine–governor system.
- The AVR may operate under the voltage or reactive power control mode. In addition, the governor may operate under the droop or active power control mode.
- Use typical parameters for the turbine, which can be found in [29].
- Use typical parameters for the excitation system, which can be found in [28].
- Tune the AVR and governor controllers to provide stable steady-state operation of the SGBMG.

b) The control mode of the SSSG has considerable impacts on its behavior and stability, especially after short-circuit faults. To quantify the SSSG stability, the critical clearing time (CCT) index can be exploited. CCT is the maximum time interval of a three-phase fault at F for which the SSSG remains stable and its angular frequency recovers to the nominal value after the fault clearance. Compare the CCT of the SSSG shown in Fig. 1.26 under the following control modes:

- Voltage and reactive power control modes for the excitation system.
- Droop and active power control modes for the governor system.

1.2 In this problem, it is assumed that the SGBMG shown in Fig. 1.26 operates at the islanded mode and the PCC breaker is at the open position. In this case, the excitation and governor systems of the SSSG should control the voltage and frequency of the islanded SGBMG, respectively. In this case, compare the performance of shunt, AREP, and PMG excitation techniques for the following scenarios:

- SSSG increased current up to 300% of nominal current during short-circuit faults and high overload conditions.
- Motor starting.
- A three-phase short-circuit fault at F, which is cleared after 250 ms.

Meanwhile, present the SGBMG voltage, field current, and voltage for each case. Moreover, for the short-circuit scenario, investigate the SGBMG stability for the mentioned excitation systems.

1.3 Using the implemented system of Problem (1.1):

a) Compare the SSSG CCT amounts for various values of governor controller parameters.
b) Compare the SSSG CCT amounts for various values of AVR controller parameters.

References

1 Z. Li, M. Shahidehpour, F. Aminifar, A. Alabdulwahab, and Y. Al-Turki, "Networked microgrids for enhancing the power system resilience," *Proc. IEEE*, vol. 105, no. 7, pp. 1289–1310, 2017.

2 M. N. Alam, S. Chakrabarti, and A. Ghosh, "Networked microgrids: State-of-the-art and future perspectives," *IEEE Trans. Ind. Inform.*, vol. 15, no. 3, pp. 1238–1250, 2018.

3 J. Yang, A. Paudel, and H. B. Gooi, "Compensation for power loss by a proof-of-stake consortium blockchain microgrid," *IEEE Trans. Ind. Inform.*, vol. 17, no. 5, pp. 3253–3262, 2021.

4 W. El-Khattam and M. M. Salama, "Distributed generation technologies, definitions and benefits," *Electr. Power Syst. Res.*, vol. 71, no. 2, pp. 119–128, 2004.

5 Zaheeruddin and M. Manas, "Analysis of design of technologies, tariff structures, and regulatory policies for sustainable growth of the smart grid," *Energy Technol. Policy*, vol. 2, no. 1, pp. 28–38, 2015.

6 M. Hemmati *et al.*, "Economic-environmental analysis of combined heat and power-based reconfigurable microgrid integrated with multiple energy storage and demand response program," *Sustain. Cities Soc.*, vol. 69, p. 102790, 2021.

7 P. K. Mianaei, M. Aliahmadi, S. Faghri, M. Ensaf, A. Ghasemi, and A. A. Abdoos, "Chance-constrained programming for optimal scheduling of combined cooling, heating, and power-based microgrid coupled with flexible technologies," *Sustain. Cities Soc.*, vol. 77, p. 103502, 2022.

8 S. Pramanik and R. V. Ravikrishna, "A review of concentrated solar power hybrid technologies," *Appl. Therm. Eng.*, vol. 127, pp. 602–637, 2017.

9 M. H. Ahmadi *et al.*, "Solar power technology for electricity generation: A critical review," *Energy Sci. Eng.*, vol. 6, no. 5, pp. 340–361, 2018.

10 T. Panda *et al.*, "Losses in bifacial PERC solar cell due to rear grid design and scope of improvement," *Sustain. Energy Techn. and Assess.*, vol. 52, p. 102280, 2022.

11 M. Sanaye-Pasand and R. Bekhradian, "Technical, economic, and environmental assessment of implementing a pilot floating solar power plant in Iran, and development of a financial model," Technical Report, Satba, Ministry of Energy, Iran, 2025. Available: https://www.satba.gov.ir/.

12 X. Wu, C. Li, L. Shao, J. Meng, L. Zhang, and G. Chen, "Is solar power renewable and carbon-neutral: Evidence from a pilot solar tower plant in China under a systems view," *Renew. Sust. Energ. Rev.*, vol. 138, p. 110655, 2021.

13 A. Dabbaghiyan, F. Fazelpour, M. D. Abnavi, and M. A. Rosen, "Evaluation of wind energy potential in province of Bushehr, Iran," *Renew. Sust. Energ. Rev.*, vol. 55, pp. 455–466, 2016.

14 J. L. Domínguez-García, O. Gomis-Bellmunt, F. D. Bianchi, and A. Sumper, "Power oscillation damping supported by wind power: A review," *Renew. Sust. Energ. Rev.*, vol. 16, no. 7, pp. 4994–5006, 2012.

15 A. Honrubia-Escribano, E. Gómez-Lázaro, J. Fortmann, P. Sørensen, and S. Martin-Martinez, "Generic dynamic wind turbine models for power system stability analysis: A comprehensive review," *Renew. Sust. Energ. Rev.*, vol. 81, pp. 1939–1952, 2018.

16 S. Wang and S. Wang, "Impacts of wind energy on environment: A review," *Renew. Sust. Energ. Rev.*, vol. 49, pp. 437–443, 2015.

17 K. Dai, A. Bergot, C. Liang, W. N. Xiang, and Z. Huang, "Environmental issues associated with wind energy–A review," *Renew. Energy*, vol. 75, pp. 911–921, 2015.

18 John W. Lund, Characteristics, development and utilization of geothermal resources, *GHC Bulletin*, 2007. PP. 1–9. https://www.researchgate.net/profile/John-Lund-6/publication/226005395_Development_and_Utilization_of_Geothermal_Resources/links/5705604b08aef745f7176af4/Development-and-Utilization-of-Geothermal-Resources.pdf.

19 Y. Situmorang, Z. Zhao, A. Yoshida, A. Abudula, and G. Guan, "Small-scale biomass gasification systems for power generation (< 200 kW class): A review," *Renew. Sust. Energ. Rev.*, vol. 117, p. 109486, 2020.

20 D. K. Okot, "Review of small hydropower technology," *Renew. Sust. Energ. Rev.*, vol. 26, pp. 515–520, 2013.

21 "Technology characterization: Reciprocating engines," Environmental Protection Agency, 2008.

22 "Technology characterization: Gas turbines," Environmental Protection Agency, 2008.

23 "Renewable energy statistics," International Renewable Energy (IRENA), 2022. Available: https://www.irena.org/publications/2022/Jul/Renewable-Energy-Statistics-2022.

24 J. Bird, "Electrical and electronics principles and technology, Chapter 37: Alternators," 2021. Available: https://nit-edu.org/wp-content/uploads/2021/09/Ch-37-Alternators.pdf.

25 LEROY-SOMER, "Excitation systems – 10 kVA to 35 MVA alternators," 2020.

26 J. K. Nøland, S. Nuzzo, A. Tessarolo, and E. F. Alves, "Excitation system technologies for wound-field synchronous machines: Survey of solutions and evolving trends," *IEEE Access*, vol. 7, pp. 109699–109718, 2019.

27 CATERPILLAR, "Excitation selections," 2020. Available: https://www.cat.com/en_ZA/by-industry/electric-power/Articles/White-papers/excitation-selections.

28 "IEEE recommended practice for excitation system models for power system stability studies," IEEE Std 421.5-2005, 2006.

29 R. Bekhradian, M. Sanaye-Pasand, M. Davarpanah, and Y. Kor, "Governor controller improvement to enhance SSSG performance in presence of electrical arc furnace," *IET Gener. Transm. Distrib.*, vol. 16, no. 18, pp. 3645–3660, 2022.

30 M. D. R. Greidanus and M. L. Heldwein, "Model-based control strategy to reduce the fault current of a gas turbine synchronous generator under short-circuit in isolated networks," *Electr. Power Syst. Res.*, vol. 204, p. 107687, 2022.

31 M. Abedini, M. Sanaye-Pasand, M. Davarpanah, H. Lesani, and M. Shahidehpour, "Predictive auto-reclosure approach to enhance transient stability of grid-connected DGs," *IET Gener. Transm. Distrib.*, vol. 13, no. 14, pp. 3011–3019, 2019.

32 "IEEE standard for interconnection and interoperability of inverter-based resources (IBRs) interconnecting with associated transmission electric power systems," IEEE Std 2800-2022, 2022.

33 "IEEE standard for interconnection and interoperability of distributed energy resources with associated electric power systems interfaces," IEEE Std 1547-2018 (Revision of IEEE Std 1547-2003), 2018.

34 D. Reimert, *Protective Relaying for Power Generation Systems*. Taylor & Francis, 2006.

35 "IEEE recommended practice for protection and coordination of industrial and commercial power systems," IEEE Standard 242, 2001.

36 "IEEE guide for AC generator protection," IEEE Standard C37. 102, 2006.

37 A. F. Zobaa, S. H. A. Aleem, S. M. Ismael, and P. F. Ribeiro, *Hosting Capacity for Smart Power Grids*. Cham: Springer International Publishing, 2020.

38 "Guidelines for the hosting capacity evaluation of distribution networks for distributed energy resources," IEC TS 63276, 2024.

39 "Distributed energy resources connection with the grid," IEC TS 62786, 2023.

40 "Assessment of power quality - Characteristics of electricity supplied by public networks," IEC TS 62749, 2020.

41 "Short-circuit currents in three-phase a.c. systems," IEC TS 60909, 2016.

2

Stability of Synchronous Generator-Based Microgrids: Fundamentals, Classification, and Assessment

2.1 Introduction

The primary goal of the protection system of a synchronous generator-based microgrid (SGBMG) can be divided into three groups: (i) fast operation during the faulty condition and preventing SGBMG unstable operation and equipment damage, (ii) high security against nonfaulty conditions and during SGBMG stable operation through proper coordination with the SGBMG control system, and (iii) preserving SGBMG stability using suitable protection schemes such as load-shedding and islanding detection. Therefore, various stability types of the modern SGBMGs should be carefully investigated to be able to properly design and adjust the protection system. Meanwhile, the configuration, architecture, and operating mode of the SGBMG control system have serious impacts on the SGBMG stability as well as the performance of the protection system. Therefore, the behavior of the control system and its impacts should also be considered for designing and adjusting the protection system.

Since considering the impact of the control system is crucial for investigating SGBMG stability and designing the protection system properly, this chapter discusses the SGBMG control system including control loops and control modes. Then, it presents applicable definitions and classifications for various types of SGBMGs stability, some of which are entirely different from those of sizeable conventional power systems. SGBMG stability types can be divided into three major groups: (i) control system stability, (ii) load stability, and (iii) power supply stability [1]. Meanwhile, the assessment methodology for some kinds of stabilities may be completely different depending on the SGBMG operation mode, i.e., grid-connected and standalone modes.

In order to study the control system stability, the impacts of the prime mover governor, automatic voltage regulator (AVR), and power-sharing system on the steady-state, dynamic, and transient behaviors of SGBMG are investigated. On the other hand, the stable operation of SGBMG local loads is crucial, and thus, if these loads are intentionally tripped, the SGBMG may no longer be able to perform its primary task and may not even remain stable. In other words, unlike conventional large power systems, the continuous stable operation of local loads should be evaluated for the stability assessment of SGBMGs. Therefore, this chapter carefully investigates the possible circumstances that can engender the local load stability and cease the SGBMG stable operation. Furthermore, it discusses the SGBMG ability to provide a stable power supply within a permissive operation range and maintain a suitable balance between the generation and demand. In this regard, voltage, frequency, and transient stability of the SGBMG have been investigated after the occurrence of perturbations, including short-circuit faults, small-scale synchronous generator (SSSG) outages, fault-initiated islanding, and unintentional loss of main. In addition, the dynamic behavior of local loads, especially induction motors (IMs), which may lead to some types of instabilities, would be studied through different simulation studies.

Due to the strong coupling between the voltage and frequency parameters and the high value of R/X parameter in microgrids (MGs), SGBMG stability assessment may differ from that of traditional large-power systems. Therefore, to cover this concern, suitable analytical methods for evaluating the SGBMG stability have been presented in this chapter. Furthermore, dynamic phenomena that may threaten the SGBMG stability, such as the unbalanced fault-induced delayed voltage recovery (FIDVR), have been introduced and thoroughly studied using simulation studies.

Traditionally, the stability types of the conventional power systems were classified into main three categories: (i) voltage stability, (ii) frequency stability, and (iii) rotor angle stability (transient stability) [2–4]. Although each instability form may finally lead to the other forms of instability in the power systems, the initiating cause of system instability is considered one of the aforementioned categories. Furthermore, for the detection of each instability form, various features have been recommended in standards and grid codes. These features are widely utilized in protection and control systems to take proper actions and prevent power system instability [5].

In modern power systems including smart MGs, the conventional forms of stability and their features may not be sufficient for stability assessment. In fact, some new aspects and forms of stability are recognized in recent years, which cannot be considered a subset of any of the abovementioned categories. Furthermore, the aforementioned categories may occur with new features in modern systems. Based on the recently published research works on modern MG stability and latest grid codes, a new stability classification is presented in [1], which is different from the traditional one. The main reasons, which result in the new forms of MG stability and changing the features of the previous stability forms, include:

- Architecture of MG control systems
- Different operation modes of MGs
- Adjacency of distributed generations (DGs) and loads
- High penetration of unbalance loads
- Behavior of loads including smart and dynamic loads
- High values of resistance to reactance ratio
- Different relationships between voltage magnitude, load angle, and active and reactive powers compared to those of the traditional power systems
- Higher uncertainty of DG operation
- Reactive power capability and voltage disturbance ride-through capability of DGs.

Based on these explanations, MG control systems, load behavior, and reactive power capability of DGs have significant impacts on the MG stability and protection system. Therefore, in this section, at first, the requirements of SGBMG design, interconnection, and operation are presented. Then, the control system architecture of MGs, including the primary and secondary controls, is explained. Finally, various stability aspects of the modern SGBMGs are described and investigated.

2.2 Control System of SGBMGs

In different countries, specific national grid codes have been provided for design, operation, and grid connection of MGs. MGs are mostly connected to the distribution system level, i.e., low- and medium-voltage networks, and thus, these grid codes are usually organized by the local electrical energy distribution companies. Moreover, several international standards and codes for the purpose of design, operation, control, and interconnections of MGs are prepared by the professional associations, committees, and task forces, such as IEEE, IEC, and CIGRE. IEEE 1547 [6] is the main international standard that provides the major instructions and requirements for designing protection and control systems of modern SGBMGs and inverter-based MGs (IBMGs). This standard has been considered by various countries for developing their national grid codes [7, 8].

For investigating the stability of modern MGs and their protection systems, the related control system structure and performance should be scrutinized. On the other hand, realizing the control system requirements of modern MGs, especially for designing MGs, is necessary for system engineers. Therefore, the main requirements of control system of modern SGBMGs are presented in this section, considering the instructions of IEEE 1547-2018 standard.

2.2.1 Overview of SGBMG Control System

SGBMGs and the included SSSGs may operate in various control modes during the grid-connected and isolated operation modes of the SGBMGs. A conceptual schematic of the SGBMG control system is illustrated in Fig. 2.1. As shown, the SGBMG control system contains two control levels: SSSG control level and upstream control level. The SSSG control level includes the excitation system (e.g., AVR) and governor systems, which have been discussed in Chapter 1, Section 1.2. Excitation system regulates the amount of output reactive power of the SSSGs, while the governor system is exploited for adjusting the SSSG output active power. Meanwhile, the upstream control level determines the suitable references for SSSG control level to fulfill the requirements of the IEEE 1547 standard including upstream system support during both steady-state operation and after the occurrence of system transients, such as voltage drop and short-circuit faults. In fact, exchanging a suitable amount of active and reactive powers between the MG and upstream grid, which are determined by the upstream control level, could help the system to maintain its voltage and frequency within the permissible ranges during both steady-state operation and system transients. Moreover, regulating the SSSG output at these suitable values could also help maintain the MG stability and prevent unwanted outages, trippings, and load sheddings.

The SSSG operation characteristic, known as "capability curve," is an important criterion for determining the suitable amount of active/reactive power to be exchanged with the upstream grid [9]. The SSSG capability curve is

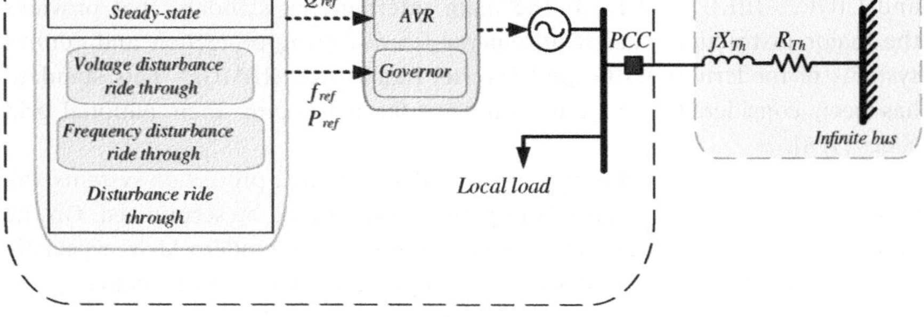

Fig. 2.1 Overview of the SGBMG control system.

commonly expressed by a region in which the permissible boundaries of the SSSG active/reactive power output are demonstrated. SSSG operation outside the mentioned boundaries may endanger the SSSG stability and cause SSSG thermal and mechanical damage. Therefore, the upstream control system should consider the permissible range of the SSSG capability curve for determining suitable active/reactive power amounts. A typical capability curve of the SSSG is depicted in Fig. 2.2 [9]. As shown, the maximum amount of reactive power that can be absorbed by the SSSG is commonly less than the maximum permissible injected reactive power to the grid due to various limitations such as stability margin and stator end iron limit. It should also be considered that the SSSG flexibility for injecting/absorbing reactive power is less than that of inverter-based resources, and thus, this limitation should be considered in the upstream control level.

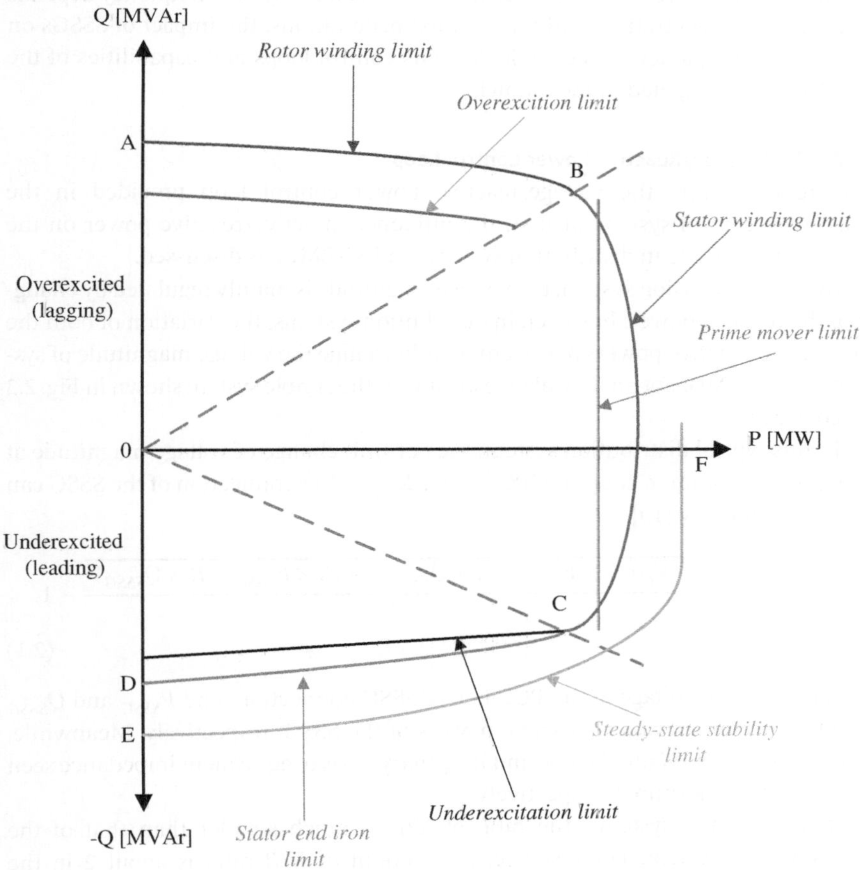

Fig. 2.2 SSSG capability curve. [9] / MDPI / CC BY 4.0.

The upstream control level may include various control loops and capabilities as presented in Table 2.1. These control loops have considerable impacts on the system stability, especially after the occurrence of a disturbance. The associated strategies, which are commonly utilized in the upstream control loops, are explained in the rest of this section.

2.2.2 Steady-State Operation

In the steady-state operation, i.e., within the normal operating range, SSSG control system can change the amount of the output active and reactive powers to respond to the system voltage changes and enhance the voltage profile of the distribution system. Meanwhile, SSSGs may also be exploited to support the frequency of the system. However, the impact of the SSSGs on the system frequency depends on their vast penetration, and for minimal penetrations, the impact of SSSGs on the system frequency is negligible. Various control loops and capabilities of the SSSGs are investigated in the following.

2.2.2.1 Voltage/Reactive Power Control Loop

Before explaining the voltage/reactive power control loop provided in the upstream control system, at first, the influence of active/reactive power on the voltage magnitude in distribution systems and SGBMGs is discussed.

In the transmission system, the voltage magnitude is mainly regulated by changing the reactive power; however, in distribution systems, the variation of both the active and reactive powers may significantly change the voltage magnitude of system and SGBMG. For analytical investigations, the simple system shown in Fig. 2.3 is considered.

In this typical distribution system, the per-unit change of voltage magnitude at the point of common coupling (PCC) (i.e., ΔV_{pu}) after connection of the SSSG can be determined by [10]

$$\Delta V_{pu} = \frac{\sqrt{(V_t + R \times P_{SSSG} + X \times Q_{SSSG})^2 + (X \times P_{SSSG} - R \times Q_{SSSG})^2}}{V_t^2} - 1 \tag{2.1}$$

where V_t is the voltage at the PCC before SSSG connection, and P_{SSSG} and Q_{SSSG} are the output active and reactive powers of the SSSG, respectively. Meanwhile, parameters R and X are the real and imaginary parts of equivalent impedance seen from the SSSG terminal, respectively.

In distribution systems, the ratio of X/R is much smaller than that of the transmission system. For instance, the amount of X/R ratio is about 2 in the distribution systems, while this ratio can be greater than 10 for the transmission

Table 2.1 Various control loops and capabilities of SSSGs.

State	Control loop	Control mode	SGBMG operation mode	Corresponded system
Normal (*steady-state*)	Voltage/reactive power	Fixed reactive power	Grid-connected	Excitation system
		Fixed power factor		
		Droop		
		Volt–VAr		
		Watt–VAr		
		Constant voltage	Islanded	
	Frequency/active power	Fixed active power	Grid-connected	Governor system
		Droop		
		Volt–Watt	Both grid-connected and islanded	
		Fixed frequency	Islanded	
Abnormal	Voltage disturbances ride-through	Dynamic voltage support	Both grid-connected and islanded	Excitation system
	Frequency disturbances ride-through	Fast frequency response	Both grid-connected and islanded	Governor system

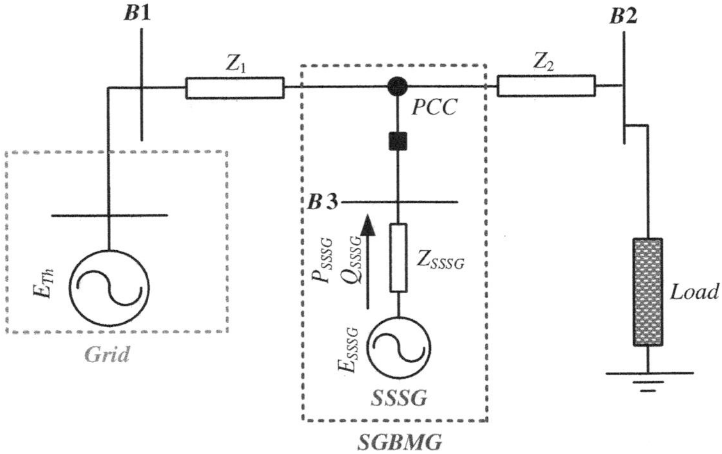

Fig. 2.3 Typical configuration of a grid-connected SSSG.

systems. Therefore, the impact of active power variations on the system voltage magnitude would significantly increase in distribution systems. On the other hand, the impact of reactive power changes on the voltage magnitude is much less than that of the transmission system. For better demonstration, the amount of voltage change for various amounts of active and reactive powers is depicted in Fig. 2.4. As shown, the sensitivity of voltage magnitude to active power is considerably increased for high values of R/X in the distribution systems. Therefore, the impact of active power changes on the voltage magnitude should be considered in the control loops of SSSGs.

In normal operation, SSSG provides voltage regulation capability by adjusting the output reactive power through its excitation system. The following control modes can be utilized in a typical SGBMG [10]:

- **Fixed reactive power mode:** In this mode, the SSSG regulates its output reactive power at a constant value without considering the amount of output active power and SSSG terminal voltage. In fact, in this mode, the SSSG does not participate in regulating the SGBMG voltage, and this voltage is adjusted by the upstream grid or other resources located in SGBMGs.
- **Constant voltage mode:** In this mode, the SSSG regulates the SGBMG voltage by changing its output reactive power and injecting or absorbing the appropriate amount of reactive power. In this condition, the SSSG may operate in both over- and underexcited modes to regulate the voltage. It should be noted that in this case, the control system will take its maximum effort to regulate the SGBMG voltage and to completely mitigate the error of voltage magnitude from its set-point value.

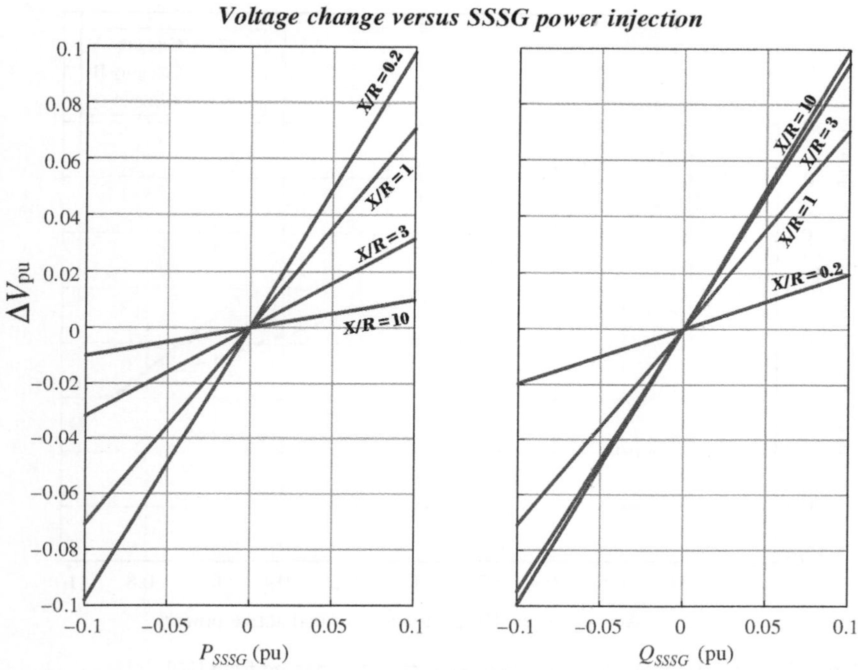

Fig. 2.4 Change of voltage at the point of SSSG connection for various amounts of SSSG output active and reactive powers. Adapted from [10].

- **Fixed power factor and Watt–VAr mode:** In this mode, the SSSG power factor should remain constant. In other words, if the amount of SSSG output active power decreases/increases, the amount of its reactive power will also be decreased/increased by the excitation system to maintain the SSSG power factor at a constant value. Watt–VAr mode is basically similar to the fixed power factor mode; however, it may include a deadband for small amounts of active power. In fact, this mode aims to compensate for the impact of the exchanged active power on the change of PCC voltage magnitude by suitable adjustment of the SGBMG reactive power. In other words, in the Watt–VAr control loop, the reactive power setpoint is not directly determined by the voltage magnitude, and the amount of active power exchange determines the setpoint of the SSSG reactive power. In such a condition, the SSSG may inject or absorb reactive power when it is not actually required for voltage regulation. Therefore, this control loop may be disabled by the SGBMG operator. Three different Watt–VAr characteristics are shown in Fig. 2.5, including category A, category B, and constant power factor equal to 0.96 lead ($pf = 0.96$ lead). As shown, in category A and category B, for active power between −0.5 and 0.5 pu, the SSSG operates

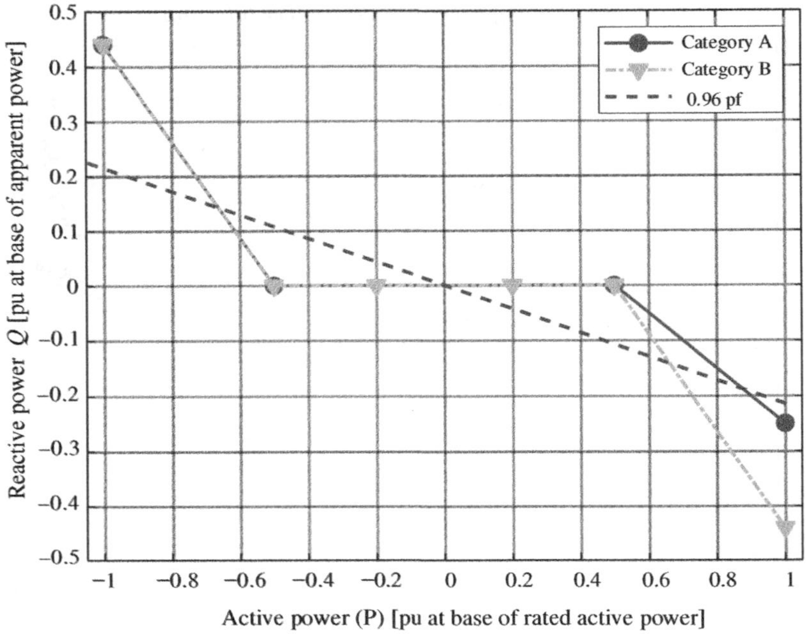

Fig. 2.5 Various characteristics of Watt–VAr control. Adapted from [10].

at unity power factor; while for $P > 0.5$ pu, the SSSG absorbs reactive power to compensate for the impact of SSSG on the system voltage [10]. Category B has a higher capability for the absorption of reactive power. As depicted, for nominal injection of active power, i.e., $P = 1$ pu (at the base of the rated active power), the amount of reactive power absorbed by category B is about 0.45 pu (at the base of the rated apparent power); while for category A, it is about 0.25 pu. In addition, for the constant power factor, the reactive power amount changes continuously in proportion to the output active power. The left side of this figure in which the active power is negative is not commonly applicable to the SSSGs. However, in inverter-based DGs (IBDGs), including battery-chargers, the active power can be absorbed, i.e., $P < 0$.

Example 2.1

A Watt–VAr control mode has been enabled for a grid-connected SSSG whose rated active and apparent powers are 10 MW and 15 MVA, respectively. This SSSG is connected to a 20 kV network. Find the voltage change at the PCC if the SSSG operates at its rated active power. Moreover, the equivalent impedance seen from the SSSG terminals is $4.5 + j5.1\ \Omega$.

a) Use category A (shown in Fig. 2.5).
b) Use category B (shown in Fig. 2.5).
c) Use category $pf = 0.96$ (shown in Fig. 2.5).

Solution

a) Considering category A, for injection with the rated active power $P_{SSSG} = 1$ pu, the reactive power absorption is 0.25 pu ($Q_{SSSG} = -0.25$ pu). Thus, based on (2.1):

$$\Delta V_{pu} = \frac{\sqrt{\left(V_t^2 + R \times P_{SSSG} + X \times Q_{SSSG}\right)^2 + (X \times P_{SSSG} - R \times Q_{SSSG})^2}}{V_t^2} - 1$$

$$= \frac{\sqrt{\begin{array}{l}((20 \times 10^3)^2 + 4.5 \times 10 \times 10^6 + 5.1 \times -0.25 \times 15 \times 10^6)^2 \\ + (5.1 \times 10 \times 10^6 - 4.5 \times -0.25 \times 15 \times 10^6)^2\end{array}}}{(20 \times 10^3)^2} - 1$$

$$= 7.81\%$$

b) Considering category B, for injection with the rated active power $P_{SSSG} = 1$ pu, the reactive power absorption is 0.25 pu ($Q_{SSSG} = -0.45$ pu). Thus, based on (2.1):

$$\Delta V_{pu} = \frac{\sqrt{\left(V_t^2 + R \times P_{SSSG} + X \times Q_{SSSG}\right)^2 + (X \times P_{SSSG} - R \times Q_{SSSG})^2}}{V_t^2} - 1$$

$$= \frac{\sqrt{\begin{array}{l}((20 \times 10^3)^2 + 4.5 \times 10 \times 10^6 + 5.1 \times -0.45 \times 15 \times 10^6)^2 \\ + (5.1 \times 10 \times 10^6 - 4.5 \times -0.45 \times 15 \times 10^6)^2\end{array}}}{(20 \times 10^3)^2} - 1$$

$$= 4.64\%$$

c) For $pf = 0.96$ lead, the reactive power is determined as

$$Q_{SSSG} = -tg(\cos^{-1}(pf)) \times P_{SSSG}$$
$$= -tg(\cos^{-1}(0.96)) \times 10MW = -2.92MVAR$$

$$\Delta V_{pu} = \frac{\sqrt{\left(V_t^2 + R \times P_{SSSG} + X \times Q_{SSSG}\right)^2 + (X \times P_{SSSG} - R \times Q_{SSSG})^2}}{V_t^2} - 1$$

$$= \frac{\sqrt{\begin{array}{l}((20 \times 10^3)^2 + 4.5 \times 10 \times 10^6 + 5.1 \times -2.92 \times 10^6)^2 \\ + (5.1 \times 10 \times 10^6 - 4.5 \times -2.92 \times 10^6)^2\end{array}}}{(20 \times 10^3)^2} - 1$$

$$= 8.72\%$$

Based on these calculations, category B has better performance in preventing overvoltage at the point of the SSSG connection.

- **Droop mode:** In the droop/Volt–VAr modes, although the SSSG participates in voltage regulation of the SGBMG, the voltage magnitude error from its setpoint value does not become zero. In the droop mode, the amount of output reactive power of the SSSG (Q_{output}) is determined by

$$Q_{output} = Q_{ref} + R \times (V_{ref} - V_t) \tag{2.2}$$

where R is the constant positive value, which denotes the droop value. Parameters Q_{ref} and V_{ref} are the references of output reactive power and voltage magnitude, respectively. Moreover, V_t is the terminal voltage magnitude. Based on this equation, if the voltage magnitude decreases, the amount of Q_{output} will be increased by the control system to participate in SGBMG voltage regulation. For larger amounts of R, the changes of Q_{output} with variation of V_t increases and the participation of SSSG in voltage regulation increases. For $R = 0$, the droop mode develops to the fixed reactive power mode, and for $R = \infty$, the droop mode converts to the constant voltage mode.

For reactive power sharing among the SSSGs, considering their rated powers, the droop coefficient should be determined considering the following equation:

$$S_i/R_i = S_j/R_j \tag{2.3}$$

where subscripts i and j correspond to the SSSG number in the SGBMG.

- **Volt–VAr mode:** This control mode is somewhat similar to the droop control mode. In this mode, a piecewise linear characteristic is utilized to adjust the amount of SSSG reactive power for various amounts of voltage magnitudes. A sample characteristic of the Volt–VAr control mode is depicted in Fig. 2.6. This piecewise characteristic commonly includes two major parts: (i) deadband range, in which the amount of Q_{output} remains constant for a range of voltage variations and (ii) control range, in which the amount of Q_{output} changes linearly with voltage variations. For instance, in this figure, for the deadband range of $0.9\,\text{pu} < V_t < 0.93\,\text{pu}$, the amount of Q_{output} is constant and about +0.43 pu. Moreover, the amount of Q_{output} for the control range of $0.93\,\text{pu} < V_t < 0.98\,\text{pu}$ changes linearly with the rate of $\Delta Q_{output}/\Delta V_t = +2.5$.

2.2.2.2 Frequency/Active Power Control Mode

The SSSG output active power is regulated by its associated turbine–governor system as explained in Chapter 1 of this book. The output active power of the SSSG could maintain the SGBMG frequency and voltage magnitude during both the steady-state operation and the occurrence of transient disturbances. Depending on

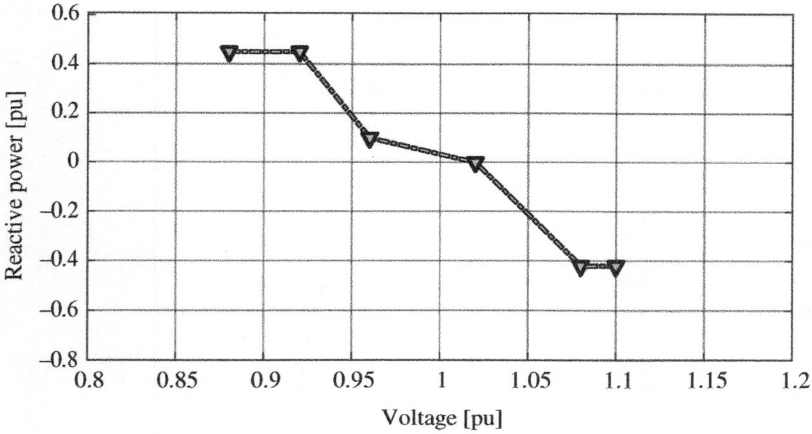

Fig. 2.6 Sample Volt–VAr characteristic of a grid-connected SSSG. Adapted from [10].

the SGBMG operation mode, i.e., grid-connected or islanded, the following control modes may be exploited in the steady-state operation [6, 10]:

- **Fixed active power mode:** In this mode, the SSSG adjusts its output active power at a fixed value without considering the system frequency or voltage magnitude. In fact, in this condition, the SSSG does not participate in regulating the SGBMG frequency. Commonly, the grid-connected SGBMGs are operated under the fixed active power mode.
- **Droop mode:** In the droop mode, the SSSG participates in the adjustment of the SGBMG frequency. Although this mode could help the system to regulate its frequency and decrease the frequency error from its setpoint value, this error does not become zero. The final amount of steady-state error depends on the frequency droop value (R). In the droop mode, the amount of output SSSG active power (P_{output}) is calculated by

$$P_{output} = P_{ref} + R \times (f_{ref} - f) \tag{2.4}$$

where parameters P_{ref} and f_{ref} are the reference setpoints of output power and SGBMG frequency, respectively. If the SGBMG frequency decreases, the amount of $f_{ref} - f$ increases, and thus, P_{output} will increase. The amount of P_{output} increment depends on both droop value and frequency error from its setpoint, i.e., $f_{ref} - f$. It is worth mentioning that for $R = 0$, the frequency droop mode would correspond to the fixed active power mode. As the frequency droop value increases, the SGBMG controller effort would also increase for adjusting the frequency at its setpoint and thus the steady-state error of the frequency will decrease.

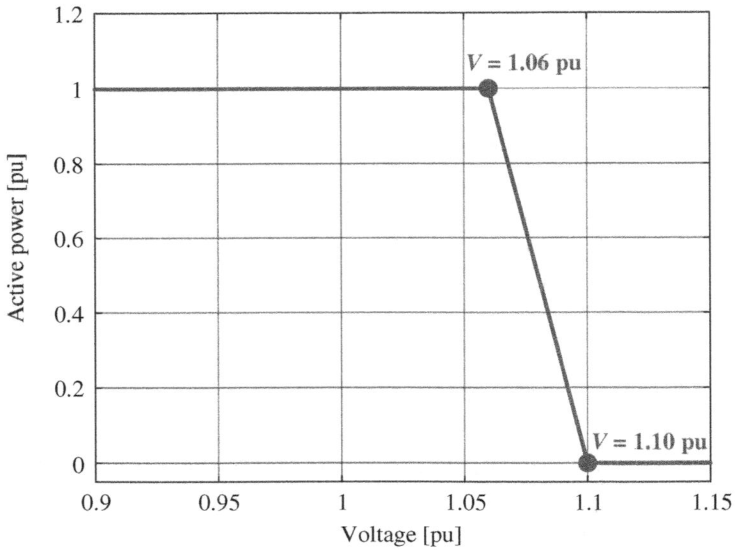

Fig. 2.7 A sample Volt–Watt characteristic. Adapted from [10].

- **Voltage-active power (Volt–Watt) mode:** The active power generation of the SSSG may have a significant impact on the PCC voltage and may lead to an overvoltage condition. Thus, this control loop is implemented in SGBMGs to resolve the voltage rise issue by limiting the amount of injected active power. A sample characteristic of a Volt–Watt loop is depicted in Fig. 2.7 [6, 10]. As shown, for $V < 1.06$ pu, the SSSG injects the rated power to the upstream network. However, for $1.06\,\text{pu} < V < 1.1$ pu, the amount of the SSSG output power should be reduced to prevent voltage rise. Moreover, to take the maximum active power capability of the SGBMG, the Volt–Watt loop shall be coordinated with the Volt–VAr control loop.
- **Fixed frequency mode:** In the islanded operation mode of the SGBMG, the frequency should be regulated by the SSSGs. To do so, the SSSG regulates its output active power at the suitable amount. It should be noted that in this case, the control system would take its maximum effort to regulate the SGBMG frequency and to completely alleviate the frequency error from its setpoint value.

2.2.3 Transient Abnormal Conditions

During transient abnormal conditions, which may occur in upstream system or SGBMG, the upstream control system and SSSGs shall respond properly to support the system, maintain the system/SGBMG stability, and prevent damage to the equipment. In this regard, two types of grid support during abnormal conditions are commonly required: (i) voltage disturbances ride-through and (ii) frequency disturbances ride-through. It should be noted that the probability of the occurrence of voltage disturbances in distribution system is considerably higher than that of the frequency disturbances, and thus, providing voltage disturbance ride-through capability is much more important.

If the grid support through the SGBMG control system is not effective, the protection system of the SSSG and SGBMG may operate to island the SGBMG, shed loads, shed generators, etc., to avoid equipment damage and retain system stability. In fact, the effectiveness of the SGBMG grid support during the transient abnormal conditions has a considerable impact on the protection system, and thus, the ride-through capability of the SGBMG should be scrutinized. In this regard, the SGBMG protection system should be properly coordinated with the required ride-through capabilities. In this section, the aforementioned required ride-through capabilities of the SGBMG are presented.

2.2.3.1 Voltage Disturbance Ride-Through

After the occurrence of a voltage disturbance, i.e., voltage drop or voltage rise in the upstream system or SGBMG, the SSSGs should not be tripped instantaneously. In such a condition, the SSSGs should remain connected to the system for a while to support the system against voltage disturbances. Thus, the voltage protection functions, i.e., over- and undervoltage, should remain stable for a prespecified duration of ride-through. The duration of voltage disturbance ride-through is dependent on the severity of the voltage drop or voltage rise. For severe voltage disturbances, the necessary duration of the required ride-through considerably decreases to prevent SSSG damage. The ride-through duration t_{RT} is commonly determined, considering the technical requirements of the upstream grid and the capability of the SSSGs. Commonly, the voltage disturbance ride-through capability includes the following regions:

- **Region 1:** For voltage disturbances in which the voltage magnitude remains in the permissible continuous operation range, the SSSG should remain stable and continue injecting active power. In such a condition, the amount of active power is determined by [6, 10]:

$$P_{SSSG} = \begin{cases} \dfrac{|V_{post}|}{|V_{nom}|} \times P_{pre} & for\ |V_{post}| < |V_{nom}| \\ P_{pre} & for\ |V_{post}| > |V_{nom}| \end{cases} \tag{2.5}$$

where P_{pre} is the output active power of the SSSG before the disturbance occurrence. Moreover, $|V_{post}|$ and $|V_{nom}|$ are postdisturbance and nominal voltage, respectively. For undervoltage disturbances, such as short-circuit faults, $|V_{post}|$

is assumed to be equal to the least voltage magnitude of the individual phase. However, for overvoltage disturbances, the greatest magnitude should be considered. Based on (2.5), the SSSG active power could be decreased during the low-voltage ride-through conditions, considering the least voltage magnitude of the individual phase. For this region, the voltage protection of the SSSG should remain stable. Thus, the voltage protection thresholds should be coordinated with this region.

- **Region 2:** This region is applicable when the voltage magnitude becomes lower or higher than the permissible range. In this region, the SSSG should remain connected to the system and maintain its synchronism with the upstream grid for a predefined duration. This capability is known as "voltage-disturbance ride-through" [6, 10]. The predefined duration of the SSSG ride-through is commonly specified based on the grid requirements and SSSG specifications. Moreover, the mentioned duration depends on the amount of voltage drop or voltage rise. For a larger amount of voltage changes, the duration of ride-through decreases to prevent possible damage to SSSGs and other equipment.

  During high-voltage ride-through, i.e., $V_{L1} > V_H$, the SSSG shall cease to inject active power to the upstream system. However, when the undervoltage ride-through capability is activated, the SSSG shall continue exchanging current with the upstream system for a prespecified voltage interval $V_{L1} < V < V_{L2}$. Meanwhile, for severe undervoltage disturbances in which the voltage becomes lower than a predefined threshold, i.e., $V < V_{L1}$, exchanging current with the upstream system is not necessary and the SSSG may cease to inject power. The amount of the exchanged active/reactive current (power) with the upstream system highly depends on the SSSG characteristics, its capability curve, and the mutual agreement between the SGBMG owner and the upstream grid operator [6, 10].
- **Region 3:** After the duration of the ride-through elapses, the SSSG should be tripped to prevent equipment damage. Therefore, the protection system, especially voltage protection functions, shall be properly coordinated with the mentioned regions. Meanwhile, over- and undervoltage functions may be utilized in SGBMGs for the detection of the islanding condition. Since the over- and under-voltage functions should be coordinated with the voltage disturbance ride-through, they may not detect the islanding condition quickly. Therefore, using other types of islanding detection functions, such as vector-shift or vector-jump, may be necessary in such a condition [6, 10].

Besides the aforementioned requirements of voltage disturbance ride-through, dynamic voltage support capability in low- and high-voltage disturbances may also be required by the grid operator [10]. The major purposes of this voltage support are to (i) prevent tripping SSSG, loads, and DGs, (ii) help mitigate the FIDVR phenomenon in the distribution system and SGBMG, (iii) improve the static voltage stability and loadability of the system, and (iv) enhance the voltage profile. Meanwhile, in distribution systems, the effectiveness of dynamic voltage support through exchanging reactive power highly depends on the R/X ratio. Thus, for large values of R/X ratio, enabling reactive current-based dynamic voltage support may be arguable, and the necessity of enabling dynamic voltage support, in this case, should be investigated by the grid operator. Furthermore, the penetration of dynamic loads, especially three-phase and single-phase IMs, should also be considered for enabling dynamic voltage support.

For enabling the dynamic voltage support function at SGBMGs, the SSSG should have the capability of fast injecting/absorbing reactive current to/from the upstream system. Commonly, the required amount of additional reactive current is calculated based on the characteristics depicted in Fig. 2.8. As shown, the required amount of additional reactive current is proportional to the amount of voltage deviation. The parameters of these characteristics are determined, considering the SSSG capability and requirement of the upstream grid. However, implementing dynamic voltage support using SSSGs may include various challenges and limitations, as highlighted in the following:

- Limitation for injection/absorbing reactive current due to the SSSG capability curve
- Limitation of excitation system for providing reactive current during low-voltage disturbances, as presented in Chapter 1
- Slow response for decreasing active current and providing the required reactive current.

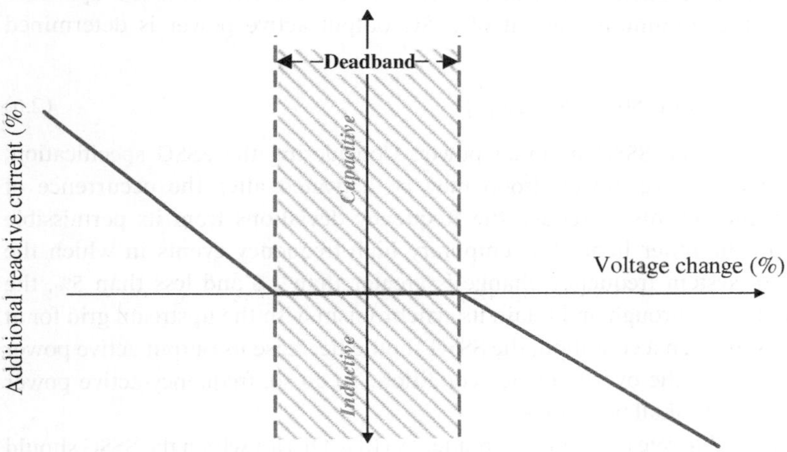

Fig. 2.8 Additional reactive current during voltage disturbances. Adapted from [10].

Therefore, as a common practice, IBDGs are commonly utilized in distribution systems for providing dynamic voltage support due to the abovementioned limitations of the SSSGs. If the voltage dynamic support is expected to be enabled for SSSGs, the performance of this function should be considered for adjusting the protection functions of the SSSG and SGBMG. Enabling voltage support capability may also help the SGBMG retain its transient stability.

2.2.3.2 Frequency Disturbance Ride-Through

Although the probability of the occurrence of frequency disturbances in grid-connected SGBMGs is considerably lower than that of voltage disturbances, the SSSGs shall also provide frequency disturbance ride-through capability. During frequency ride-through, the SSSG shall remain connected to the system for a predefined duration. Therefore, the SSSG and its associated prime mover should be designed to provide such a frequency ride-through capability without impairment. After the duration of the frequency ride-through elapses, the SSSG should be tripped to prevent mechanical damage. Therefore, the frequency ride-through capability shall be considered for adjustment of over- and under-frequency functions. If the frequency change is less than 2% of the nominal frequency and the per-unit ratio of voltage/frequency remains less than 1.1, the SSSG shall remain connected to the system and inject active power at least equal to the predisturbance value for an infinite time. However, when the frequency changes are greater than 2%, the duration of SSSG ride-through is limited. For temporary low-frequency events in which the system frequency drops below 0.98 pu and remains greater than 0.95 pu, the SSSG shall ride-through and retain its synchronism with the upstream grid for at least 299 s. In such a condition, the SSSG shall exchange both active and reactive currents with the upstream grid, and the minimum amount of SSSG output active power is determined by [6]

$$P_{SSSG} = \min\{80\% \times S_{nom}, P_{pre}\} \tag{2.6}$$

where S_{nom} is the SSSG nominal power. Considering the SSSG specification, the frequency-active power droop may be activated after the occurrence of low-frequency events to reduce the frequency deviations from its permissible range. On the other hand, for temporary high-frequency events in which the amount of system frequency change is greater than 2% and less than 5%, the SSSG shall ride-through and retain its synchronism with the upstream grid for at least 299 s. In such a condition, the SSSG should decrease its output active power to help mitigate the overfrequency condition. To do so, frequency-active power droop capability shall be enabled.

The acceptable rate of change of frequency (ROCOF) for which the SSSG should remain connected to the system is determined based on the SSSG specifications and capability. The minimum amount of acceptable ROCOF for which the SSSG should retain its synchronism is 0.833 pu/s. Moreover, if the displacement of positive-sequence voltage and individual phase voltage at PCC are, respectively, less than 20 and 60 electrical degrees, the SSSG should remain connected to the upstream grid and retain its stability [6].

As will be presented in Chapter 5 of this book, various frequency-based protection functions are utilized for unintentional islanding detection including underfrequency, overfrequency, ROCOF, and vector shift [11]. The operation of these functions shall not restrict the SSSG ride-through capability during nonislanding frequency disturbances. In this regard, the setting parameters of these functions shall be adjusted properly to prevent their maloperation for nonislanding frequency disturbances. The possibility of maloperation of frequency-based islanding detection approaches during frequency disturbances is discussed in Chapter 5.

2.3 SGBMG Stability Classification

2.3.1 Control System Stability

The main control systems of SGBMGs include the excitation and governor systems of SSSGs. As mentioned in the previous sections of this chapter, these control systems should cover various control loops and may operate in various operational modes. In addition, considering the impact of control system behavior on protection system during both the steady-state condition and transient disturbances is crucial. In fact, without scrutinizing the dynamics of the control system, utilizing proper protection schemes and adjustment of protection functions may not be possible. Therefore, in the rest of this section, the issues related to the control system dynamic response and its stability in SGBMGs are discussed.

2.3.1.1 Poorly Tuned Controllers

Tuning of the parameters of the SGBMG controllers, including the SSSGs' controller parameters, is a vital task, which should be properly performed by SGBMG operators. Poor-tuning of these controllers may lead to serious issues in SGBMG operation, including (i) insufficient damping torque coefficient (K_D); (ii) insufficient synchronizing torque coefficient (K_S); (iii) decreasing critical clearing time (CCT) and increasing the possibility of transient instability; (iv) poor dynamic response, i.e., significant overshoot and long settling time; and (v) increasing mechanical fatigue of the generator shaft, especially in adjacent to nonlinear varying loads.

Various methods have been proposed in the literature for optimum selection of controller parameters to investigate and diminish the abovementioned issues and enhance the controller stability of the SGBMG in both the grid-connected and islanded operation modes [12]. The first two issues, items (i) and (ii), have been widely investigated in conventional power systems, which may lead to local instability under small disturbances. Considering a single machine to infinite bus model for a grid-connected SSSG, its characteristic equation can be described by

$$s^2 + \frac{K_D}{2H}s + \frac{K_S\omega_0}{2H} = 0 \tag{2.7}$$

where parameter H is the SSSG inertia constant and ω_0 is the SSSG rated electrical speed. Based on this equation, the lack of synchronizing torque coefficient, i.e., $K_S = 0$, results in an aperiodic increase in the SSSG rotor angle due to the positive amount of the real part of eigenvalues. Moreover, when the damping torque coefficient is not sufficient ($K_D \approx 0$), the real part of the eigenvalues and damping ratio become very small, and thus, the SSSG rotor angle behavior will become oscillatory. Moreover, negative amount of damping torque coefficient ($K_D < 0$) leads to undamped oscillations. Therefore, appropriate values for controller parameters shall be selected to guarantee sufficient amounts for synchronizing and damping torques. The dynamic response of SSSG under small-signal disturbances for various amounts of K_D and K_S is presented in Table 2.2 [2].

Item (iii), i.e., transient stability of SSSG, is important especially for grid-connected SGBMGs. Since the SSSG output active power considerably decreases during close-in faults, the SSSG rotor speed significantly increases based on the generator swing equation, and thus, the SSSG may become unstable. Meanwhile, this issue is more severe for SSSGs than large-scale SGs due to lower inertia constant of SSSGs and also high fault-clearing time (FCT) in distribution systems. In addition, under the fixed active power control mode of SSSGs, the difference between the active power setpoint and the output active power injected to the governor PI controller increases, and thus, the mechanical input power would also be increased by the governor system. This leads to more increment of SSSG rotor speed. The parameters of the governor controller, especially the proportional controller, have a considerable impact on the mechanical power changes during short-circuit faults. Some literatures have proposed the smallest permissible value for proportional controller to reduce the amount of mechanical power increment during faults. For instance, Fig. 2.9 illustrates the obtained CCTs for various PI controller parameters calculated for a three-phase short-circuit at 1 km from a sample SSSG [12]. As shown in Fig. 2.9(a) and (b), reducing K_I and K_p parameters both modifies the CCT value. K_p directly affects CCT, while K_I operates based on an integrator whose value gradually increases. Therefore, K_p influence on CCT is more than K_I effect during the short-fault period.

Furthermore, some research works have proposed to block the steady-state power control loop in such a condition [12]. It is worth mentioning that, for the short-circuit faults at the ending parts of the feeders, the acceleration of the

Table 2.2 SSSG dynamic response for various amounts of K_D and K_S [2].

State	Dynamic response	Electromechanical eigenvalues
Stable	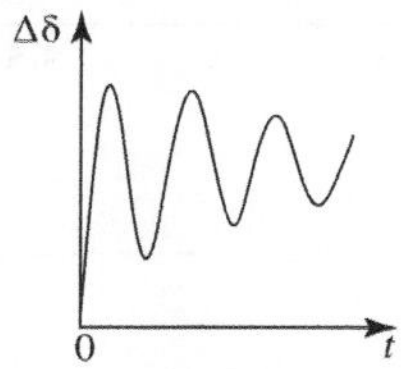 $K_S > 0$ and $K_D > 0$	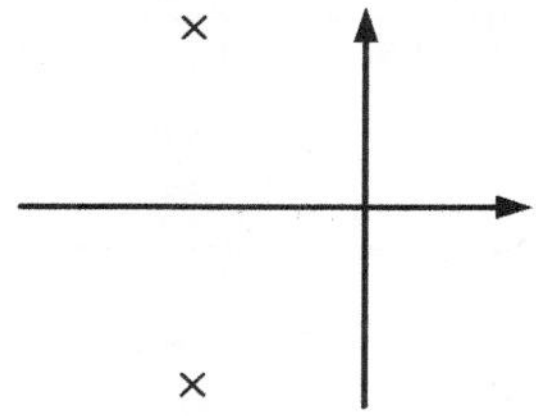
Unstable	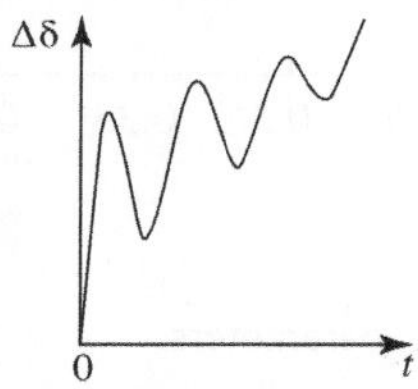 Nonoscillatory instability under constant field voltage $K_S < 0$ and $K_D > 0$	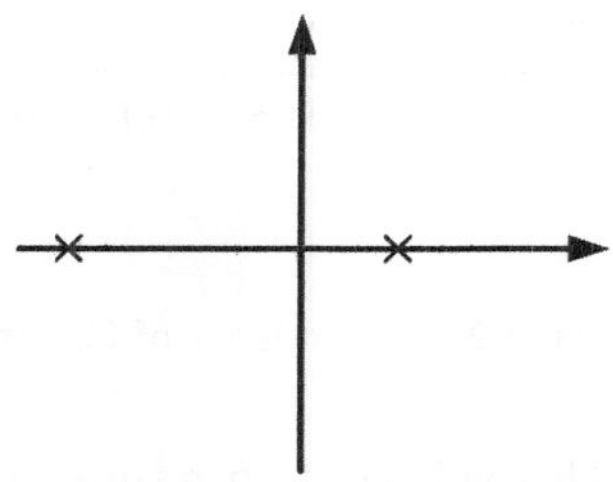
	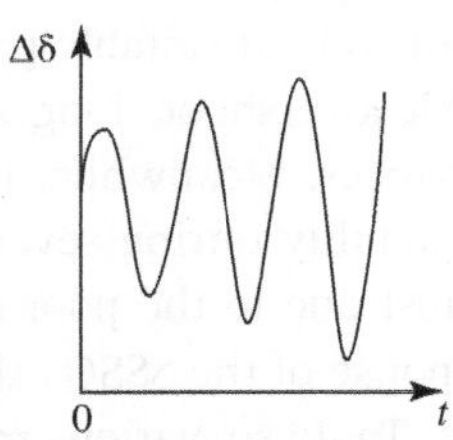 Nonoscillatory instability under field voltage control $K_S > 0$ and $K_D < 0$	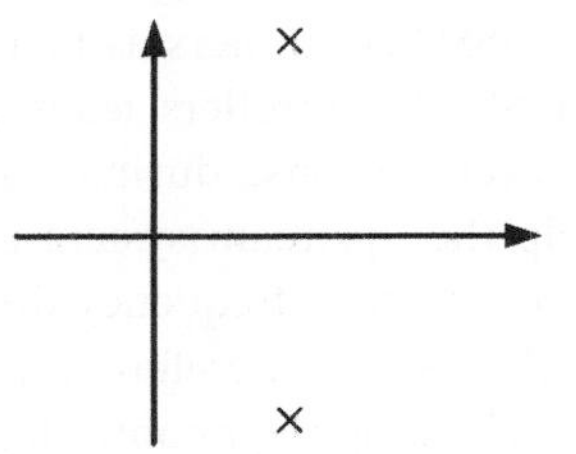

rotor speed may not be considerable, because the injected output active power of the SSSG during such faults is absorbed due to the resistance nature of the SGBMGs and distribution systems. In addition, the behavior of dynamic loads such as IMs decreases the SSSG transient stability margin. For instance, the speed of IM loads reduces during a fault, and after the fault clearance, these loads absorb a considerable amount of reactive power for reaccelerating. Therefore, the voltage at SSSG terminals may remain low for a duration after the fault clearance.

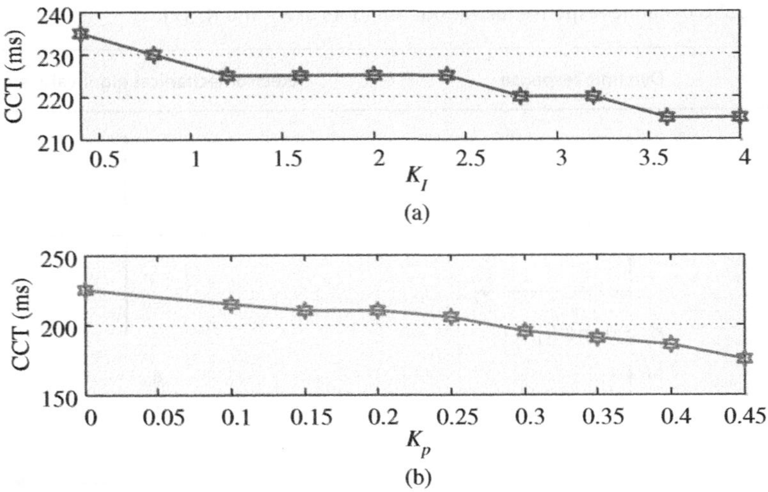

Fig. 2.9 Dependency of CCT on (a) K_p and (b) K_I parameters.

This voltage drop results in reducing the transient stability of SSSG after-fault clearance, as fully discussed in the following sections of this chapter.

The dynamic response of the SSSG highly depends on the parameters of the SSSG controllers as mentioned in item (iv). Unsuitable parameter selection for SSSG controllers leads to considerable overshoot, long settling time, and delayed response during SGBMG disturbances. Meanwhile, the opportunity to help the upstream system and SGBMG stability during severe events, such as severe voltage/frequency drops, may be lost due to the poor dynamic response of the SSSGs. Therefore, the dynamic response of the SSSGs shall be considered for adjusting their controllers' parameters. To do so, various methods have been proposed in the literature such as using eigenvalues of the linearized system and off-line simulation studies for various cases [12, 13]. In [12], to quantize the SSSG dynamic response for any sudden system change, a proper generator performance index (GPI) has been introduced as

$$\mathrm{GPI} = \int_{t_0}^{\infty} \left[(P_{ref} - P_e)^2 + K_u \times u^2\right] \tag{2.8}$$

where t_0 is the instant of SGBMG disturbance, and P_{ref} and P_e are the SSSG reference and active powers, respectively. The PI controller output of the SSSG governor system is denoted by u. This parameter should also be considered in the proposed GPI value to include the controller efforts in the system dynamic performance. Parameter K_u is a constant coefficient to determine contribution of the controller efforts on GPI. After changing P_{ref} in typical SSSGs, the obtained GPI values for

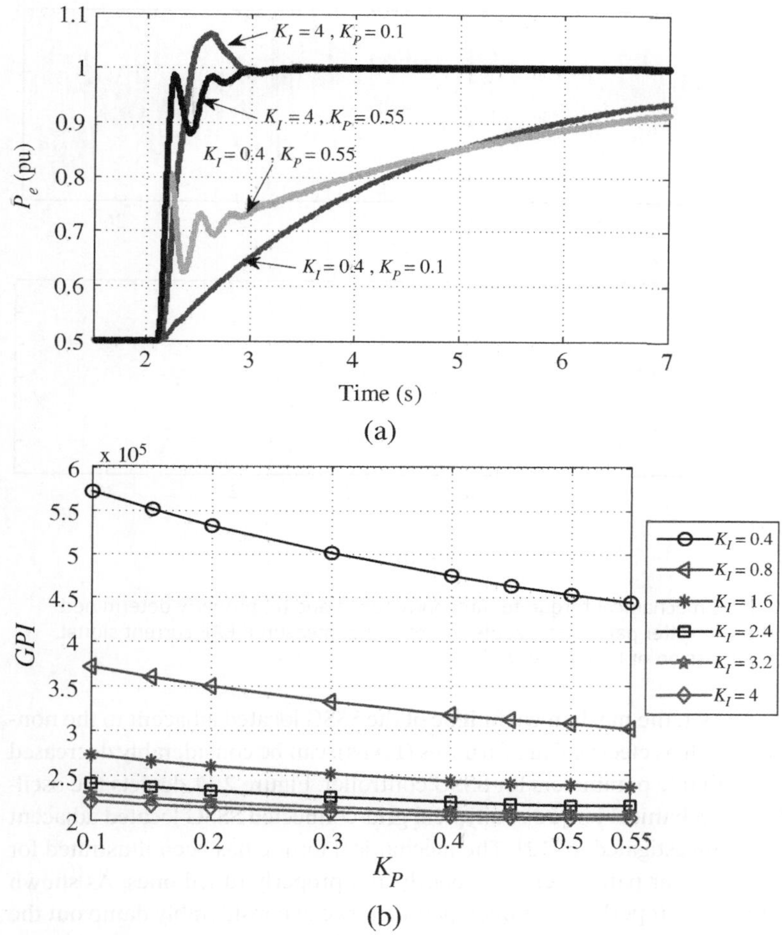

Fig. 2.10 SSSG dynamic performance, (a) generator output power for different PI controller parameters, (b) obtained GPI for various amounts of K_I and K_P parameters. [12] / with permission of John Wiley & Sons.

various amounts of integral (K_I) and proportional (K_P) gains of the governor PI controller are depicted in Fig. 2.10. As shown, for various controller parameters, the dynamic response of the SSSG and the PI controller effort would considerably change.

Meanwhile, nonlinear time-varying loads, especially in an islanded SGBMG, have negative impacts on the rotating machines, including SSSGs. In such a condition, the output power and other electrical parameters of the SSSG considerably change, and thus, the SSSG mechanical torque varies. The SSSG controller parameters have significant impacts on the amount of SSSG mechanical torque

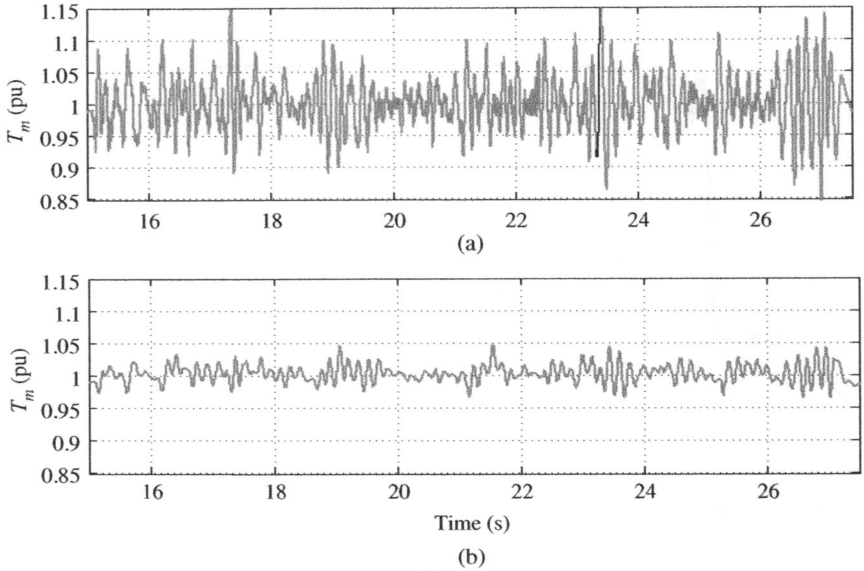

Fig. 2.11 SSSG mechanical torque for (a) poorly tuned and (b) properly determined values of the controller parameters subjected to actual measured EAF current signal. [12] / with permission of John Wiley & Sons.

variations. In fact, the mechanical fatigue of the SSSG located adjacent to the nonlinear loads, such as electrical arc furnaces (EAFs), can be considerably decreased by utilizing suitable parameters for SSSG controller. Figure 2.11 depicts the oscillation of the mechanical torque of a typical grid-connected SSSG located adjacent to an EAF, as investigated in [12]. The mechanical torque has been illustrated for two sets of governor parameters, i.e., poorly and properly tuned ones. As shown in this figure, the properly determined parameters can considerably damp out the mechanical torque oscillations.

2.3.1.2 Miscoordination

Coordination between various control loops of the SSSG and also between the control system of SSSG, other DGs, as well as energy storage systems and VAr compensators equipment is necessary to take the maximum capability of the SSSGs and prevent overlaps, especially during disturbances. For instance, the miscoordination between control loops may limit the SSSG capability for injection/absorption of the reactive power during voltage disturbances, which could threaten the voltage stability of SGBMG. In this regard, one of the main required coordinations is between the Volt–VAr and Volt–Watt control loops by ensuring that the voltage response ranges do not overlap. Moreover, for synchronous type DGs, i.e., SSSG, both the Volt–VAr and Volt–Watt characteristics shall be coordinated with the

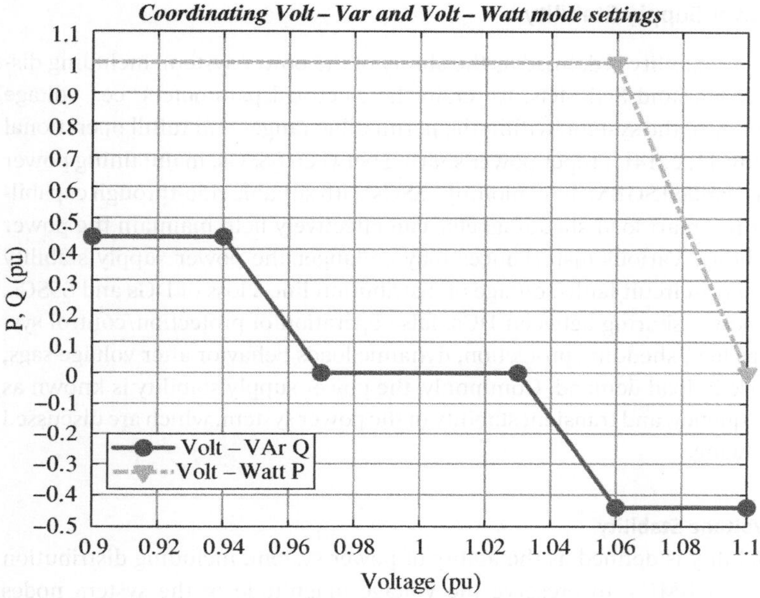

Fig. 2.12 Coordination of Volt–VAr and Volt–Watt control loops. Adapted from [10].

generator capability curve to prevent working the generator outside the permitted region.

Figure 2.12 depicts an example for coordination between the voltage response range of Vol–VAr and Volt–Watt control loops [10]. As shown, considering the Volt–Watt characteristic, the SSSG maintains its rated active power up to 1.06 pu of PCC voltage, where the SSSG capability for absorbing reactive power reaches to its maximum permissible value; then, the Volt–Watt control loop curtails the active power for the PCC voltage above this value. By this proper coordination, the maximum reactive power absorption capability of the SSSG could be exploited to prevent SGBMG damage and voltage instability.

The main portions and subsystems of the SGBMG that should be considered for coordination include:

- Control loops and characteristics of each SSSG
- Control system and protection system of each SSSG
- Control/protection systems of various SSSGs located in the SGBMG
- Control systems of SSSGs and other adjacent controllable equipment, including voltage regulators, capacitor banks, on-load tap changers, and energy storage systems
- Considering the requirements of upstream grid operator.

2.3.2 Power Supply Stability

Power supply stability is defined as the ability of the power system, including distribution system and SGBMGs, to retain the electrical parameters, i.e., voltage and frequency of the system, within the permissible ranges and fulfill operational requirements [1, 2, 14]. Proper power sharing between SSSGs, maintaining power balance in islanded SGBMGs, exploiting SSSGs with suitable ride-through capabilities, utilizing smart load-shedding, etc., can effectively help maintain the power supply stability. Various disturbances may endanger the power supply stability including short-circuit faults, outage of distribution lines, loss of DGs and SSSGs, improper power sharing between DGs, false operation of protection/control system such as load-shedding protection, dynamic loads behavior after voltage sags, and increase in load demand. Commonly, the power supply stability is known as voltage, frequency, and transient stability in the power system, which are discussed in the following.

2.3.2.1 Voltage Stability

Voltage stability is defined as the ability of power system, including distribution system and SGBMGs, to preserve the voltage magnitude of the system nodes within the permissible range during the steady-state condition and after the occurrence of a disturbance. In conventional power grids in which $X \gg R$, the main circumstance of voltage instability is the lack of required amount of reactive power. However, this argument may not be valid in modern distribution systems and SGBMGs due to their small values of X/R. Traditionally, the static voltage instability in conventional power systems was studied and analyzed through the P–V and Q–V curves in each node of the system, as depicted in Fig. 2.13. As shown, by increasing the amount of active power P, the operation point moves toward the knee point of the P–V characteristic and becomes closer to the critical point. At the knee point, the voltage considerably drops by a small increase in the load active power P, and the power-flow solution methods fail to converge beyond this point. In fact, beyond such a limit, the system does not have any stable operation point, and thus, the voltage dramatically decreases. This knee point is commonly considered the critical voltage collapse point and denotes the limit for the voltage stability [15].

The static voltage stability may also be studied using Q–V curves. In stable operation of the system, the voltage magnitude of the nodes increases with the injection of the reactive power (Q). In other words, the amount of dQ/dV is positive in all nodes of the system. However, after voltage instability, the amount of dQ/dV becomes negative. Therefore, the bottom point of the Q–V curve, where $dQ/dV = 0$, is considered the voltage stability limit. In fact, operation of the

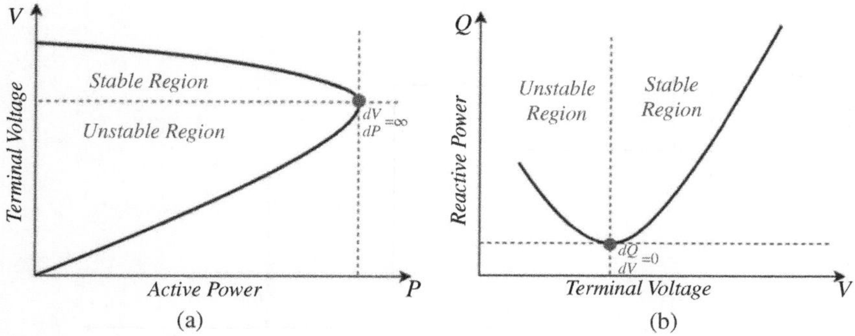

Fig. 2.13 Voltage stability investigation using (a) $P-V$ curve and (b) $Q-V$ curve [15].

system on the right side of this point is stable, while it is unstable on the left side. Consequently, dQ/dV is considered a suitable index for evaluating the voltage stability condition for each node of the system.

These two curves can be effectively utilized in conventional power systems for determining the static voltage limits and identifying the voltage instability, which causes from:

- Increment of load demands, especially those located at the end of long transmission lines
- Insufficient reactive power compensation
- Weak-voltage source
- Long electrical distance between voltage source and load centers.

Although the abovementioned curves could help the system engineers for determining the operational limits of the system, they can be only used for evaluating the static voltage stability limit of conventional transmission systems. In fact, they cannot identify dynamic voltage instability, especially in distribution systems and SGBMGs. In the aforementioned classical methods, the system loads are modeled as constant static loads, and the behavior of dynamic loads, such as IM loads, is neglected. However, in distribution systems, since the loads are very close to the system nodes, the dynamic behavior of the loads should be considered for voltage stability assessment. Moreover, in real distribution systems, after sudden voltage changes, for example, after voltage sags, the load consumption power has a dynamic variation, as shown in Fig. 2.14, and does not change abruptly. In this regard, various load models are developed to consider the transient behavior of the loads and to investigate dynamic voltage stability. For instance, in [16, 17], the transient load pattern is mathematically modeled

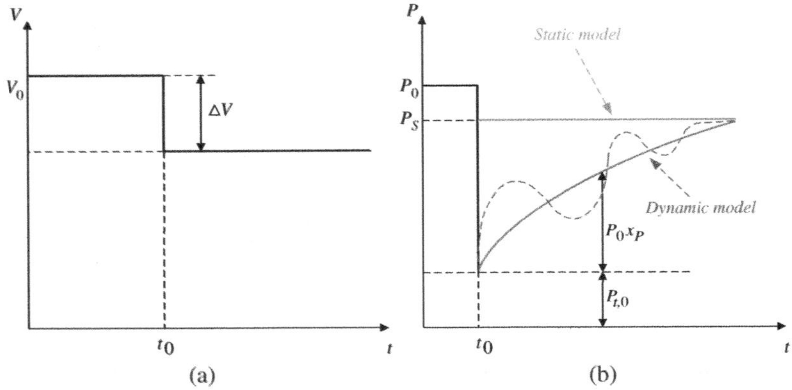

Fig. 2.14 Dynamic response of loads after sudden voltage drop: (a) voltage drop curve and (b) dynamic response of the load [16].

through algebraic-differential equations as

$$\tau_P \frac{dx_P}{dt} = \frac{1}{P_0}(P_s - P_t) \tag{2.9}$$

$$\tau_Q \frac{dx_Q}{dt} = \frac{1}{Q_0}(Q_s - Q_t) \tag{2.10}$$

where P_0 and Q_0 represent the load active and reactive power consumptions, respectively, under the nominal operation state of the system, i.e., nominal voltage V_0. Parameters P_s and Q_s are the static active and reactive power consumptions at the system voltage V, correspondingly. Meanwhile, parameters P_t and Q_t are the transient active and reactive power consumptions, respectively. To describe the transient behavior, first-order differential equations have been exploited in which time constants τ_P and τ_Q represent the dynamic behavior of load restoration, and variables x_p and x_Q are the generic dimensionless state variables. P_s and Q_s can be modeled by

$$P_s = P_0 \cdot \left(\frac{V}{V_0}\right)^{\alpha_P}, \quad Q_s = Q_0 \cdot \left(\frac{V}{V_0}\right)^{\alpha_Q} \tag{2.11}$$

where exponents α_P and α_Q are constant parameters, which represent the dependency of active and reactive power consumptions on the voltage magnitude for the static behavior of loads, respectively. These parameters are selected based on the type of system loads and off-line studies as described in [2, 17]. It should be noted that for grid-connected SGBMGs, the frequency dependency is neglected, while for islanded SGBMGs in which the frequency deviations may be quite large, the frequency dependency should also be considered. Moreover, based on the firs-order differential equation, which represents the dynamic process of the load during transient condition, two models for P_t and Q_t can be exploited as [17]

$$\begin{cases} \textit{Additive generic model: } P_t = P_0 \cdot \left[\left(\frac{V}{V_0}\right)^{\alpha_{P,t}} + x_P\right], \quad Q_t = Q_0 \cdot \left[\left(\frac{V}{V_0}\right)^{\alpha_{Q,t}} + x_Q\right] \\ \textit{Multiplicative model: } P_t = x_P \cdot P_0 \cdot \left(\frac{V}{V_0}\right)^{\alpha_{P,t}}, \quad Q_t = x_Q \cdot Q_0 \cdot \left(\frac{V}{V_0}\right)^{\alpha_{Q,t}} \end{cases} \tag{2.12}$$

where $\alpha_{P,t}$ and $\alpha_{Q,t}$ are constant parameters for considering the dependency of the load consumption power on the voltage during transient periods. The constant parameters of these models can be determined using online measurements of smart meters.

In [18], another load model has been developed, which is suitable for the investigation of voltage stability and FIDVR phenomena. The mathematical model is suggested to be a mixture of aggregated static and aggregated IM loads:

$$P_s = \alpha_1 \times P_{ZIP} + \beta_1 \times P_{IM} \tag{2.13}$$

$$Q_s = \alpha_2 \times Q_{ZIP} + \beta_2 \times Q_{IM} \tag{2.14}$$

where α and β are the percentages of the aggregated static loads and dynamic loads, respectively. Moreover, P_{IM} and Q_{IM} represent the active and reactive powers of aggregated IM load, respectively. Moreover, P_{ZIP} and Q_{ZIP} represent the active and reactive powers of aggregated static load, respectively. The static ZIP load can be modeled by (2.15). To determine P_{IM} and Q_{IM}, state-space equations of the aggregated IM model are exploited, which are presented as follows [18]:

$$\begin{cases} \frac{dE'_d}{dt} = -\frac{1}{T'}\left[E'_d + (X - X')I_q\right] - (\omega - 1)E'_q \\ \frac{dE'_q}{dt} = -\frac{1}{T'}\left[E'_q - (X - X')I_d\right] + (\omega - 1)E'_d \\ \frac{d\omega}{dt} = -\frac{1}{2H}\left[T_0(A\omega^2 + B\omega + C + D\omega^E) - (E'_d I_d + E'_q I_q)\right] \\ I_d = \frac{1}{R_s^2 + X'^2}\left[R_s\left(U_d - E'_d\right) + X'\left(U_q - E'_q\right)\right] \\ I_q = \frac{1}{R_s^2 + X'^2}\left[R_s\left(U_q - E'_q\right) - X'\left(U_d - E'_d\right)\right] \end{cases} \tag{2.15}$$

where X' is the transient reactance, X is the open-circuit (no load) reactance, R_s is the stator resistance, H is the rotor inertia constant, ω is the mechanical speed, E'_d and E'_q are the d-axis and q-axis transient EMFs, respectively, U_d

and U_q are the d-axis and q-axis terminal voltages, respectively, and I_d and I_q are the d-axis and q-axis stator currents, respectively. Furthermore, the term $T_0(A\omega^2 + B\omega + C + D\omega^E)$ represents the load torque (T_L) in which parameters T_0, A, B, C, D, and E are constant. To determine the constant parameters, a parameter identification approach can be exploited using online measurements of smart meters.

Besides the importance of load model for the investigation of the dynamic voltage stability, the impact of SSSGs on the system voltage and other automatic reactive power compensation equipment should also be considered, which are not typically included in classical methods. On the other hand, some of the common causes for static voltage instability in conventional power systems may not be a serious challenge for modern distribution systems and SGBMGs. For instance, in SGBMGs, since the loads are located adjacent to the SSSGs and the feeders are relatively short, the large electrical distance between the loads and the voltage source may not be an issue. Moreover, proper reactive power resources are considered in the design stage of the SGBMG for the highest load demand to consider the static voltage stability limits. On the other hand, the dynamic load behavior due to various disturbances is not commonly considered in the design stage, and in such a condition, the limited reactive power resources of SGBMGs may not be enough. Therefore, lack of enough reactive power supply due to the limited reactive power capability of the SSSGs may lead to low-magnitude voltage conditions in short- and long-term situations. Such a low-voltage condition cannot be identified by classic indices of voltage instability.

One of the main conditions that may finally lead to dynamic voltage instability and undesired tripping of the healthy feeders is the FIDVR phenomenon. FIDVR is a temporary condition where the system voltage remains low due to the IM behavior after the fault clearance instant [19]. IM speed reduces during a severe voltage dip subsequent to a fault event. After the fault clearance instant, IMs may reaccelerate, decelerate, or stall depending on the system voltage level, FCT, and IM characteristic along with its associated mechanical load parameters [20]. At the instant of fault clearance, the terminal voltage of IM increases, but it may not return to its prefault value due to the reacceleration of the adjacent IMs and their significant amount of consumed reactive power. Moreover, reacceleration current of the motor under study may intensify the voltage drop at the motor terminal, which finally leads to FIDVR. This large reactive current of inductive loads, which passes through the healthy feeder for a long duration, may result in maloperation of feeder protection relays.

FIDVR phenomenon commonly occurs in SGBMGs with high penetration of IMs, especially after a fault-initiated islanding event. Fault-initiated islanding is a condition in which the PCC relay disconnects the SGBMG from the upstream system due to a system short-circuit fault. Since the islanded SGBMG has limited resources of reactive power, the IM cannot reaccelerate properly after the fault clearance, which leads to a considerable voltage drop at the SGBMG buses. In such a condition, large reactive currents will pass through the SGBMG overcurrent relays for a long period, which may finally result in relay maloperation for healthy feeders. Commonly, two types of FIDVR, i.e., balanced and unbalanced, may occur in distribution systems, which are investigated in the following:

2.3.2.1.1 Balanced FIDVR Balanced FIDVR occurs symmetrically in all three phases of the distribution system. In other words, zero and negative sequences of voltage and current signals are relatively small. Depending on the SGBMG configuration, balanced FIDVRs commonly take place after two- and three-phase faults. However, the chance of balanced FIDVR occurrence after single-phase faults is very low.

For further investigations, a real balanced FIDVR case is investigated. To do so, a portion of a real distribution network is simulated using the PSCAD/EMTDC software environment [21]. Figure 2.15 illustrates this system, which includes the upstream subtransmission 63/20 kV substation (USS), two SSSGs, single-phase motors (SPMs), three-phase IMs, static loads, and a capacitor bank. Dynamic components such as SPMs and IMs are modeled and simulated using their transient differential equations as presented in [22]. The other system elements are modeled appropriately and with careful details, and the related data are given in [21]. As shown, the SSSGs are connected to an SGBMG, which can operate in both islanded and grid-connected modes. In addition, the 63 kV network is modeled by its Thevenin's equivalent as an external grid. Meanwhile, it should be considered that the SSSG excitation systems have a considerable impact on the FIDVR occurrence. Therefore, the excitation system should be carefully modeled for studying the FIDVR phenomenon occurring in SGBMGs. In this regard, the excitation system has been implemented using the AC5A dynamic model provided by [23]. Moreover, it is assumed that SSSG1 operates in the constant active/reactive power control mode. To consider the worst case, it is assumed that SSSG2 and the capacitor bank are out of service. Furthermore, the amount of SGBMG local load consumption is equal to the SSSG1 generation.

In the considered scenario, a three-phase short-circuit occurs on feeder 1 at $t = 8$ s. After 100 ms, the OFR1 relay operates as the main protection and clears the fault. Figure 2.16 depicts the rotor speeds of IMs. A three-phase short-circuit results in a symmetrical voltage dip, and thus, the speeds of SPMs and IMs, which are connected to the adjacent healthy feeders, reduce during the fault. After the fault clearance, the speeds of IM1 and IM3 recover to their nominal values. However, stall condition occurs for IM2 and SPMs due to their large electrical distances from the main bus of USS. The current magnitudes of feeder 1, feeder 2, and feeder 3 are illustrated in Fig. 2.17. As shown in this figure, the current magnitude of

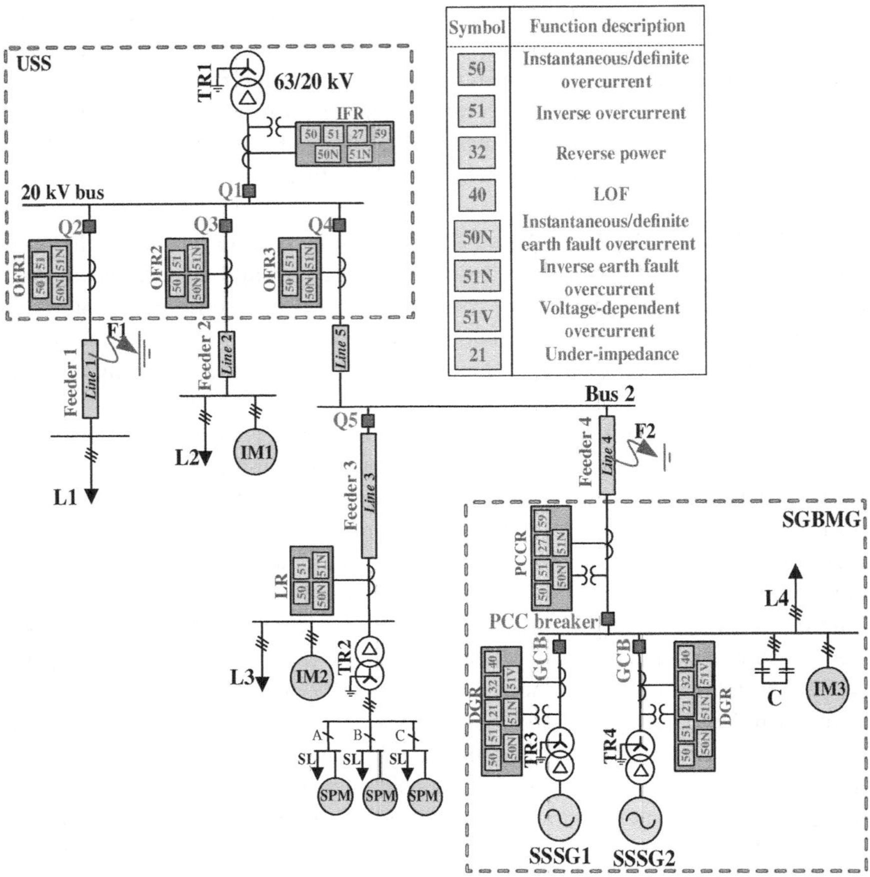

Symbol	Function description
50	Instantaneous/definite overcurrent
51	Inverse overcurrent
32	Reverse power
40	LOF
50N	Instantaneous/definite earth fault overcurrent
51N	Inverse earth fault overcurrent
51V	Voltage-dependent overcurrent
21	Under-impedance

Fig. 2.15 Sample test system. Adapted from [21].

the faulty feeder (feeder 1) increases after the fault occurrence. Albeit, the current magnitude of adjacent feeders decreases for a short duration after the fault occurrence due to the severe voltage drop. In this case, the symmetrical current of Fig. 2.17 and voltage magnitude shown in Fig. 2.18 represent a balanced FIDVR phenomenon.

2.3.2.1.2 Unbalanced FIDVR Unbalanced FIDVR is defined as a postfault condition in which the increment of current magnitude is different for each phase. This phenomenon usually occurs due to two reasons: (i) unbalanced fault occurrence in a system including high penetration of SPMs and (ii) fault occurrence in a system with different penetration of SPMs in each phase. Although unbalanced FIDVRs

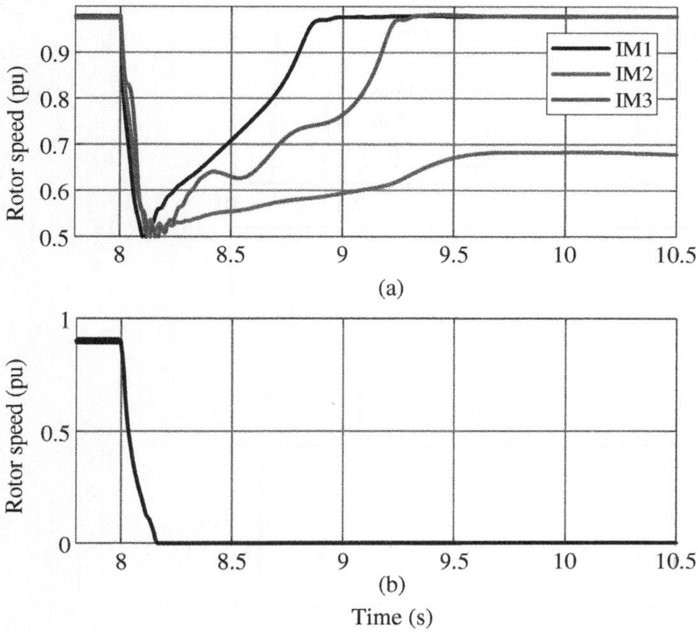

Fig. 2.16 Rotor speeds of induction motors: (a) three-phase IMs and (b) single-phase IMs.

usually occur after unsymmetrical short-circuit faults, they can also occur after three-phase symmetrical faults when the penetration of SPMs is not similar in each phase of the system. During the unbalanced FIDVR condition, motor reacceleration, deceleration, or stalling occurs differently in each phase, and thus, some negative sequence current may pass through the feeders. In such a condition, the operation of overcurrent protection in each phase is different, and overcurrent relays may operate only for one or two phases.

In order to investigate the unbalanced FIDVR condition, a two-phase short-circuit event is simulated. In this case, a two-phase B to C fault occurs on feeder 1 at $t = 8\,\text{s}$ and is cleared after 400 ms. Voltage magnitudes at the SPM terminals are depicted in Fig. 2.19. Since the connection of distribution transformers is Dyn, the amount of voltage drop at the LV side is different from that of the HV side. In other words, the voltage magnitude of phase B at the LV side considerably decreases during the phase B to C fault event. However, the voltage magnitudes of phases A and C do not change significantly, as shown in Fig. 2.19. Thus, only single-phase IMs, which are connected to phase B, are stalled during the fault, as depicted in Fig. 2.20. Meanwhile, phases A and C IMs, which are not stalled, reaccelerate after the fault clearance. Severe voltage

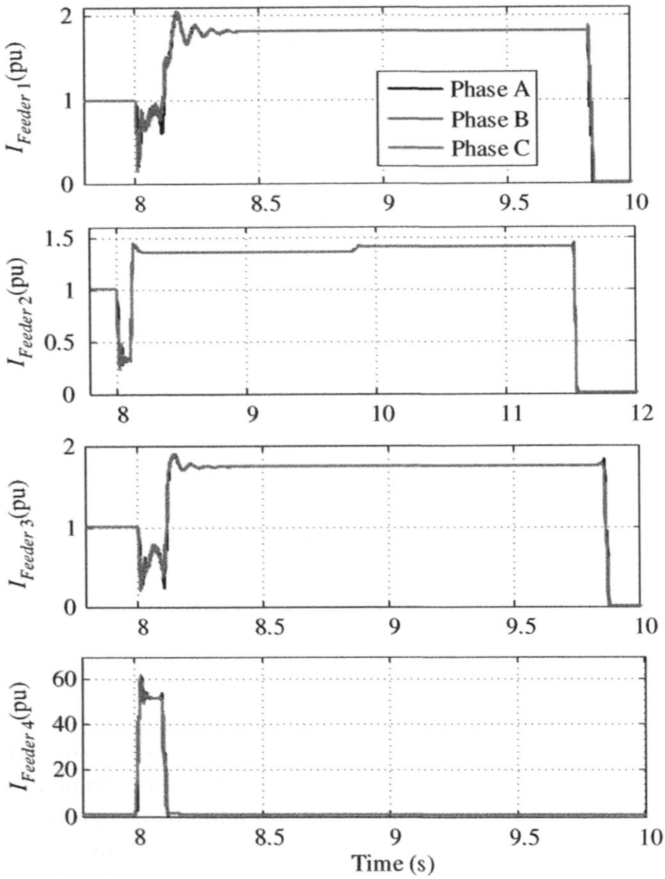

Fig. 2.17 Current magnitudes of the distribution feeders.

drops in phases B and C at the HV side of the distribution transformer, i.e., 20 kV voltage level, result in the speed reduction of three-phase IMs, as shown in Fig. 2.20. Current signal of each feeder is shown in Fig. 2.21. According to this figure, the current magnitude of healthy feeders in each phase is different from that of the other phases after the fault clearance, which represents an unbalanced FIDVR condition. After the fault clearance, the unbalanced FIDVR results in overcurrent function 51 incorrect operation of healthy feeders in one phase. Furthermore, a high-magnitude negative sequence current passes through the feeders, which may cause maloperation of the feeder broken conductor protection function.

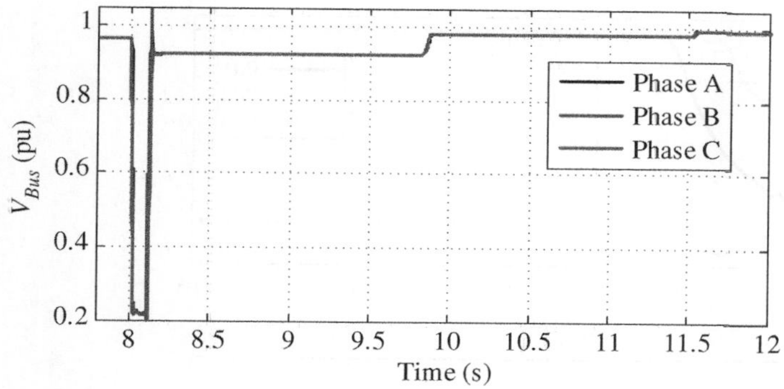

Fig. 2.18 Voltage magnitudes during balanced FIDVR.

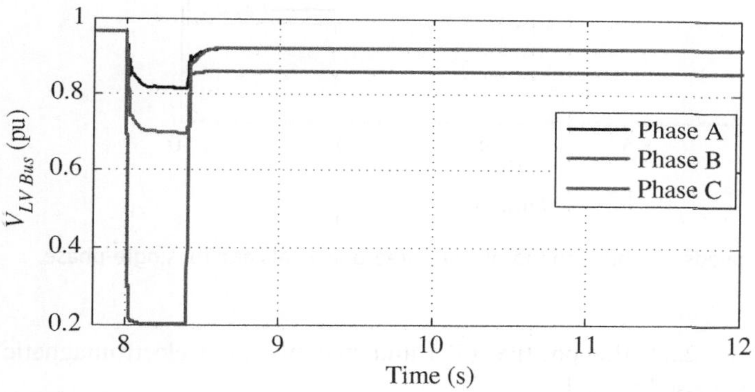

Fig. 2.19 Voltage magnitudes during unbalanced FIDVR.

The main factor, which results in FIDVR and voltage degradation conditions in SGBMGs, is IMs behavior and their instability following a short-circuit fault. This condition may have significant impact on the protection system, especially over-current, undervoltage, and load-shedding relays. Thus, for proper adjustment of the mentioned functions, the performance of IMs should be carefully investigated. To do so, in the following, the IM behavior is analytically securitized.

For analytical investigations, all IMs are modeled as an equivalent aggregated IM. Positive and negative sequence circuits of the IM aggregate model are shown in Fig. 2.22. Meanwhile, the negative sequence circuit should only be considered for an unbalanced voltage condition, and thus, the amounts of V_n and I_n are equal to zero for a balanced voltage sag. Parameters R_s, X_s, R'_r, X'_r, and s represent stator resistance, stator reactance, rotor resistance, rotor reactance, and rotor slip, respectively. Moreover, it is assumed that the negative and positive sequence parameters are equal to each other, and magnetizing reactance has been neglected.

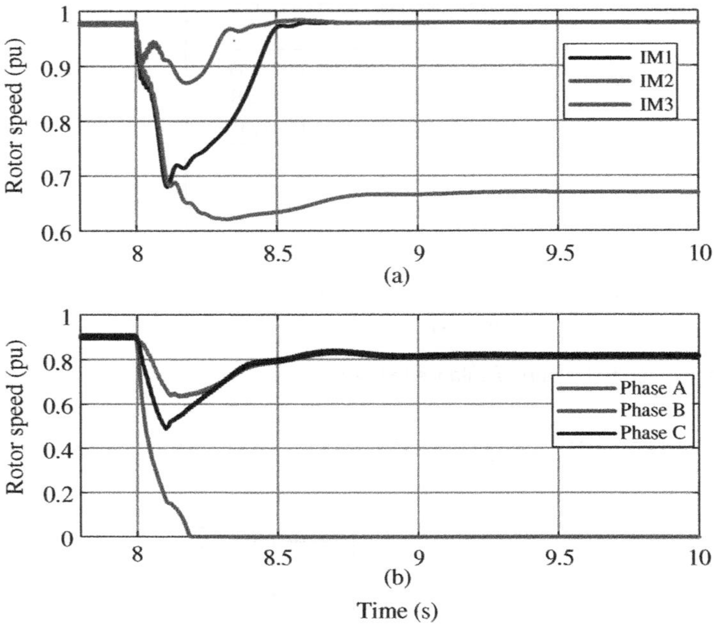

Fig. 2.20 Rotor speeds of induction motors: (a) three-phase IMs and (b) single-phase IMs.

Considering Fig. 2.22, the positive (T_p) and negative (T_n) electromagnetic torques can be expressed as [24]

$$T_p = \frac{3I_p^2 R_r'}{s\omega_s} = \frac{3V_p^2 R_r'}{s\omega_s\left[\left(R_s + \frac{R_r'}{s}\right)^2 + (X_s + X_r')^2\right]} \tag{2.16}$$

$$T_n = -\frac{3I_n^2 R_r'}{(2-s)\omega_s} = -\frac{3V_n^2 R_r'}{(2-s)\omega_s\left[\left(R_s + \frac{R_r'}{2-s}\right)^2 + (X_s + X_r')^2\right]} \tag{2.17}$$

where ω_s is the synchronous speed, and I_p and I_n are the motor positive and negative sequence currents, respectively. T_p and T_n can be simplified as follows:

$$T_p = 3V_p^2 s / \left(R_r'\omega_s\right) \tag{2.18}$$

$$T_n = -3V_n^2 R_r'(2-s) \Big/ \left(\omega_s \times \left[\left(2R_s + R_r'\right)^2 + 4\left(X_s + X_r'\right)^2\right]\right) \tag{2.19}$$

The total electromagnetic torque T_e can be calculated as

$$T_e = T_p + T_n \tag{2.20}$$

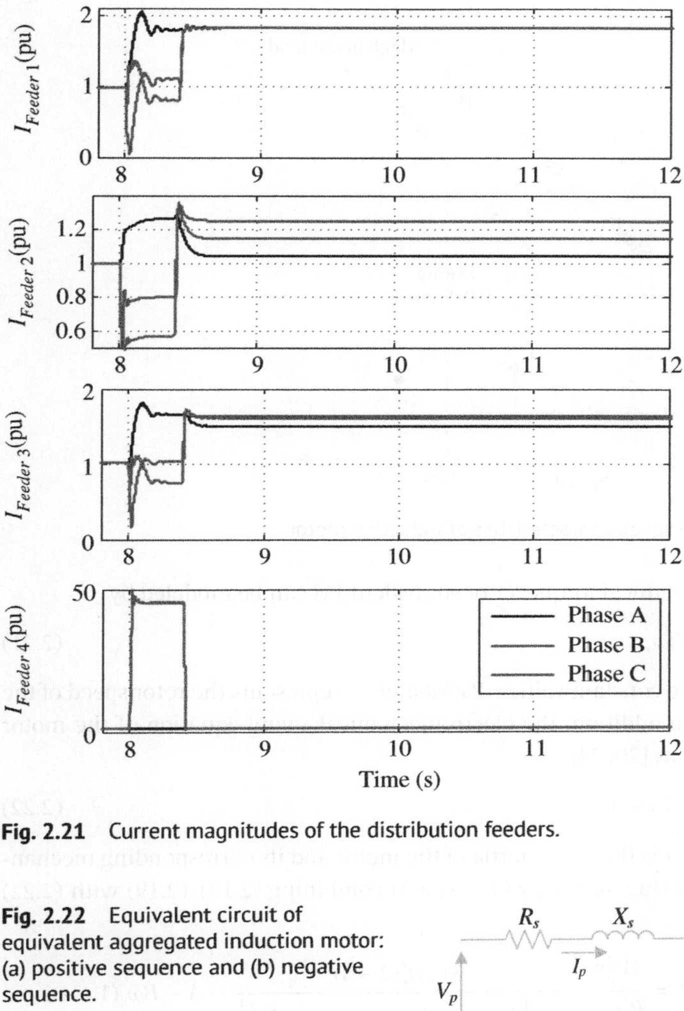

Fig. 2.21 Current magnitudes of the distribution feeders.

Fig. 2.22 Equivalent circuit of equivalent aggregated induction motor: (a) positive sequence and (b) negative sequence.

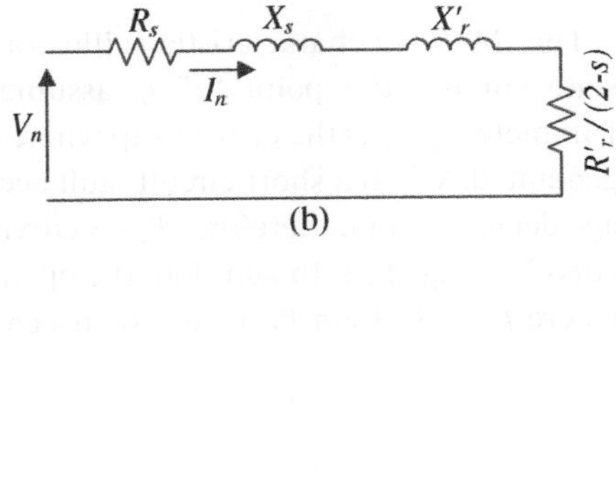

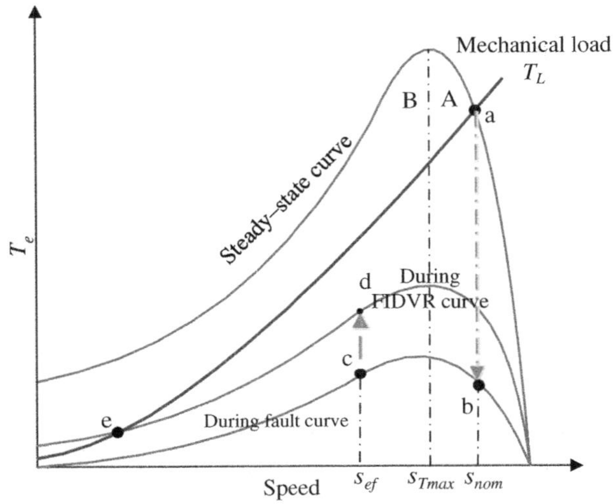

Fig. 2.23 Torque–speed characteristics of induction motor.

Moreover, mechanical torque T_L of equivalent IM can be modeled by

$$T_L = A + B\omega_r \tag{2.21}$$

where A and B are constant values. Parameter ω_r represents the rotor speed of the aggregated IM. In addition, the electromechanical swing equation of the motor can be expressed as [20, 24]

$$Jd\omega_r/dt = T_e - T_L \tag{2.22}$$

where parameter J is the total inertia of the motor and its corresponding mechanical load. Considering $\omega_r = \omega_s \times (1 - s)$ and combining (2.18)–(2.19) with (2.22) results in

$$-J\omega_s ds/dt = \frac{3V_p^2 s}{R_r'\omega_s} - \frac{3V_n^2 R_r'(2-s)}{\omega_s\left[(2R_s + R_r')^2 + 4(X_s + X_r')^2\right]} - A - B\omega_s(1-s) \tag{2.23}$$

The IM $T_e - s$ characteristic is illustrated in Fig. 2.23, which is called "steady-state curve," and point "a" is assumed as the steady-state operating point. Parameter s_{Tmax} is the critical slip where the maximum electromagnetic torque is generated. When a short-circuit fault occurs on the system, the IM terminal voltage decreases and, therefore, $T_e - s$ curve changes, which is named "during-fault curve" in Fig. 2.23. In addition, the operating point suddenly jumps to point "b," where T_e is less than T_L. Under such a condition, according to (2.22), the machine speed decreases and its slip increases. Therefore, the operating point moves to the left on the "during-fault curve," whose rate of change depends on the total machine inertia. At the instant of fault clearance, the terminal voltage of IM increases but does not usually return to its prefault condition, as the IM should reaccelerate and thus draws a significant amount of reactive power. Moreover, motor reacceleration current leads to some voltage drop at the motor terminal. In such a condition, the FIDVR phenomenon may occur. Hence, once the fault is cleared, $T_e - s$ characteristic changes to a new curve, which is named "during FIDVR curve."

By solving (2.23), the aggregated IM slip after the fault clearance instance $t = t_1$ is derived as [25]

$$s(V_p, V_n, t) = F(V_p, V_n) + [s_{ef} - F(V_p, V_n)] \times e^{-M(V_p, V_n) \times (t - t_1)} \tag{2.24}$$

$$F(V_p, V_n) = \frac{A + B\omega_s + 2k_n V_n^2}{k_n V_n^2 + k_p V_p^2 + B\omega_s}, \quad M(V_p, V_n) = \frac{k_n V_n^2 + k_p V_p^2 + B\omega_s}{J\omega_s} \tag{2.25}$$

$$k_p = \frac{3}{R'_r \omega_s}, k_n = \frac{3R'_r}{\omega_s \left[(2R_s + R'_r)^2 + 4(X_s + X'_r)^2 \right]} \tag{2.26}$$

where V_p and V_n are the voltage positive and negative sequences after the fault clearance, respectively. Meanwhile, s_{ef} is the IM slip at the fault clearance instant t_1. If the intersection of "during FIDVR curve" and T_L is less than s_{Tmax}, as shown in Fig. 2.23, the IM speed will become less than the normal condition. In this condition, IM consumes more reactive power, and consequently, its terminal voltage would decrease. Meanwhile, more voltage drop leads to further reduction of T_e and thus machine speed decreases, which may result in motor stalling. However, if T_e is greater than T_L after the fault clearance instant, the IM will accelerate, even when the fault-clearing instant slip is less than s_{Tmax}.

Based on (2.24), the reacceleration time t_R of aggregated IM for both the proper reaccelerated and stalled conditions can be obtained from [25]

$$t_R = \frac{1}{M(V_p, V_n)} \times \ln \frac{F(V_p, V_n) - s_{ef}}{F(V_p, V_n) - s_{nom}} \tag{2.27}$$

Although s_{ef} can be computed using (2.24), it is assumed to be equal to s_{Tmax} for simplification. Thus, the reacceleration time between the maximum torque slip and the operating point after the fault clearance (assumed to be s_{nom}) is derived as follows:

$$t_R = \frac{1}{M(V_p, V_n)} \times \ln \left(1 + \frac{\Delta s}{s_{nom} - F(V_p, V_n)} \right) \tag{2.28}$$

where $\Delta s = s_{ef} - s_{nom}$, which is always a positive number. Since $\ln\left(1 + \frac{\Delta s}{s_{nom} - F(V_p, V_n)}\right) \approx \frac{\Delta s}{s_{nom} - F(V_p, V_n)}$, (2.28) can be simplified as

$$t_R = \frac{J\omega_s \Delta s}{k_p s_{nom} V_p^2 - (2 - s_{nom})k_n V_n^2 - (A + B\omega_s - s_{nom} B\omega_s)} \tag{2.29}$$

This equation can be rewritten as

$$t_R = \text{TMS}/\left[\left({}^{V_p}/_{V_{Th}^p}\right)^2 - \left({}^{V_n}/_{V_{Th}^n}\right)^2 - 1\right] \tag{2.30}$$

where

$$\text{TMS} = J\omega_s \Delta s/(A + B(1 - s_{nom})\omega_s) \tag{2.31}$$

$$V_{Th}^n = \sqrt{\frac{A + B(1 - s_{nom})\omega_s}{k_n(2 - s_{nom})}}, \quad V_{Th}^p = \sqrt{\frac{A + B(1 - s_{nom})\omega_s}{s_{nom} k_p}} \tag{2.32}$$

For the stable FIDVR condition, IM speed and system voltage recover to their nominal values before t_R seconds without any control action. However, IM speed does not recover during the unstable FIDVR phenomenon. Therefore, for adjusting the time delay of the protection functions of SGBMG, the proper value of t_R should be considered to prevent maloperation of the protection system during the stable FIDVR condition. Moreover, if the IM speed does not recover to its nominal value within t_R seconds, the load-shedding function should operate and disconnect the unstable IM.

Example 2.2

A three-phase short-circuit fault occurs in a grid-connected SGBMG, which contains SSSGs and IMs and is cleared after 200 ms. Determine the IM speed after the fault clearance. The parameters of the aggregated IM model are $J = 3$ s, $R_r' = 0.15\,\Omega$, and $\omega_s = 314$ rad/s. The nominal terminal voltage is 6 kV, and the load torque is expressed as $T_L = 510$ N.m. The nominal slip is considered 5%.

Solution

Considering the worst condition, the voltages, both V_p and V_n at the IM terminal, become zero during the three-phase fault. Therefore, T_e is equal to zero during the short-circuit fault. Therefore, (2.22) can be written as

$$Jd\omega_r/dt = -T_L$$

where $J = 0.5$ s and $T_L = 510$ N.m. Therefore, the rate of speed reduction can be determined by

$$|\Delta\omega_r/\Delta t| = T_L/J = \frac{510}{3} = 170 \text{ rad/s}^2$$

Therefore, during the short-circuit fault, i.e., 200 ms, the amount of speed reduction can be determined as

$$|\Delta\omega_r| = 170\ \text{rad/s}^2 \times 0.2\ \text{s} = 34\ \text{rad/s}$$

The initial rotor speed is equal to

$$\omega_{r\,(0)} = (1 - s) \times \omega_s = (1 - 0.05) \times 314\ \text{rad/s} = 298.3\ \text{rad/s}$$

Therefore, the speed of IM after the fault clearance is

$$\omega_{r\,(final)} = \omega_{r\,(0)} - |\Delta\omega_r| = 298.3\ \text{rad/s} - 34\ \text{rad/s} = 264.3\ \text{rad/s}$$

and the slip increases to 15.82%.

Example 2.3

In the previous example, calculate the aggregated IM recovery time. It is assumed that the voltage reaches to 0.8 pu after the fault clearance.

Solution

After the fault clearance $V_p = \frac{0.8\times 6\ \text{kV}}{\sqrt{3}} = 2.77\ \text{kV}$, $V_n = 0$. Meanwhile, k_p is calculated as

$$k_p = \frac{3}{R'_r\omega_s} = \frac{3}{0.15 \times 314} = 0.0637$$

Therefore, $F(V_p, V_n)$ and $M(V_p, V_n)$ are determined according to (2.25) as follows:

$$F(V_p, V_n) = \frac{A + B\omega_s + 2k_n V_n^2}{k_n V_n^2 + k_p V_p^2 + B\omega_s} = \frac{510}{0.0637 \times \left(\frac{6000}{\sqrt{3}} \times 0.8\right)^2} = 0.001$$

$$M(V_p, V_n) = \frac{k_n V_n^2 + k_p V_p^2 + B\omega_s}{J\omega_s} = \frac{0.0637 \times \left(\frac{6000}{\sqrt{3}} \times 0.8\right)^2}{3 \times 314} = 519.36$$

Meanwhile, $\Delta s = s_{ef} - s_{nom} = 15.82\% - 5\% = 10.82\%$. Considering (2.28), the recovery time can be calculated as

$$t_R = \frac{1}{M(V_p, V_n)} \times \ln\left(1 + \frac{\Delta s}{s_{nom} - F(V_p, V_n)}\right)$$

$$= \frac{1}{519.36} \times \ln\left(1 + \frac{10.82\%}{5\% - 0.001}\right) = 2.24\ \text{ms}$$

2.3.2.2 Frequency Stability

Frequency stability is the ability of the power network to preserve the system frequency within the desired ranges during steady-state operation and after the occurrence of disturbances [1, 14]. Traditionally, the frequency instability was a major concern in the transmission system, and fast frequency response and other control actions for retaining the frequency stability were mainly taken at the higher voltage levels. Nowadays, due to the increased penetration of DGs, frequency regulation has become important, even in isolated SGBMGs; however, it is not a serious concern for the grid-connected SGBMGs. Frequency stability in the isolated SGBMGs has the following challenges compared to that of conventional power systems:

- Limited power resources and lack of spinning reserves
- Large excursions and rapid changes of SGBMG frequency due to low-inertia generation units
- High risk of frequency instability after the outage of generation units due to their low number
- Loss of active/reactive power balance due to the unintentional islanding of the SGBMG such as fault-initiated islanding
- Power imbalance in the isolated SGBMG due to load dynamic behavior.

Therefore, the frequency stability in the isolated SGBMGs requires careful attention, especially after unintentional islanding events. In the grid-connected operation of an SGBMG, the distribution system preserves the SGBMG frequency and guarantees its frequency stability, and therefore, the possibility of frequency instability decreases. In this condition, the SGBMG resources, especially SSSGs, can help the upstream system during frequency change events and provide fast frequency response. Unlike the inverter-based resources, SSSGs provide "inherent inertial response" after frequency events due to the inherent mechanical inertia of the SSSG and its associated prime mover. In fact, without using any specific control action, SSSGs can provide inertial response in both the grid-connected and isolated SGBMGs, exploiting the kinetic energy stored in the SSSG rotating masses. In such a condition, the SSSG active power will be changed in proportion to the ROCOF. The active power change (ΔP) of an SSSG due to frequency changes can be simply determined by

$$\Delta P = \Delta S \times H \times \mathrm{ROCOF}/30 \tag{2.33}$$

where S is the SSSG rating in MVA, H is the inertia constant in s, ROCOF is the rate of change of frequency in Hz/s, and ΔP gives the active power change in MW. Based on this equation, for instance, after a low-frequency event in the SGBMG or upstream grid, the ROCOF is negative and thus ΔP is positive, and the active power output of the SSSG increases. It is worth mentioning that the inertial response of the SG is commonly referred to as "fast frequency response," which is a more aggressive and faster response compared to the governor PI controller response, known as the primary frequency control [10].

2.3.2.3 Transient Stability

Generally, transient stability is known as the ability of the power system to preserve its synchronism after severe transient events, including short-circuit faults, loss of generation, transmission line outage, etc. [16, 26]. In the grid-connected SGBMGs, the transient stability is defined as the ability of SSSGs to preserve their synchronism with the upstream system after large disturbances. Meanwhile, in an islanded SGBMG, the transient stability is referred to as the ability of SSSGs to maintain their synchronism with the other SSSGs located in the islanded SGBMG after the occurrence of large events. Due to smaller inertia constant of SSSGs, longer FCT, and higher possibility of fault occurrence in distribution feeders, the possibility of transient instability of SGBMGs is higher than that of large-scale generators, and thus, this issue requires special care. The major events that endanger the SGBMG transient stability include:

- Close severe short-circuit faults
- Fault-initiated islanding events
- SSSG faults such as the loss of field
- Distribution line outages
- Changes of operation modes.

To quantify the transient stability margin of SSSGs, various indices have been developed. However, the most suitable index, which is widely utilized in both transmission and distribution levels, is CCT. CCT is the maximum time duration between the initiation and clearance of a short-circuit fault for which the generator/system remains stable and its speed recovers [2, 26]. Various techniques have been utilized for determining CCT and evaluating the transient stability margin, including (i) equal-area criterion, (ii) time-domain electromagnetic simulation-based techniques, (iii) Lyapunov techniques, and (iv) hardware-in-the-loop (HIL) methods. The principles of these methods are briefly discussed in the following sections.

The operation mode of the SSSG and the SSSG ride-through capability have considerable impacts on the transient stability of the SSSG and SGBMG. When the SSSG participates in voltage regulation during the faults and increases its reactive power, the SSSG internal and terminal voltages increase. Therefore, the amount of the SSSG active power increases compared to that of a control mode in which the SSSG does not participate in voltage regulation during the fault, and therefore, its field current relatively remains constant. The CCT values for a grid-connected SSSG for various operation modes are depicted in Fig. 2.24 [27]. As shown, for the voltage control mode in which the SSSG participates in voltage regulation during the fault, the CCT is much higher than that of the other control modes. However, for no voltage regulation during the fault, the CCT is decreased. Furthermore, the location of fault has a significant impact on the transient stability of the SSSG.

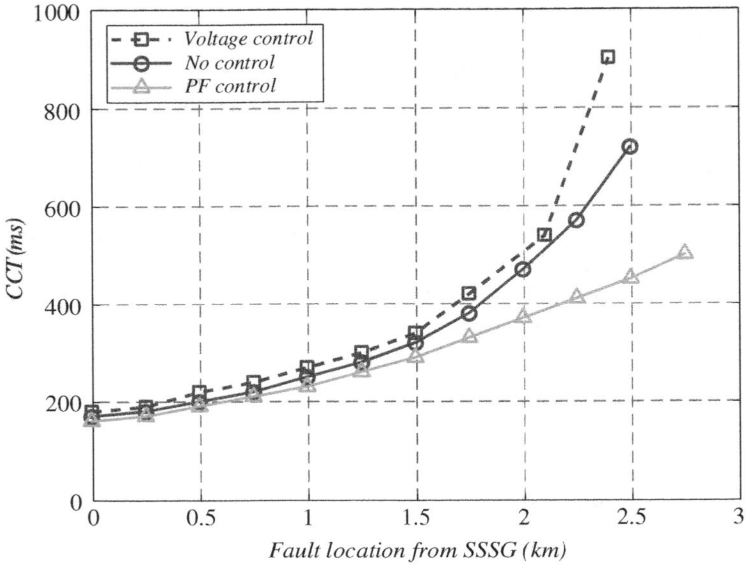

Fig. 2.24 CCT values for various fault locations under different control modes. Adapted from [27].

As depicted, for remote faults, the CCT amount is considerably greater than that for the close-in faults.

To improve the transient stability of SGBMGs, various techniques are exploited, including:

- Using proper control modes
- Exploiting proper ride-through capability
- Decreasing FCT using a suitable protection system
- Using a mechanical flywheel for increasing the inertia time constant
- Fast detection of unintentional islanding events.

Unlike large-power systems in which SGs are connected to the transmission/subtransmission voltage levels, the SSSGs are connected to distribution systems and operate in adjacent to the system loads. Therefore, the behavior of dynamic loads, especially IMs, could have much more impact on the transient stability of SSSGs, and thus, this should be considered in stability studies. For instance, as mentioned before, the penetration of IMs in distribution systems has been significantly increased. IM slip increases during the short-circuit faults. This may lead to FIDVR phenomenon where the system voltage remains low for a period due to the large slip of IMs and their significant reactive power consumptions after the fault clearance. This condition could jeopardize the SSSG transient stability, as investigated in this section. In this regard, a typical

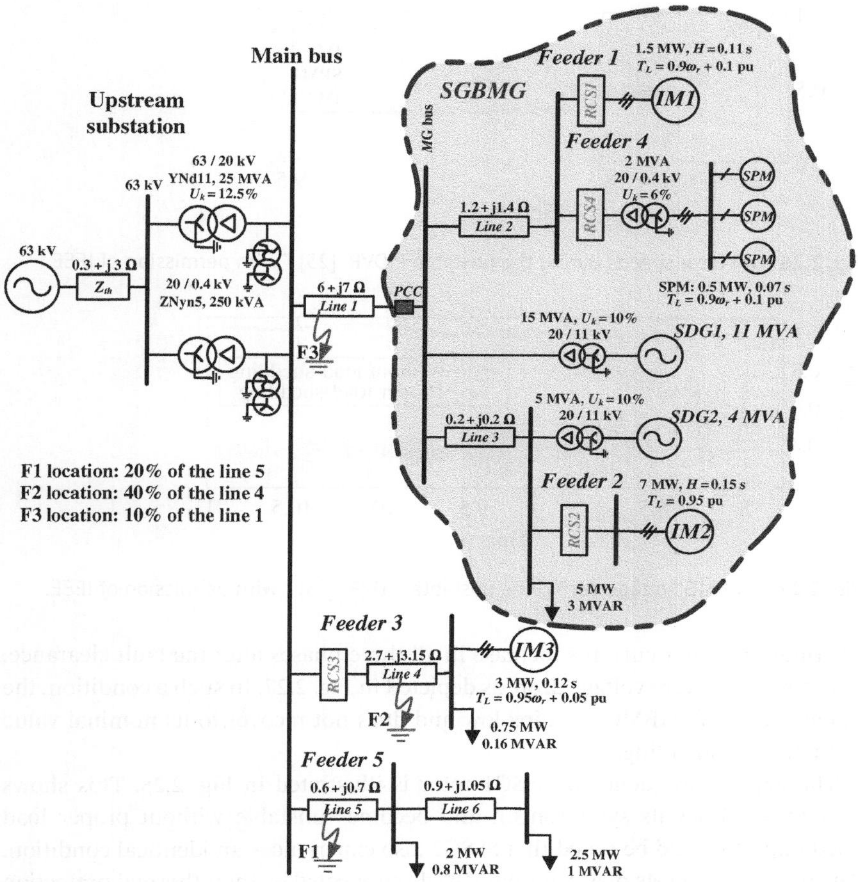

Fig. 2.25 Single line diagram of a real distribution network. [25] / with permission of IEEE.

distribution system for FIDVR studies, as shown in Fig. 2.25, is investigated [25]. This system contains two SSSGs, two aggregated three-phase IMs (IM1 and IM2), a set of three aggregated single-phase IMs (SPMs), aggregated static loads, and remote-control switches (RCSs). To study the impact of FIDVR on the SSSG transient stability, various short-circuit scenarios are simulated as explained in the following paragraphs.

Figure 2.26 illustrates the speeds of IMs for a three-phase short-circuit fault occurring at F1 and cleared after 200 ms by the feeder protection relay. It shows that IM2 rotor stalls due to its large mechanical load torque (T_L = 0.95 pu). Moreover, the stall condition takes place for SPMs because of their associated small inertias (0.07 s). Due to the large values of motor slips at the fault-clearing instant,

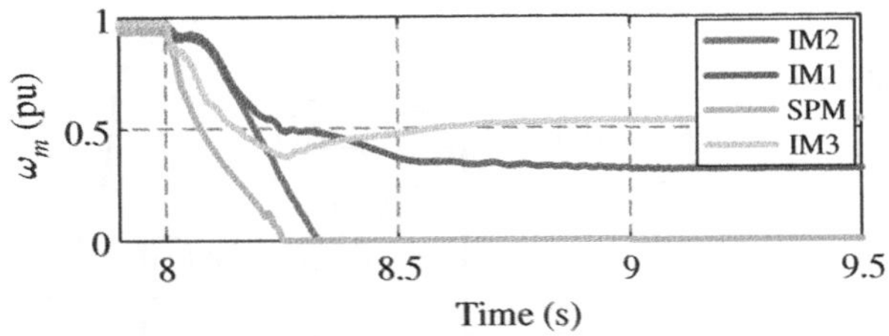

Fig. 2.26 IM rotor speeds during the unstable FIDVR. [25] / with permission of IEEE.

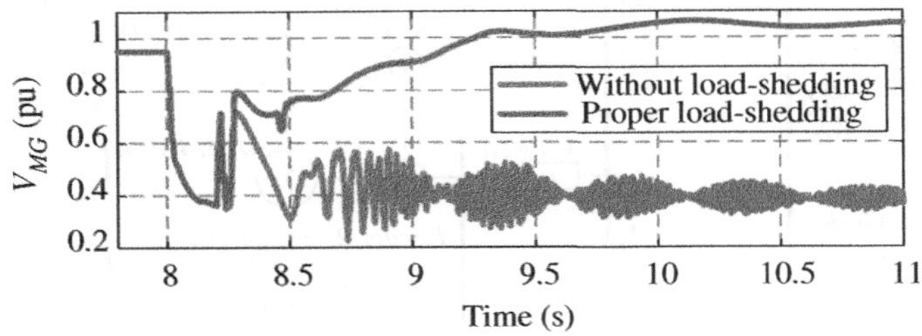

Fig. 2.27 SGBMG voltage during the unstable FIDVR. [25] / with permission of IEEE.

distribution feeder currents increase in all three phases after the fault clearance, resulting in a severe voltage drop, as depicted in Fig. 2.27. In such a condition, the voltage level of SGBMG remains low and does not recover to its nominal value without load shedding.

The angular frequency of SSSG1 rotor is illustrated in Fig. 2.28. This shows that SSSG1 loses its synchronism and becomes unstable without proper load shedding. It should be noted that SSSG2 also experiences an identical condition. The unrecovered IMs may eventually be disconnected by their thermal protection relays. However, the thermal relay usually has a large operation delay, which may lead to transient instability of the SGBMG. Therefore, for the unrecovered FIDVR

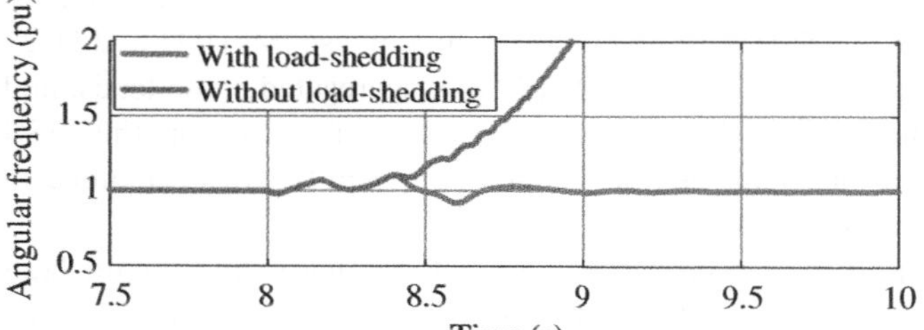

Fig. 2.28 Angular frequency of SSSG1 [25].

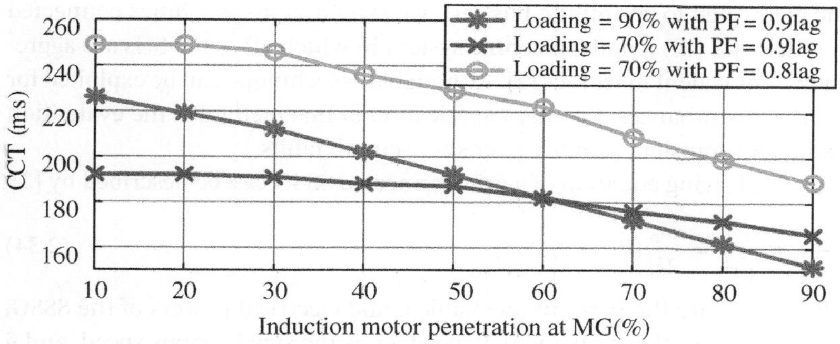

Fig. 2.29 CCT values of SSSG1 for various amounts of IM penetration [25].

phenomenon, a fast load-shedding algorithm is required to detect unrecovered feeders and disconnect them from the system to prevent the SGBMG and SSSG instability. For instance, the stability of SSSGs can be retained after interrupting the IM loads connected to the SGBMG at $t = 8.38$ s, as shown in Fig. 2.28 [25].

For further investigations, the influence of IM penetration level on the CCT is investigated. To do so, the ratio of the SGBMG IM power to its total load is varied from 0% to 100%. In such a condition, the CCT amount of SSSG1 for a three-phase short-circuit fault at F1 is depicted in Fig. 2.29 for various cases. As shown, the amount of CCT significantly reduces with increasing the IM penetration. Moreover, the CCT is influenced by various real-time factors, including FCT, fault location, and SSSG loading conditions. Therefore, the load-shedding strategy to tackle the transient instability problem should be designed considering such parameters.

2.4 SGBMG Stability Assessment

2.4.1 Large Disturbances

For evaluating the SGBMG stability during and after a large disturbance, various techniques and algorithms have been proposed. The nonlinear behavior of the SSSG control systems and dynamic loads shall be considered for analyzing large perturbations. Appropriate techniques, which can be exploited for examining the SGBMG stability following sizable disturbances, are explained as follows.

2.4.1.1 Equal-Area Criterion Technique

Equal-area criterion is a classic graphical method widely exploited for the quick prediction of the transient first-swing stability of SGs, including the SSSGs. By utilizing this technique, the stability of a grid-connected SSSG can be analyzed. It should be noted that this method is only applicable to (i) a single synchronous machine connected to an infinite bus, (ii) two synchronous machines connected to each other, and (iii) a multi-machine system in which all of the SGs are aggregated in two equivalent machines [2]. Although this technique can be exploited for various large disturbances, the main application of this method is the evaluation of the generator transient stability after short-circuit faults.

The so-called swing equation of a grid-connected SSSG can be described by [2]

$$\frac{d\omega}{dt} = \frac{d^2\delta}{dt^2} = \frac{\omega_0}{2H}(P_m - P_e) \tag{2.34}$$

where P_m and P_e are the per-unit mechanical and electrical powers of the SSSG, respectively. ω is the electrical angular speed, ω_0 is the synchronous speed, and δ is the electrical power angle. Both sides of the swing equation can be multiplied by $2 \times d\delta/dt$ as

$$2\frac{d\delta}{dt}\frac{d^2\delta}{dt^2} = \frac{\omega_0}{2H}(P_m - P_e)\frac{d\delta}{dt} \tag{2.35}$$

This can be rewritten as

$$\frac{d}{dt}\left[\left(\frac{d\delta}{dt}\right)^2\right] = \frac{2\pi f_0}{H}(P_m - P_e)\frac{d\delta}{dt} \tag{2.36}$$

Then,

$$d\left[\left(\frac{d\delta}{dt}\right)^2\right] = \frac{2\pi f_0}{H}(P_m - P_e)d\delta \tag{2.37}$$

Consequently, ${}^{d\delta}/_{dt}$ can be calculated as follows by integration of both sides of (2.37):

$$\frac{d\delta}{dt} = \sqrt{\frac{2\pi f_0}{H}\int_{\delta_0}^{\delta}(P_m - P_e)d\delta} \tag{2.38}$$

If the machine remains stable, the amount of ${}^{d\delta}/_{dt}$ and the relative speed with respect to the synchronous speed should eventually become zero after the disturbance. In this regard, the transient stability criterion can be written as

$$\int_{\delta_0}^{\delta}(P_m - P_e)d\delta = 0 \tag{2.39}$$

In fact, based on (2.39), the change of the SSSG total energy after the disturbance should become zero. For better explanation of this method and the abovementioned criterion, the SGBMG shown in Fig. 2.30 is considered. This figure illustrates the single line diagram of a distribution system containing an SSSG and its associated transformer. In addition, Fig. 2.31 illustrates the (P_e – δ) curve of the grid-connected SSSG. When a short-circuit fault occurs on the upstream system,

Fig. 2.30 A typical grid-connected SSSG.

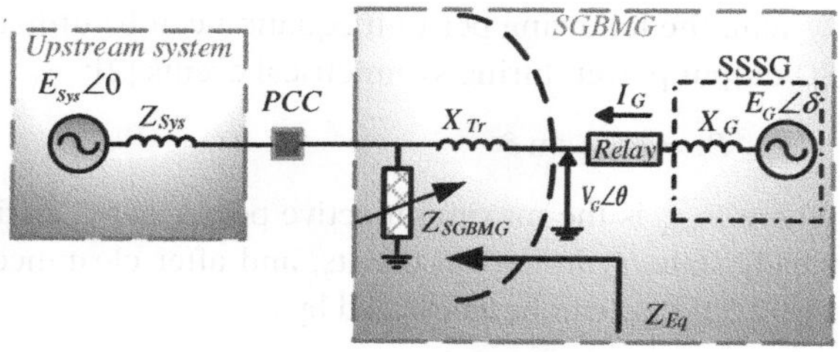

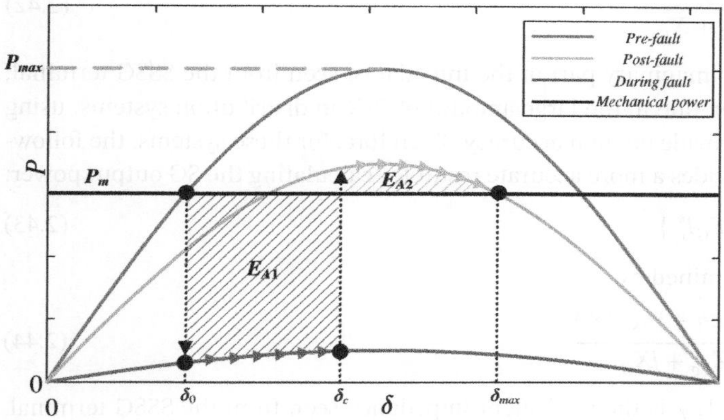

Fig. 2.31 P–δ curve of the grid-connected SSSG.

the operating point of SSSG at δ_0 moves downward suddenly as shown in Fig. 2.31. In this condition, the SSSG rotor total energy E_{total} can be calculated as

$$E_{total} = E_{A1} - E_{A2} = \int_{\delta_0}^{\delta_c} (P_m - P_{e-A1})d\delta + \int_{\delta_c}^{\delta_{max}} (P_m - P_{e-A2})d\delta \tag{2.40}$$

where δ_{max} is the maximum amount of SSSG rotor angle after the short-circuit event for which the SSSG remains stable. Meanwhile, the mechanical input power P_m is commonly considered constant in this method. P_{e-A1} and P_{e-A2} are the electrical SSSG powers obtained using the during-fault and after-fault clearance curves, respectively. Moreover, P_m is the constant SSSG mechanical power. In fact, E_{A1} results in positive acceleration during the fault, which increases the SSSG kinetic energy. While, E_{A2} decrease the generator rotor speed and its kinetic energy. Based on (2.40), if E_{total} becomes zero before the SSSG rotor angle reaches to its maximum value, the SSSG remains stable.

One of the main challenges of transient stability assessment using equal-area criterion technique is to determine the amount of SSSG output power during various disturbances, especially unsymmetrical ones such as two-phase and single-phase short-circuit faults. In the transmission and subtransmission levels of the power system, the following per-unit equation can be utilized for the calculation of the SG output power during symmetrical events [2]:

$$P_e = P_{max} \sin \delta \tag{2.41}$$

where P_{max} is the maximum active power of SG during various conditions, e.g., steady-state, symmetrical events, and after clearance of the event. Considering Fig. 2.30, P_{max} can be calculated by

$$P_{max} = \frac{E_{Sys} E_G}{X_G + X_{Eq}} \tag{2.42}$$

where X_{Eq} is the imaginary part of the impedance seen from the SSSG terminal, i.e., Z_{Eq}. However, due to the large amount of R/X in distribution systems, using (2.42) may not provide enough accuracy. Therefore, for these systems, the following equation provides a more accurate result for calculating the SG output power:

$$P_e = \mathbf{Re}\left\{E_G I_G^*\right\} \tag{2.43}$$

where I_G is determined by

$$I_G = \frac{|E_G|\angle\delta - |E_{Sys}|\angle 0}{Z_{Eq} + jX_G} \tag{2.44}$$

where $Z_{Eq} = |Z_{Eq}|\angle\gamma$ is the equivalent impedance seen from the SSSG terminal. Considering the system shown in Fig. 2.30, Z_{Eq} can be determined by

$$Z_{Eq} = Z_{SGBMG} \| Z_{Sys} \tag{2.45}$$

Combining (2.43) and (2.44) results in

$$P_e = \mathbf{Re}\left\{ (|E_G|\angle\delta) \times \left(\frac{|E_G|\angle\delta - |E_{Sys}|\angle 0}{|Z_{Eq}|\angle\gamma} \right)^* \right\} \tag{2.46}$$

This equation can be rewritten as

$$P_e = \frac{|E_G|^2}{|Z_{Eq}|} \cos\gamma - \frac{|E_G||E_{Sys}|}{|Z_{Eq}|} \cos(\delta + \gamma) \tag{2.47}$$

For high ratio of X/R, such as transmission and subtransmission systems, γ can be considered $\pi/2$. However, in distribution systems whose X/R ratio is not large, (2.47) can be utilized instead of (2.41) for more accuracy. It should be noted that considering $\gamma = \pi/2$ for distribution systems may lead to a more conservative result.

Equal-area criterion technique has been widely exploited for off-line studies in distribution systems. However, its utilization for online applications, such as out-of-step and load-shedding functions, is recently suggested by various research works [27]. For online applications, the amount of equivalent impedance seen by the SSSG, i.e., Z_{Eq}, and the rotor angle should also be estimated to examine the transient stability margin. However, determining the SSSG rotor angle using local measurements may be difficult and not be practical in real applications. Therefore, instead of the rotor angle, the load angle parameter can be exploited in the equal-area criterion for investigating SSSG transient stability. Load angle is the phase difference between the SSSG internal voltage ($E_G \angle \delta$) and its terminal voltage ($V_G \angle \theta$), which is defined as $\varphi = \delta - \theta$ and can be easily determined using local measurements. In this regard, P_e could be described based on load angle, and thus, (2.47) should be revised accordingly. To do so, the typical grid-connected SGBMG shown in Fig. 2.30 is considered. For this system, the SSSG terminal voltage is determined by

$$V_G \angle \theta = E_G \angle \delta - jX_G I_G \tag{2.48}$$

Combining (2.44) and (2.48) results in

$$V_G \angle \theta = E_G \angle \delta - \left(\frac{jX_G}{Z_{Eq} + jX_G} \right) (E_G \angle \delta - E_{sys} \angle 0) \tag{2.49}$$

Assuming $\frac{jX_G}{Z_{Eq}+jX_G} = k + jm$, (2.49) can be rewritten as

$$V(\cos\theta + j\sin\theta) = E_G \cos\delta + jE_G \sin\delta - (k+jm)(E_G \cos\delta + jE_G \sin\delta - E_{Sys}) \tag{2.50}$$

Thus, the following equation is derived:

$$\cot\theta = \frac{\cos\delta(1-k) + m\sin\delta + k\lambda}{\sin\delta(1-k) - m\cos\delta + m\lambda} \tag{2.51}$$

where $\lambda = E_{Sys}/E_G$, which could be considered constant for transient stability assessment. Considering $\theta = \delta - \varphi$, the amount of load angle φ can be expressed as

$$\cot(\varphi) = f(\delta) = \frac{\cot(\delta) + \dfrac{\sin\delta(1-k) - m\cos\delta + m\lambda}{\cos\delta(1-k) + m\sin\delta + k\lambda}}{1 - \cot(\delta)\dfrac{\sin\delta(1-k) - m\cos\delta + m\lambda}{\cos\delta(1-k) + m\sin\delta + k\lambda}} \tag{2.52}$$

Consequently,

$$\delta = f^{-1}(\cot(\varphi)) \tag{2.53}$$

Therefore, by combining (2.53) and (2.47), P_e can be expressed as a function of load angle as

$$P_e = \frac{|E_G|^2}{|Z_{Eq}|}\cos\gamma - \frac{|E_G||E_{Sys}|}{|Z_{Eq}|}\cos(f^{-1}(\cot(\varphi)) + \gamma) \tag{2.54}$$

Moreover, using (2.54), the P_e–φ curve (instead of the P_e–δ curve) can be obtained. Although using load angle instead of rotor angle may resolve difficulties in online measurement of rotor angle, other limitations and shortcomings of the equal-area criterion technique cannot be resolved. The serious weaknesses and restrictions of this classical method include:

- For large-scale SGs, changing the position of mechanical valves includes considerable delays and, thus, the parameter P_m can be considered relatively constant for studying the transient stability issue in which the time interval is less than 500 ms. However, for SSSGs, considering the constant value for P_m may be arguable. For these small generators, the mentioned delays may become less than 500 ms, and thus, the amount of mechanical power can vary during the time interval of the transient stability study.
- Using P–δ curve for studying the unbalanced conditions, including asymmetrical short-circuit faults, could be troublesome.
- In SGBMGs, especially islanded SGBMGs, the stability subject following a large disturbance may be associated with various aspects of stability issues, such as transient stability and voltage stability. In other words, the stability phenomenon after large disturbances may not be directly related to the transient stability problem.
- This approach could be applied to multi-machine systems only by aggregation of the generators into two equivalent machines. This can be considered a major limitation of this method.

2.4.1.2 Lyapunov Technique

Various stability assessment methods have been developed in different research works based on the Lyapunov technique. Two major types of commonly exploited methods based on this technique are direct and indirect methods, which are discussed in the following.

1) **Direct Lyapunov technique:** The direct Lyapunov technique is exploited for direct assessment of SGBMG stability without requiring to solve the nonlinear differential equations associated with a dynamic system. The equations of an autonomous nonlinear system can be represented by [28, 29]

$$\dot{X} = F(X) \tag{2.55}$$

where $X = (x_1, x_2, x_3, \dots, x_n)$ is the state vector of the nonlinear system F. Furthermore, it is assumed that $X^* = (0, 0, \dots, 0)$ is the equilibrium point of the system for which $f(X^*) = 0$. By using the Lyapunov direct method, the stability of the equilibrium point can be analyzed. Point X^* is referred to as a "stable" equilibrium point if for any amount of $\varepsilon > 0$, there exists $\rho = \rho(\varepsilon)$

such that

$$\|\Phi(0)\| < \rho \Rightarrow \|\Phi(t)\| < \varepsilon \tag{2.56}$$

where $\Phi(t)$ is the solution of $\dot{X} = F(X)$ at time $t \geq 0$. In fact, when X^* is referred to as stable, it means that if an operation point starts close to X^*, then for $t \geq 0$, it stays close to this point. The equilibrium point is not stable if (2.56) does not hold. Moreover, the equilibrium point is called "asymptotically stable," if the following condition and (2.56) are fulfilled:

$$\|\Phi(0)\| < \rho \Rightarrow \lim_{t \to \infty}\{\Phi(t)\} = X^* \tag{2.57}$$

In fact, asymptotic stability is a condition in which the operation point finally settles at X^* and, thus, it is stronger than the stable equilibrium point (SEP) condition in which only (2.56) is fulfilled.

For evaluating the stability of X^* of the autonomous nonlinear system defined in (2.55) through the direct Lyapunov technique, at first a continuous definite positive Lyapunov function U with continuous partial derivatives shall be determined in the closed region of Ω around X^*, as follows [28, 29]:

$$U\{X = 0\} = 0 \text{ and } U\{X \neq 0\} > 0 \tag{2.58}$$

Then, based on the Lyapunov technique, point X^* is asymptotically stable if the derivative of the Lyapunov function with respect to time, dU/dt would be a definite negative function in the closed region of Ω. The main challenge of this technique is construction of a suitable Lyapunov function, which may be complicated especially for nonlinear systems. Some approaches have been proposed in the literature for constructing and determining an appropriate Lyapunov function. In the linear and linearized systems, one of the well-known techniques for determining U is utilizing the following quadratic function:

$$U = X^T P X \tag{2.59}$$

where matrix P can be determined by

$$\begin{cases} A^T P + PA = Q \\ \dot{U} = A^T Q A \end{cases} \tag{2.60}$$

Matrix A is defined by the state-space equations of the linearized/linear system as

$$\dot{X} = AX \tag{2.61}$$

The main challenge of this approach is determining the matrix Q and solving (2.60). For power system applications, which the associated state equations cannot be linearized after large perturbations, the best Lyapunov function is suggested to be determined using physical interpretation. It is worth mentioning that for a typical SGBMG including SSSGs, the nonlinear system equations can be presented by [17]

$$\omega_i = \frac{d\delta_i}{dt} \tag{2.62}$$

$$J_i \frac{d\omega_i}{dt} = P_{mi} - P_{ei} \tag{2.63}$$

where

$$P_{ei} = E_i^2 Y_{ii} \sin \beta_{ii} + \sum_{\substack{j=1 \\ j \neq i}}^{n} E_i E_j Y_{ij} \sin(\delta_i - \delta_j - \beta_{ij}) \tag{2.64}$$

where Y_{ii} and Y_{ij} are the diagonal and nondiagonal elements of the SGBMG admittance matrix, respectively. Moreover, β_{ii} and β_{ij} are the angle of Y_{ii} and Y_{ij}, respectively. Meanwhile, E_i is voltage of bus i, E_j is the voltage of bus j, δ_i is the voltage angle of bus i, and δ_j is the voltage angle of bus j. Various methods have recommended to use energy function as the Lyapunov function for SGBMGs [17, 30]. As such, for SGBMGs and SG-based power systems, the following energy-based Lyapunov function is widely recommended for evaluating the transient stability of synchronous-based systems such as SGBMGs:

$$U = K(\Delta\omega) + E(\delta) \tag{2.65}$$

where $K(\omega)$ corresponds to the kinetic energy and $E(\delta)$ represents the potential energy. For SGBMGs in which only SG-based generations exist, $K(\omega)$ and $E(\delta)$ are calculated as follows:

$$K(\omega) = \sum_{i=1}^{n} H_i \times (\omega - \omega_0)^2$$

$$\begin{aligned} E(\delta) = -\Bigg[& \sum_{i=1}^{n} \left(P_{mi} - E_i^2 Y_{ii} \sin \beta_{ii}\right)(\delta_i - \delta_{0i}) \\ & + \sum_{i=1}^{n} \sum_{j=2}^{i} E_i E_j Y_{ij} (\cos(\delta_i - \delta_j) - \cos(\delta_{0i} - \delta_{0j})) \Bigg] \end{aligned} \tag{2.66}$$

By using the mentioned energy-based Lyapunov function, the boundary of the stable region, which is referred to as attraction region, can be determined. Since the mentioned energy-based Lyapunov function results in nonlinear algebraic equations, determining the boundary of attraction region may be difficult or impossible through analytical techniques. Thus, to determine the attraction region, the commonly simulation-based analysis or repetitive numerical-based approaches such as Newton–Raphson method are exploited.

Based on the direct Lyapunov technique and exploiting the energy-based Lyapunov function, the stability of an SGBMG after large disturbances can

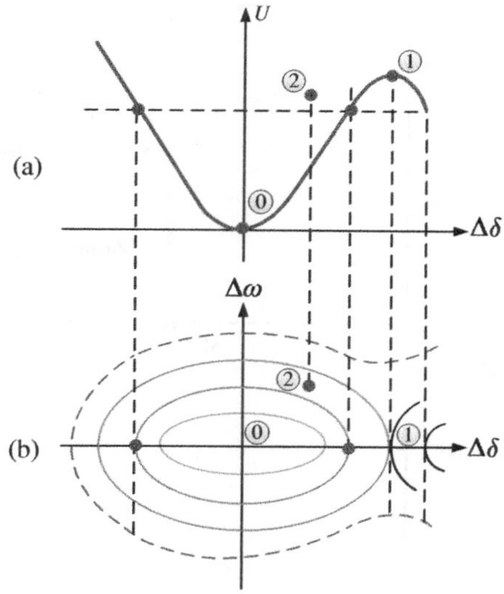

Fig. 2.32 Stability assessment of a grid-connected SSSG using the Lyapunov technique: (a) $U - \Delta\delta$ and (b) $\Delta\omega - \Delta\delta$ [17].

be evaluated. As an example, for evaluating the transient stability of a typical grid-connected SGBMG, parameters $\Delta\omega$ and δ are considered space variables. Figure 2.32 depicts the graphical visualization of Lyapunov direct technique for such a second-order system [17]. Figure 2.32(a) illustrates the Lyapunov function versus $\Delta\delta$, and Fig. 2.32(b) depicts $\Delta\omega$ versus $\Delta\delta$. For point "0," the amount of U and $dU/d\Delta\omega$ are zero, and thus, this initial point is stable based on the Lyapunov technique. However, the saddle point "1" represents the boundary of the attraction region in which the system remains stable. In other words, for the point beyond the saddle point "1," the system becomes unstable. In fact, if the energy level of the system becomes greater than U_1, the system becomes unstable, while for $U < U_1$, the system is stable. Based on these explanations, the attraction region is depicted in Fig. 2.32(b). As shown, for points $(\Delta\delta, \Delta\omega)$, which are located in the attraction region, the energy level U is less than U_1 and the system is stable. However, for the points located outside the mentioned region, the SGBMG becomes unstable.

The direct Lyapunov technique can be utilized for real-time protection schemes and online evaluation of the SSSG transient stability. For instance, in [30], the attraction region is determined using the energy-based function. Furthermore, the boundary of the attraction region has been computed through the simulation-based analysis. The SEP, unstable equilibrium point (UEP), and computed region of attraction (ROA) are shown in Fig. 2.33. Based on this figure, the state variables, i.e., δ and $\Delta\omega$, are monitored using the real-time protection scheme, and the related trajectory is mapped to the attraction region. The unstable

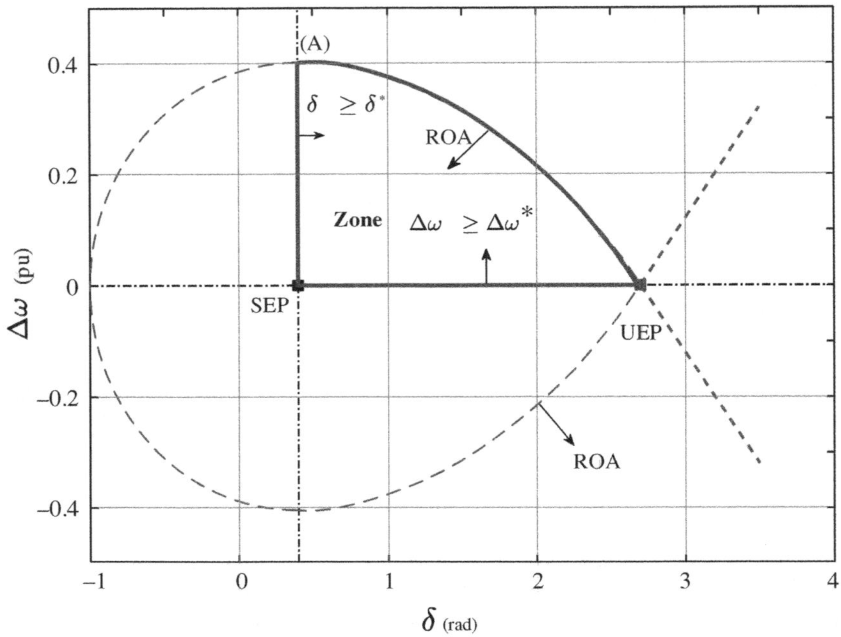

Fig. 2.33 Attraction region and tripping zone. [30] / with permission of IEEE.

operation of the SGBMG will be realized when the trajectory violates the ROA. The application of direct Lyapunov technique for online evaluation of the SSSG transient stability is investigated in the following example.

Example 2.4

A grid-connected SGBMG is shown in Fig. 2.34. The three-phase and single-phase short-circuit levels at the PCC are 10 and 3 kA, respectively. Investigate the transient stability of the SGBMG using the direct Lyapunov technique in the following two conditions and illustrate the trajectory of state variables e.g., $\Delta\omega$ and $\Delta\delta$ using the simulation-based analysis:

a) **Case I:** Three-phase short-circuit fault at PCC whose FCT is 100 ms.
b) **Case II:** Three-phase short-circuit fault at PCC whose FCT is 200 ms.

Solution

At first, the boundary of the attraction region is computed through the simulation-based analysis. To do so, three-phase short-circuit faults with various FCTs are simulated at the PCC. Then, the boundary is determined based on the worst-case three-phase short-circuit fault with the maximum FCT, for

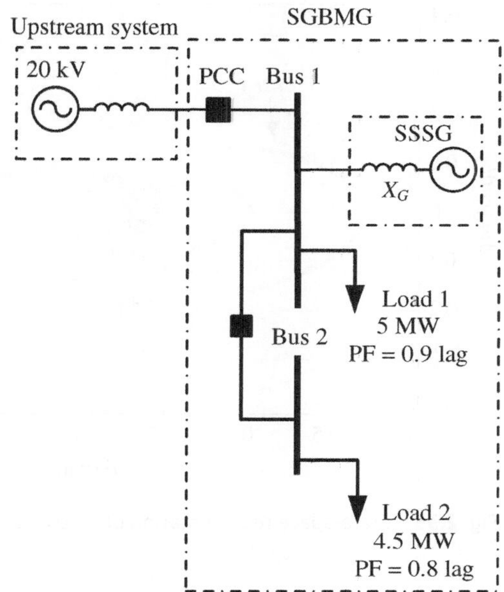

Fig. 2.34 Typical grid-connected SGBMG.

which the SSSG remains stable. The trajectory of the state variables for both cases is shown in Fig. 2.35. As shown, for case I, the trajectory of state variables remains inside the attraction region, and thus, the MG remains stable. However, for the second case, the trajectory of the state variables is not inside the attraction region, and thus, the system becomes unstable. Moreover, the results of the time-domain simulation are depicted in Fig. 2.36. As shown, for case I, the SSSG remains stable, while for case II, the SSSG loses its stability. These results are completely in accordance with the results of the Lyapunov technique.

2. **Indirect Lyapunov technique:** In the indirect Lyapunov method, the nonlinear equations of the system, i.e., $\dot{x} = f(x)$, will be linearized as

$$\dot{x} = Ax \tag{2.67}$$

where A is the Jacobian matrix around the equilibrium point determined by [30]

$$\mathbf{A} = \left.\frac{\partial f}{\partial \mathbf{x}}\right|_{\mathbf{x}=\mathbf{x}^*} = \begin{bmatrix} \frac{\partial f_1}{\partial x_1} & \frac{\partial f_1}{\partial x_2} & \cdots & \frac{\partial f_1}{\partial x_n} \\ \vdots & & \ddots & \vdots \\ \frac{\partial f_n}{\partial x_1} & \frac{\partial f_n}{\partial x_2} & \cdots & \frac{\partial f_n}{\partial x_n} \end{bmatrix}_{\mathbf{x}=\mathbf{x}^*} \tag{2.68}$$

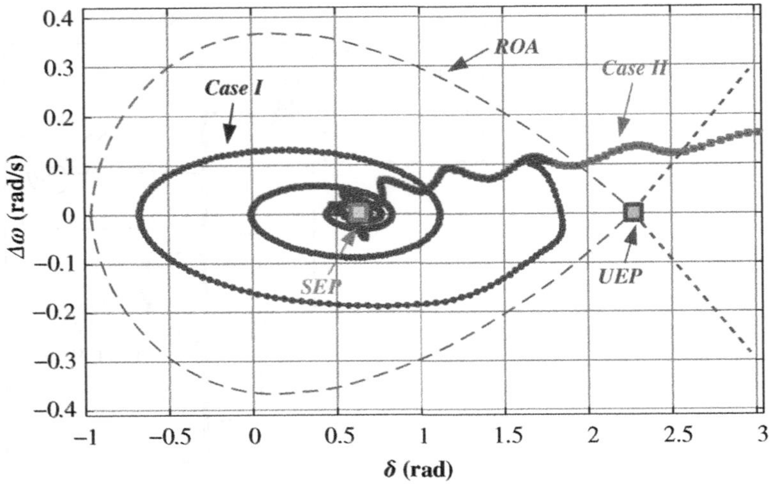

Fig. 2.35 State-space representation of cases I and II.

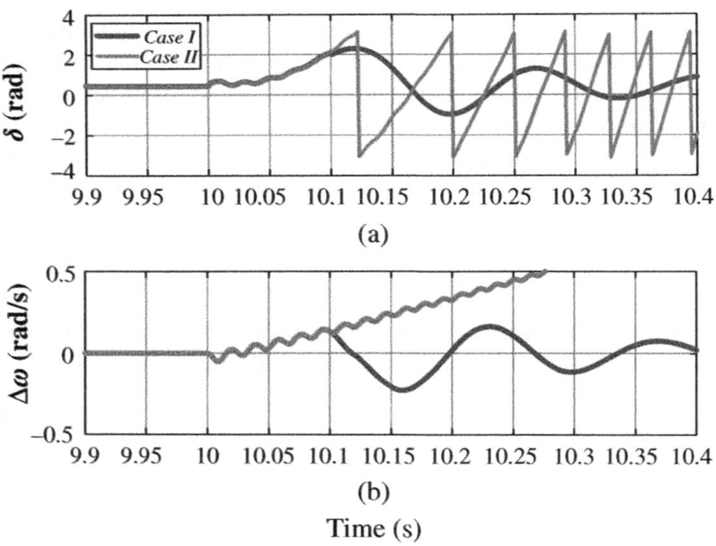

Fig. 2.36 Time-domain simulation results for cases I and II.

Point X^* is asymptotically stable if all of the eigenvalues associated with A are located on the left-half plane. However, the associated equilibrium point X^* is unstable if A has some eigenvalues on the right-half plane. Various literatures have exploited the indirect Lyapunov method to extract the system eigenvalues and evaluate the system stability after large disturbances [30]. It should be noted that after large disturbances, the linearization around the equilibrium point, especially for SSSG equations, may not have enough accuracy; thus, the direct Lyapunov technique is usually preferred for such disturbances.

2.4.1.3 Simulation-Based Techniques

One of the widely used techniques for the evaluation of the stability of power systems, including MGs, is the time-domain simulation study. Various hardware and software tools have been developed for time-domain simulations. These tools can provide accurate electromechanical models of MG components including SSSGs and its associated controllers, dynamic loads such as three-phase and single-phase IMs, transformers, and static loads. Some major software environments for time-domain simulation of MGs include PSCAD/EMTDC, DIgSILENT, and ETAP.

The accuracy of the time-domain simulation-based techniques is significantly higher than that of the aforementioned methods [31]. In fact, the SGBMG stability boundaries derived exploiting the simulation-based techniques have higher validity and accuracy. To do so, each component of the system shall be modeled in detail, which increases the computation complexity, especially for large MGs. Therefore, depending on the hardware capability of the utilized simulation-based tools, simulating the MGs behavior and evaluating their stability may take very long time using this technique, especially for distribution systems including large numbers of nodes and buses. Furthermore, evaluating large distribution systems whose number of state variables and differential equations are relatively high requires heavy computations, which may not be feasible for real-time applications. In addition, for large systems, the required tool for simulation may become very expensive, and it may not be cost-effective compared to the other methods [1].

Various methods have been proposed in research works to resolve the computation complexity issues. Electromechanical-based simulation tools in which only the electromechanical equations have been considered are widely exploited for investigating the transient stability in traditional systems [1]. However, for unbalanced systems, this approach is not suitable. To address such a condition, network reduction techniques are widely utilized. Various techniques have been proposed for network reduction including utilizing aggregate model of loads, especially dynamic loads, to decrease the number of system differential equations.

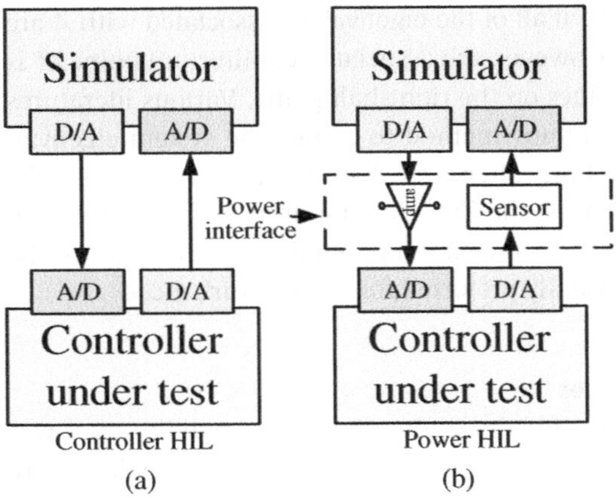

Fig. 2.37 Schematic of a hardware in the loop, (a) controller in the loop, and (b) power device in the loop. Adapted from [32].

Moreover, to investigate the stability of grid-connected SGBMGs, the upstream network can be modeled through its Thevenin's equivalent [1].

One of the widely used types of time-domain simulations is the real-time simulation in which the simulation is performed at the same period of time as the actual clock time [32, 33]. The real-time simulation tools can be effectively utilized for evaluating the performance of the protection/control equipment and power devices. In fact, the shortcomings and defects of these devices can be extracted through a cost-effective real-time method. HIL is the common technique used for real-time simulation of the power systems, including SGBMGs. Two common types of HIL are depicted in Fig. 2.37 [32]. Figure 2.37(a) illustrates a protection/control equipment in the loop, in which the real equipment is connected to an MG and the MG is completely modeled and simulated through a real-time digital simulator (RTDS). Then, the whole system will be simulated in real-time and the protection/control equipment sends/receives signals to/from the RTDS. Figure 2.37(b) depicts a power device in the loop, which is connected to the MG through a power interface. This interface includes an amplifier for providing the required voltage for the power device. Moreover, the power interface contains the required sensors for measuring the signals and transferring them to the MG. Similar to the previous case, the MG is simulated using RTDS [1, 32].

2.4.2 Small-Signal Disturbances

Small disturbances are referred to as the perturbations with small magnitude that do not substantially change the operation point of the system. Traditionally, to evaluate the MG stability after small perturbations, the eigenvalues of the system are investigated using the linearized state-space model of the system. Various techniques have been proposed to derive the state-space model of the MG and its components, including DGs, SSSGs, dynamic loads, static loads, etc. The eigenvalues of the system should be calculated for various operating conditions. These eigenvalues considerably depend on the control strategy and SSSG loading. For example, the system eigenvalues for a PQ control mode are different than those of a droop control mode. Moreover, changes in active/reactive power of the SSSG can change the system eigenvalues and its dynamic indices, such as the damping ratio.

For small-signal stability assessment of SGBMGs, the state differential equations including equations of the SSSG, related to both electrical and mechanical parts, should be linearized. To do so, the eighth-order differential equations of the SSSG could be utilized, including the sixth-order model for the electrical portion and the second-order model for the mechanical portion. As a common practice, in the electrical portion, the SG variables are transformed to the rotor reference frame for simplicity. In fact, the *d*–*q* frame of the model rotates at the angular speed of rotor ω_r. Considering these explanations, the sixth-order model of the SSSG electrical portion, including stator winding, field winding, and damper windings, could be described by [22]

$$
\begin{aligned}
v_{qs}^r &= -r_s i_{qs}^r + \frac{\omega_r}{\omega_b}\psi_{ds}^r + \frac{1}{\omega_b}\frac{d\psi_{qs}^r}{dt} \\
v_{ds}^r &= -r_s i_{ds}^r - \frac{\omega_r}{\omega_b}\psi_{qs}^r + \frac{1}{\omega_b}\frac{d\psi_{ds}^r}{dt} \\
v'^r_{kq1} &= r_{kq1} i'^r_{kq1} + \frac{1}{\omega_b}\frac{d\psi'^r_{kq1}}{dt} \\
v'^r_{kq2} &= r_{kq2} i'^r_{kq2} + \frac{1}{\omega_b}\frac{d\psi'^r_{kq2}}{dt} \\
v'^r_{fd} &= r_{fd} i'^r_{fd} + \frac{1}{\omega_b}\frac{d\psi'^r_{fd}}{dt} \\
v'^r_{kd} &= r_{kd} i'^r_{kd} + \frac{1}{\omega_b}\frac{d\psi'^r_{kd}}{dt}
\end{aligned}
\tag{2.69}
$$

where superscript "′" denotes the amount of rotor variables, which are referred to the stator windings. Furthermore, subscripts *s*, *kq*1, *kq*2, *fd*, and *kd*, respectively, denote stator winding, first *q*-axis damper winding, second *q*-axis damper winding, field winding, and *d*-axis damper winding. In addition, ω_b, r, i, v, and ψ are base rotor speed, winding resistance, current, voltage, and flux linkages, respectively. The flux linkages can be calculated as follows:

$$
\begin{aligned}
\psi_{qs}^r &= -X_{ls_{qs}}^r + X_{mq}\left(-i_{qs}^r + i'^r_{kq} + i'^r_{kq}\right) \\
\psi_{ds}^r &= -X_{ls_{ds}}^r + X_{md}\left(-i_{ds}^r + i'^r_{fd} + i'^r_{kd}\right) \\
\psi_{0s} &= -X_{ls} i_{0s}
\end{aligned}
$$

$$
\begin{aligned}
\psi'^{r}_{kq1} &= X'_{lkq} i'' i'^{r}_{k1} + X_{mq} \left(-i^{r}_{qs} + i'^{r}_{kq1} + i'^{r}_{kq2}\right) \\
\psi'^{r}_{kq2} &= X'_{lkq} i'_{kq2} + X_{mq} \left(-i^{r}_{qs} + i'^{r}_{kq1} + i'^{r}_{kq2}\right) \\
\psi'^{r}_{fd} &= X''_{ldd} i^{r}_{fd} + X_{md} \left(-i^{r}_{dd} + i'^{r}_{fd} + i'^{r}_{kd}\right) \\
\psi'^{r}_{kd} &= X'_{lkd} i^{r}_{kd} + X_{md} \left(-i^{r}_{ds} + i'^{r}_{fd} + i'^{r}_{kd}\right)
\end{aligned}
\tag{2.70}
$$

where X_{ls}, X_{lkq1}, X_{lkq2}, X_{lfd}, and X_{lkd} are leakage reactances of stator, first q-axis damper winding, second q-axis damper winding, field winding, and d-axis damper winding, respectively. Furthermore, X_{mq} and X_{md} are the magnetizing reactance of q-axis and d-axis, respectively. Combining (2.69) and (2.70) results in [22]

$$
\begin{bmatrix} v^{r}_{qs} \\ v^{r}_{ds} \\ v_{0s} \\ v'^{r}_{kq1} \\ v'^{r}_{kq2} \\ e'^{r}_{xfd} \\ v'^{r}_{kd} \end{bmatrix} =
\begin{bmatrix}
-r_s - \frac{p}{\omega_b} X_q & -\frac{\omega_r}{\omega_b} X_d & 0 & \frac{p}{\omega_b} X_{mq} & \frac{p}{\omega_b} X_{mq} & \frac{\omega_r}{\omega_b} X_{md} & \frac{\omega_r}{\omega_b} X_{md} \\
\frac{\omega_r}{\omega_b} X_q & -r_s - \frac{p}{\omega_b} X_d & 0 & -\frac{\omega_r}{\omega_b} X_{mq} & -\frac{\omega_r}{\omega_b} X_{mq} & \frac{p}{\omega_b} X_{md} & \frac{p}{\omega_b} X_{md} \\
0 & 0 & -r_s - \frac{p}{\omega_b} X_{ls} & 0 & 0 & 0 & 0 \\
-\frac{p}{\omega_b} X_{mq} & 0 & 0 & i'_{kq1} + \frac{p}{\omega_b} X'_{kq1} & \frac{p}{\omega_b} X_{mq} & 0 & 0 \\
-\frac{p}{\omega_b} X_{mq} & 0 & 0 & \frac{p}{\omega_b} X_{mq} & i'_{kq2} + \frac{p}{\omega_b} X'_{kq2} & 0 & 0 \\
0 & -\frac{X_{md}}{r_{ls}} \left(\frac{p}{\omega_b} X_{md}\right) & 0 & 0 & 0 & \frac{X_{md}}{r'_{ls}} \left(i'_{fd} + \frac{p}{\omega_b} X'_{fd}\right) & -\frac{X_{md}}{r'_{ls}} \left(\frac{p}{\omega_b} X_{md}\right) \\
0 & -\frac{p}{\omega_b} X_{md} & 0 & 0 & 0 & \frac{p}{\omega_b} X_{md} & i'_{kd} + \frac{p}{\omega_b} X'_{kd}
\end{bmatrix}
\begin{bmatrix} i^{r}_{qs} \\ i^{r}_{ds} \\ i_{0s} \\ i'^{r}_{kq1} \\ i'^{r}_{kq2} \\ i'^{r}_{xfd} \\ i'^{r}_{kd} \end{bmatrix}
\tag{2.71}
$$

where

$$
\begin{aligned}
X_q &= X_{ls} + X_{mq} \\
X_d &= X_{ls} + X_{md} \\
X'_{kq1} &= X'_{lkq1} + X_{mq}
\end{aligned}
$$

$$X'_{kq2} = X'_{lkq2} + X_{mq}$$
$$X'_{fd} = X'_{lfd} + X_{md}$$
$$X'_{kd} = X'_{lkd} + X_{md}$$

$$e'^{r}_{xfd} = \frac{X_{md}}{r'_{fd}} \left(r'_{fd} f'^{r}_{fd} + \frac{p}{\omega_b} \psi'^{r}_{fd} \right) \tag{2.72}$$

In addition, the following equations are utilized to model the mechanical portion of the SSSG [2, 22]:

$$T_e = -2H \frac{1}{\omega_b} \frac{d\omega_r}{dt} + T_m$$

$$\theta_r = \int \omega_r dt; \quad \theta_e = \int \omega_e dt$$

$$\delta = \theta_r - \theta_e$$

$$T_e = \left(\psi^{r}_{ds} i^{r}_{qs} - \psi^{r}_{qs} i^{r}_{ds} \right) \tag{2.73}$$

where T_e and T_m are the per-unit electrical and mechanical torques of the SSSG, respectively. Meanwhile, ω_r and ω_e are the angular frequencies of SSSG and upstream system, respectively. In addition, δ is the rotor angle. The SGBMG equations, including the SSSG equations, should be linearized for small-signal stability. In this regard, each equation of the SSSG, i.e., g, could be expanded about its equilibrium point, i.e., f_{i0}, using the Taylor series as [22]

$$g(f_i) = g(f_{io}) + g'(f_{io})\Delta f_i + \frac{g''(f_{io})}{2!}\Delta f_i^2 + \cdots \tag{2.74}$$

where Δf_i is given by

$$f_i = f_{io} + \Delta f_i \tag{2.75}$$

For small changes in equilibrium points, i.e., small amount of Δf_i, the second- and higher-order terms of Taylor series can be neglected and thus

$$g(f_i) \approx g(f_{io}) + g'(f_{io})\Delta f_i \tag{2.76}$$

This equation can be rewritten as

$$\Delta g(f_i) = g'(f_{io})\Delta f_i \tag{2.77}$$

where $\Delta g(f_i) = g(f_i)–g(f_{i0})$. Using (2.77), the equations of the SSSG, i.e., equations (2.71)–(2.73) are linearized as

$$E \times \dot{\Delta x} = Fx + u \tag{2.78}$$

where

$$
\begin{aligned}
(\mathbf{x})^T &= \left[\Delta i_{qs}^r \; \Delta i_{ds}^r \; \Delta i'^r_{kq1} \; \Delta i'^r_{kq2} \; \Delta i'^r_{fd} \; \Delta i'^r_{kd} \; \Delta\omega_r/\omega_b \; \Delta\delta\right] \\
(\mathbf{u})^T &= \left[\Delta v_{qs}^r \; \Delta v_{ds}^r \; \Delta v'^r_{kq1} \; \Delta v'^r_{kq2} \; \Delta e'^r_{xfd} \; \Delta v'^r_{kd} \; \Delta T_l \; 0\right]
\end{aligned}
\tag{2.79}
$$

and

$$
\mathbf{E} = \frac{1}{\omega_b}\begin{bmatrix}
-X_q & 0 & -X_{mq} & X_{mq} & 0 & 0 & 0 & 0 \\
0 & -X_d & 0 & 0 & X_{md} & X_{md} & 0 & 0 \\
-X_{mq} & 0 & X'_{kq1} & X_{mq} & 0 & 0 & 0 & 0 \\
-X_{mq} & 0 & X_{mq} & X'_{kq2} & 0 & 0 & 0 & 0 \\
0 & -\frac{X_{md}^2}{f_{dd}} & 0 & 0 & \frac{X_{mm}X'_{dq}}{T_{dd}} & \frac{X_{md}^2}{T_{rd}} & 0 & 0 \\
0 & -X_{md} & 0 & 0 & X_{md} & X'_{kd} & 0 & 0 \\
0 & 0 & 0 & 0 & 0 & 0 & 2H\omega_b & 0 \\
0 & 0 & 0 & 0 & 0 & 0 & 0 & \omega_b
\end{bmatrix}
$$

$$
\mathbf{F} = -\begin{bmatrix}
-r_s & -\frac{\omega_e}{\omega_b}X_d & 0 & 0 & \frac{\omega_e}{\omega_b}X_{md} & \frac{\omega_e}{\omega_b}X_{md} & -X_d i_{dso}^r + X_{md} i'^r_{fdo} & 0 \\
\frac{\omega_e}{\omega_b}X_q & -r_s & -\frac{\omega_e}{\omega_b}X_{mq} & -\frac{\omega_e}{\omega_b}X_{mq} & 0 & 0 & X_q i_{qso}^r & 0 \\
0 & 0 & i'_{kq1} & 0 & 0 & 0 & 0 & 0 \\
0 & 0 & 0 & i'_{kq2} & 0 & 0 & 0 & 0 \\
0 & 0 & 0 & 0 & X_{md} & 0 & 0 & 0 \\
0 & 0 & 0 & 0 & 0 & i'_{kd} & 0 & 0 \\
X_{mq} i_{dso}^r - X_{md}\left(i_{dso}^r - i'^r_{fdo}\right) & -X_{md} i_{qso}^r + X_{mq} i_{qso}^r & -X_{mq} i_{qso}^r & -X_{mq} i_{qso}^r & X_{md} i_{qso}^r & X_{md} i_{qso}^r & 0 & 0 \\
0 & 0 & 0 & 0 & 0 & 0 & -\omega_b & 0
\end{bmatrix}
\tag{2.80}
$$

These equations can only be exploited for an islanded SGBMG in which an SSSG operates autonomously and supplies local passive loads. For a grid-connected SSSG, variations in the upstream system voltage lead to a change in the SSSG terminal voltage. Therefore, the terminal voltage of the grid-connected SSSG should be determined considering the upstream system voltage. It should be noted that the upstream system voltage is expressed in the synchronously referenced frame, while the SSSG equations are represented in the rotor-reference frame. Thus, the voltage at the SSSG terminal should be determined using the KVL rule in the synchronously referenced frame and then it is transformed to the SSSG rotor-reference frame using [22]

$$\underbrace{\begin{bmatrix}\Delta f_{qs}^{e}\\ \Delta f_{ds}^{e}\end{bmatrix}}_{\Delta \mathbf{f}_{qds}^{r}} = \underbrace{\begin{bmatrix}\cos\delta_o & \sin\delta_o\\ -\sin\delta_o & \cos\delta_o\end{bmatrix}}_{\mathbf{T}}\underbrace{\begin{bmatrix}\Delta f_{qs}^{r}\\ \Delta f_{ds}^{r}\end{bmatrix}}_{\Delta \mathbf{f}_{qds}^{e}} + \underbrace{\begin{bmatrix}f_{dso}^{e}\\ -f_{qso}^{e}\end{bmatrix}}_{\mathbf{F}^{r}}\Delta\delta \tag{2.81}$$

where these equations can be written as

$$\Delta \mathbf{f}_{qds}^{r} = \mathbf{T}\Delta \mathbf{f}_{qds}^{e} + \mathbf{F}^{r}\Delta\delta$$

$$\Delta \mathbf{f}_{qds}^{e} = (\mathbf{T})^{-1}\Delta \mathbf{f}_{qds}^{r} + \mathbf{F}^{e}\Delta\delta \tag{2.82}$$

Therefore, for a grid-connected SSSG, Δv_{qs}^{r} and Δv_{ds}^{r} can be expressed as

$$\begin{aligned}\Delta v_{qs}^{r} &= \cos\delta_0\Delta v_{qs}^{e} - \sin\delta_0\Delta v_{ds}^{e} + v_{ds0}^{e}\Delta\delta\\ \Delta v_{ds}^{r} &= -\sin\delta_0\Delta v_{qs}^{e} + \cos\delta_0\Delta v_{ds}^{e} - v_{qs0}^{e}\Delta\delta\end{aligned} \tag{2.83}$$

Replacing (2.83) into (2.71) results in

$$E\times \Delta\dot{x} = Fx + u \tag{2.84}$$

where [22], [34]

$$\begin{aligned}
(\mathbf{x})^{T} &= \left[\left(\Delta \mathbf{i}_{qds}^{e}\right)^{T} \ (\Delta \mathbf{i}_{rr})^{T}\right]\\
&= \left[\Delta i_{qs}^{e}\ \Delta i_{ds}^{e}\ \Delta i'^{r}_{kq1}\ \Delta i'^{r}_{kq2}\ \Delta i'^{r}_{fd}\ \Delta i'^{r}_{kd}\ \tfrac{\Delta\omega_r}{\omega_b}\ \Delta\delta\right]\\
(\mathbf{u})^{T} &= \left[\left(\Delta \mathbf{v}_{qds}^{e}\right)^{T} \ (\Delta \mathbf{v}_{rr})^{T}\right]\\
&= \left[\Delta v_{qs}^{e}\ \Delta v_{ds}^{e}\ \Delta v'^{r}_{kq1}\ \Delta v'^{r}_{kq2}\ \Delta e'^{r}_{xfd}\ \Delta v'^{r}_{kd}\ \Delta T_I\ 0\right]\\
\mathbf{E} &= \begin{bmatrix}(\mathbf{T})^{-1}\mathbf{W}_p\mathbf{T} & (\mathbf{T})^{-1}\mathbf{Y}_p\\ \mathbf{Q}_p\mathbf{T} & \mathbf{S}_p\end{bmatrix}\\
\mathbf{F} &= -\begin{bmatrix}(\mathbf{T})^{-1}\mathbf{W}_k\mathbf{T} & (\mathbf{T})^{-1}\mathbf{Y}_k\\ \mathbf{Q}_k\mathbf{T} & \mathbf{S}_k\end{bmatrix}\\
\mathbf{W}_p &= \frac{1}{\omega_b}\begin{bmatrix}-X_q & 0\\ 0 & -X_d\end{bmatrix}\\
\mathbf{Y}_p &= \frac{1}{\omega_b}\begin{bmatrix}X_{mq} & X_{mq} & 0 & 0 & 0 & X_q i_{dso}^{r}\\ 0 & 0 & X_{md} & X_{md} & 0 & -X_{diq}^{ir}\end{bmatrix}\\
(\mathbf{Q}_p)^{T} &= \frac{1}{\omega_b}\begin{bmatrix}-X_{mq} & -X_{mq} & 0 & 0 & 0 & 0\\ 0 & 0 & -\frac{X_{md}^{2}}{r_{sd}} & -X_{md} & 0 & 0\end{bmatrix}
\end{aligned}$$

$$\mathbf{S}_p = \frac{1}{\omega_b}\begin{bmatrix} X'_{kq1} & \mathrm{X_{mq}} & 0 & 0 & 0 & X_{mq}i^r_{dso} \\ X_{mq} & X'_{kq2} & 0 & 0 & 0 & X_{mq}i^r_{dso} \\ 0 & 0 & \frac{X_{mm}X'_{\mu s}}{r'_{/s}} & \frac{X^2_{gd}}{r_{sd}} & 0 & -\frac{X^2_{md}}{r_{js}}r^r_{qso} \\ 0 & 0 & X_{md} & X'_{kd} & 0 & -X_{md}i^r_{qso} \\ 0 & 0 & 0 & 0 & 2H\omega_b & 0 \\ 0 & 0 & 0 & 0 & 0 & 1 \end{bmatrix}$$

$$\mathbf{W}_k = \begin{bmatrix} -r_s & -\frac{\omega_e}{\omega_b}X_d \\ \frac{\omega_e}{\omega_b}X_q & -r_s \end{bmatrix}$$

$$\mathbf{Y}_k = \begin{bmatrix} 0 & 0 & \frac{\omega_e}{\omega_b}X_{md} & \frac{\omega_e}{\omega_b}X_{md} & -X_d i^r_{dso} + X_{md}i^r_{ddo} & r_s i^r_{dso} - \frac{\omega_e}{\omega_b}X_d i^r_{qso} + v^r_{dso} \\ -\frac{\omega_e}{\omega_b}X_{mq} & -\frac{\omega_e}{\omega_b}X_{mq} & 0 & 0 & X_q q^r_{qso} & -r_s i^r_{qso} - \frac{\omega_e}{\omega_b}X_q i^r_{dso} - v^r_{qso} \end{bmatrix}$$

$$(\mathbf{Q}_k)^T = \begin{bmatrix} 0\ 0\ 0\ 0\ X_{mq}i^r_{dso} - X_{md}\left(i^r_{dso} - i'^r_{fdo}\right)\ 0 \\ 0\ 0\ 0\ 0\ \quad -X_{md}i^r_{qso} + X_{mq}i^r_{qso} \quad\ 0 \end{bmatrix}$$

$$\mathbf{S}_k = \begin{bmatrix} r'_{kq1} & 0 & 0 & 0 & 0 & 0 \\ 0 & r'_{kq2} & 0 & 0 & 0 & 0 \\ 0 & 0 & X_{md} & 0 & 0 & 0 \\ 0 & 0 & 0 & r'_{kq} & 0 & 0 \\ -X_{mq}i^r_{dso} & -X_{mq}i^r_{dso} & X_{md}i^r_{qso} & X_{md}i^r_{qso} & 0 & \begin{matrix} -i^r_{dso}\left[X_{mq}i^r_{dso} \right. \\ \left. -X_{md}\left(i^r_{dso} - i^r_{\mathrm{fdo}}\right)\right] \\ -i^r_{qso}\left(X_m i^r_{qso} - X_{mi}i^r_{qso}\right) \end{matrix} \\ 0 & 0 & 0 & 0 & 0 & 0 \end{bmatrix} \tag{2.85}$$

So far, the dynamics of the associated AVR and governor, i.e., active/reactive power control loops, have not been considered. In the rest of this section, the SSSG models and equations are modified to consider the dynamics of its components in the grid-connected operation of the SSSG.

During the steady-state operation of a grid-connected SGBMG, two major control modes are commonly utilized, i.e., droop and constant active/reactive power control modes. The control structure of a grid-connected SSSG is depicted in Fig. 2.38 [34]. For the droop control mode, the reference of the SGBMG angular frequency and its terminal voltage are determined according to the following equations:

$$\omega_{r_i}^* = \omega_{nl} - m_{p_i} P_i \tag{2.86}$$

$$v_{tr_i}^* = V_{nl} - n_{q_i} Q_i \tag{2.87}$$

where ω_r^* denotes the reference angular frequency, while ω_{nl} represents the angular frequency under no-load conditions. The parameter m_{p_i} is the active power droop gain of the ith DG. Similarly, $v_{tr_i}^*$ refers to the reference value of the terminal voltage magnitude, and V_{nl} is the terminal voltage setpoint under the no-load condition. The parameter n_{q_i} represents the reactive power droop gain of the ith DG. Furthermore, P_i and Q_i are the active and reactive powers of the ith DG, respectively. Meanwhile, the reference of the SSSG governor $\left(P_{Gov}^*\right)$ is determined using the PI controller as

$$P_{Gov}^* = K_{pp}\left(\omega_r^* - \omega_r\right) + K_{ip} \int \left(\omega_r^* - \omega_r\right) dt \tag{2.88}$$

where K_{pp} and K_{ip} are the proportional and integral gains of the governor PI controller, respectively. Moreover, ω_r is the measured angular frequency of the SSSG. For the constant active power control mode, P_{Gov}^* is considered constant, i.e., $P_{Gov}^* = P_{cte}$. Moreover, the reference of excitation system and AVR $\left(v_A^*\right)$ is determined using the PI controller as [34]

$$v_A^* = K_p\left(v_t^* - v_{tr}\right) + K_i \int \left(v_t^* - v_{tr}\right) dt \tag{2.89}$$

where K_p and K_i are the proportional and integral gains of the AVR PI controller. Additionally, v_{tr} is the SSSG terminal voltage. The state variables of droop control equations (2.88) and (2.89) are considered X_{PPI} and X_{QPI}, respectively. The dynamics of the SSSG excitation system noticeably depends on the excitation system type as mentioned in Chapter 1. However, in this chapter, a simplified model of the SSSG excitation system is utilized, which can be developed for each excitation system type.

The SSSG prime mover, which is considered a diesel engine as a widely used practice, provides the input mechanical torque, and it can be appropriately modeled using a first-order transfer function with the time constant of T_e. Moreover, the governor controller is also modeled using a first-order transfer function with the time constant of T_g. Thus, the state-space equation of the prime mover and its associated governor system is given by [34]

$$\begin{aligned} \dot{\lambda}_g &= -\frac{1}{T_g}\lambda_g + P_{Gov}^*; \quad P_{En}^* = \frac{1}{T_g}\lambda_g \\ \dot{\lambda}_{En} &= -\frac{1}{T_e}\lambda_{En} + P_{En}^*; \quad P_{En} = \frac{1}{T_{En}}\lambda_{En} \end{aligned} \tag{2.90}$$

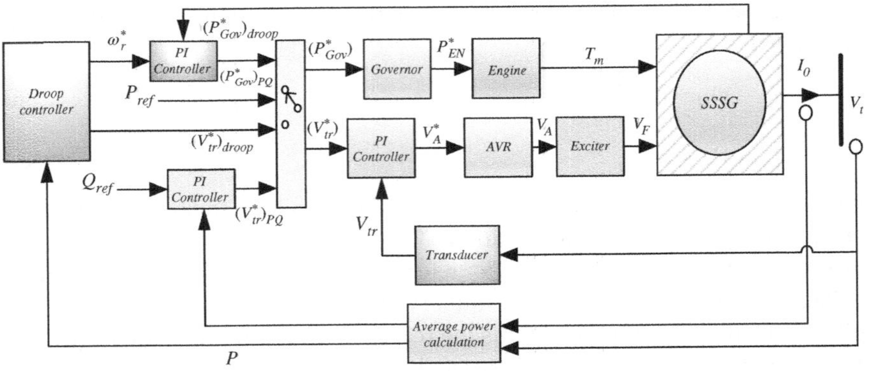

Fig. 2.38 Control structure of a typical grid-connected SSSG. Adapted from [34].

where λ_g and λ_{En} represent the state variables of the governor and the prime mover, respectively. Furthermore, variables P^*_{Gov}, P^*_{En}, and P_{En} are the governor reference power, prime mover reference power, and the prime mover output power, respectively. Meanwhile, the input mechanical torque of the SSSG is a nonlinear function of output mechanical power of the prime mover (P_{En}) and SSSG angular frequency (ω_r), which can be calculated as

$$T_m = \frac{P_{En}}{\omega_r} \tag{2.91}$$

This nonlinear equation can be linearized as follows:

$$\Delta T_m = \frac{1}{\omega_r^0}\Delta P_T - \frac{P^0_{En}}{\omega_r^0} \tag{2.92}$$

Meanwhile, the dynamics of excitation system including AVR, exciter, and corresponding transducers can be presented by the following differential equations:

$$\dot{\lambda}_A = -\frac{1}{T_A}\lambda_A + v_A^*; \quad v_F^* = \frac{K_A}{T_A}\lambda_A$$
$$\dot{\lambda}_E = -\frac{1}{T_E}\lambda_E + v_F^*; \quad v_F = \frac{1}{T_E}\lambda_E$$
$$\dot{\lambda}_r = -\frac{1}{T_r}\lambda_r + v_t$$
$$v_{tr} = \frac{1}{T_r}\lambda_r \tag{2.93}$$

where λ_r, λ_{VPI}, λ_A, and λ_E are the state variables of the transducers, voltage regulation PI controller, AVR, and exciter, respectively. The SSSG terminal voltage v_t can be determined by its d-axis and q-axis components, i.e., v_{qs} and v_{ds}, as [34]

$$v_t = \sqrt{v_{qs}^2 + v_{ds}^2} \tag{2.94}$$

This nonlinear equation can be linearized as follows:

$$\Delta v_t = \frac{V_{qs}^0}{V_t^0}\Delta v_{qs} + \frac{V_{ds}^0}{V_t^0}\Delta v_{ds} \tag{2.95}$$

These equations are derived for a grid-connected SSSG in the droop control mode. However, for the constant active/reactive power control mode (PQ), the reference of governor active power (P^*_{Gov}) is set constant, i.e., $P^*_{Gov} = P_{set}$. Moreover, for the AVR control loop, the output reactive power of the SSSG (Q_{SSSG}) should be set at a constant value of Q_{set}. Therefore, the error of the reactive power, i.e., $Q_{set} - Q_{SSSG}$, is injected to the voltage PI controller in which the proportional and integral gains are considered k_{pQ} and k_{iQ}, respectively. Based on these explanations, the complete linearized state-space model of a grid-connected SSSG and its associated controllers can be written as [22, 34]

$$\begin{aligned}
&\Delta \dot{x}_{SDG} = A_{SDG}\Delta x_{SDG} + B_{SDG}\begin{bmatrix}\Delta v_{qs}\\ \Delta v_{ds}\end{bmatrix} + B_{\omega SDG}\Delta\omega_e\\
&\Delta x_{SDG_{droop}} = \left[\Delta x_{SM}^T\ \lambda_{PPI}\ \lambda_{Gov}\ \lambda_{En}\ \lambda_r\ \lambda_{VPI}\ \lambda_A\ \lambda_E\right]^T\\
&\Delta x_{SDG_{PQ}} = \left[\Delta x_{SM}^T\ \lambda_{Gov}\ \lambda_{En}\ \lambda_{QPI}\ \lambda_A\ \lambda_E\right]^T\\
&A_{SDG} = \begin{bmatrix}A_{SDG11} & A_{SDG12}\\ & A_{SDG2}\end{bmatrix}\\
&A_{SDG11_{droop}} = A_{SM} - \frac{T_m^0}{\omega_0}\left[0_{8\times6}\ B_{TmSM}\ 0_{8\times1}\right]\\
&A_{SDG12_{droop}} = \frac{1}{\omega_0}\left[0_{8\times2}\ B_{TmSM}\ 0_{8\times3}\ B_{xfd}\right]\\
&A_{SDG12_{PQ}} = \frac{1}{\omega_0}\left[0_{8\times1}\ B_{TmSM}\ 0_{8\times2}\ B_{xfd}\right]\\
&B_{SDG} = \begin{bmatrix}B_{VSM}\\ B_{SDG2}\end{bmatrix};\quad B_{\omega SDG_{droop}} = \begin{bmatrix}B_{\omega SM}\\ 0_{7\times1}\end{bmatrix};\quad B_{\omega SDGPQ} = \begin{bmatrix}B_{\omega SM}\\ 0_{5\times1}\end{bmatrix}\\
&C_{SDG_{droop}} = \begin{bmatrix}0 & 1 & 0_{1\times13}\\ 1 & 0 & 0_{1\times13}\end{bmatrix};\quad C_{\Delta\omega\, com_{droop}} = \left[0_{1\times6}\ 1\ 0_{1\times8}\right]\\
&C_{SDGPQ} = \begin{bmatrix}0 & 1 & 0_{1\times11}\\ 1 & 0 & 0_{1\times11}\end{bmatrix};\quad C_{\Delta\omega\, com_{PQ}} = \left[0_{1\times6}\ 1\ 0_{1\times6}\right]
\end{aligned} \tag{2.96}$$

where

$$\begin{aligned}
&B_{SM} = E^{-1}, A_{SM} = E^{-1}F, B_{VSM} = B_{SM}(:, 1:2), B_{exfd} = B_{SM}(:, 5),\\
&B_{TmSM} = B_{SM}(:, 7), B_{\omega SM}(:, 8)
\end{aligned} \tag{2.97}$$

The parameters of active power/frequency control loop and reactive power/voltage control loop may considerably change the location of SGBMG eigenvalues. For example, the SGBMG eigenvalues move to the left side by increasing the active power droop gain, which enhances the small-signal stability of the SGBMG. However, the SGBMG eigenvalues move to the right side by increasing the reactive power droop coefficient, which results in a weaker small-signal stability condition. Meanwhile, the sensitivity of the system eigenvalues to these coefficients considerably depends on the operation point and loading. Therefore, to guarantee the appropriate dynamic performance, the damping ratio of the eigenvalues should be determined using the linearized model of the SSSG to adjust the controller parameters accordingly.

2.5 Stability Enhancement

In this section, several straightforward and innovative approaches for enhancing the stability of SGBMGS, as proposed in various research studies, are presented and discussed. These methods aim to address key operational and control challenges, thereby improving the overall reliability and resilience of MG systems. The following subsections elaborate on these approaches in detail, highlighting their underlying principles and implementation aspects.

2.5.1 Transient Stability Prediction

Recently, various approaches are proposed by the researchers to foresee the stability or instability of SGBMGs following a large disturbance. In [35, 36], a predictive approach is developed for prediction of transient stability of a grid-connected SSSG. Using this approach, a predictive method is suggested to enhance the transient stability of grid-connected DGs. To do so, a dynamic model of the SSSG, its associated prime mover, excitation system, and controllers are developed for real-time analysis of the SSSG unit. The framework of this approach has been depicted in Fig. 2.39. The mathematical model of the SSSG is described through a set of first-order differential equations as follows whose parameters are defined in [35]:

$$\frac{dx}{dt} = f(\mathbf{x}, \mathbf{u}) \tag{2.98}$$

where f is the set of nonlinear functions of state variables x and input vectors u:

$$\mathbf{x} = \left[\psi_d, \psi_q, E_d', \psi_{2q}, E_q', \psi_{1d}, \delta, \omega, E_{fd}, R_f, V_R, T_m, P_{Fu}, P_V\right]$$

$$\mathbf{u} = [Q_{ref}, P_{ref}, V_t] \tag{2.99}$$

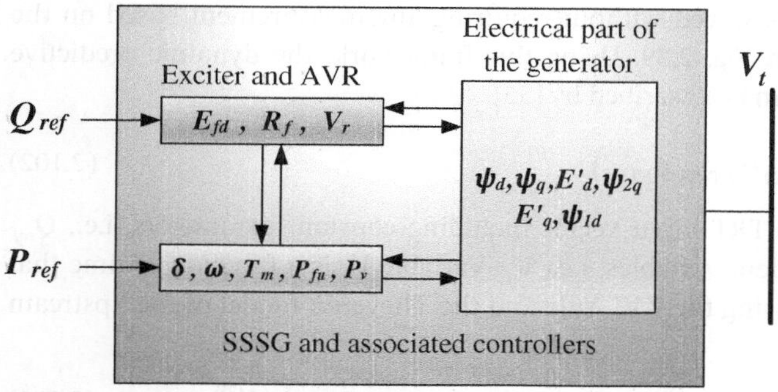

Fig. 2.39 Framework of the SSSG transient stability prediction approach. [36] / with permission of John Wiley & Sons.

Since the measured values are discrete in real-time analysis of the SSSG state variables, the forward Euler technique is employed as

$$x_k = x_{k-1} + f(x_{k-1}, u_{k-1}) \times \Delta t \tag{2.100}$$

Nonpredictive real-time schemes do not assess the forthcoming stability of the SSSG; they merely monitor its current status and state variables in real time. In contrast, predictive techniques, presented by [35, 36], evaluate the SSSG condition without waiting for new measurement samples. By solving the state-space equations of the SSSG faster than real-time (FTRT), these methods offer a significant advantage in anticipating potential instability.

To predict the SSSG condition without relying on direct measurements, the terminal voltage of the SSSG should first be estimated. For this purpose, the Thevenin parameters of the upstream grid, as seen from the SSSG terminal, need to be determined. It is generally assumed that these parameters remain relatively constant following a large disturbance. However, this assumption becomes invalid during-fault-initiated islanding events, where the SGBMG disconnects from the upstream grid after a fault occurrence. Based on these considerations, the Thevenin parameters can instead be estimated using the steady-state active and reactive power outputs of the SSSG, as expressed by the following equations:

$$P_t = \frac{V_t E_{Th}}{X_{Th}} \sin\theta_{Th}$$

$$Q_t = \frac{V_t^2}{X_{Th}} - \frac{V_t E_{Th}}{X_{Th}} \cos\theta_{Th} \tag{2.101}$$

where V_t is the SSSG terminal voltage, E_{Th} is the upstream system Thevenin voltage, and X_{Th} is the upstream system impedance. The parameter X_{Th} can be determined by an off-line short-circuit study of the system. Then, using (2.101), parameters E_{Th} and θ_{Th} are determined. Afterward, the state-space dynamic model of the SSSG is solved without requiring any measurement based on the framework shown in Fig. 2.39. Using this framework, the dynamic predictive model of the SSSG can be described by [35]

$$\frac{dx_{FTRT}}{dt} = f_{FTRT}(x_{FTRT}, u_{FTRT}) \tag{2.102}$$

where u_{FTRT} is the FTRT input vector including constant parameters, i.e., Q_{ref} and P_{ref}, and dependent variables, i.e., V_t. Variable V_t is a function of time that can be determined using the KVL rule and the Thevenin model of the upstream system as

$$V_t = (V_d + jV_q)e^{j(\delta-(\pi/2))} = E_{Th}\angle\theta_{Th} + X_{Th}(I_d + jI_q)e^{j(\delta-(\pi/2))} \tag{2.103}$$

where V_d and V_q as well as I_d and I_q are the SSSG stator voltages and currents in the d–q rotating frame. I_d and I_q can be calculated by

$$\begin{cases} I_d = \left[-\psi_d + \dfrac{(X''_d - X_{ls})}{(X'_d - X_{ls})}E'_q + \dfrac{(X'_d - X''_d)}{(X'_d - X_{ls})}\psi_{1d}\right] / X''_d \\ I_q = \left[-\psi_q + \dfrac{(X''_q - X_{ls})}{(X'_q - X_{ls})}E'_d + \dfrac{(X'_q - X''_q)}{(X'_q - X_{ls})}\psi_{2q}\right] / X''_q \end{cases} \tag{2.104}$$

Therefore, the input vector can be determined without requiring any online signals measurement. In fact, the input vector can be considered as a function of E_{Th} and θ_{Th} instead of V_t. Consequently, the equation set for predicting the state variables of the grid-connected SSSG can be described as

$$\begin{aligned} x_{FTRT} &= \left[\psi_d, \psi_q, E'_d, \psi_{2q}, E'_q, \psi_{1d}, \delta, \omega, E_{fd}\,, R_f, V_R, T_m, P_{Fu}, P_V\right] \\ u_{FTRT} &= [Q_{ref}, P_{ref}, E_{Th}, X_{Th}, \theta_{Th}] \end{aligned} \tag{2.105}$$

This technique predicts the condition of the SSSG, including its transient stability, after a fault occurrence without relying on real-time measurements. The key distinction between the real-time and predictive approaches lies in the use of online measurements. In real-time techniques, the terminal voltage V_t must be directly measured, whereas in predictive methods, V_t is estimated using the Thevenin equivalent model of the upstream system.

2.5.2 IM Instability Detection

After a fault clearance, the occurrence of FIDVR and the related degraded voltage condition in SGBMGs may endanger the SSSG transient stability. Therefore, detection of unstable IMs and disconnecting their feeders could improve the SGBMG stability, especially in the islanded-mode operation. In the following, an effective index that could be exploited for the detection of the IM instability is explained [25].

SGBMG loads can be considered as a combination of static and dynamic loads. Meanwhile, static loads are represented by a mixture of constant power (P), constant current (I), and constant impedance (Z) loads. In addition, dynamic loads are modeled by an IM simplified circuit diagram. Power consumption of static loads can be expressed as a function of the terminal voltage as follows:

$$P_{load} = P_0\left(\frac{V}{V_0}\right)^{\alpha}, \quad Q_{load} = Q_0\left(\frac{V}{V_0}\right)^{\beta} \tag{2.106}$$

where V is the system voltage, and P_0 and Q_0 are the active and reactive load powers at the rated voltage V_0, respectively. In addition, α and β are constant values. The static load model can be represented by an equivalent impedance consisting of an equivalent resistance R_{load} parallel to an equivalent reactance X_{laod}. Based on (2.106), R_{load} and X_{laod} can be determined by

$$R_{load} = R_0\, V^{2-\alpha}, \quad X_{load} = X_0\, V^{2-\beta} \tag{2.107}$$

where $R_0 = V_0^{\alpha}/P_0$ and $X_0 = V_0^{\beta}/Q_0$. The equivalent static load admittance (Y_{load}) considering the parallel combination of R_{load} and X_{load} can be calculated as [27]

$$Y_{load} = \frac{1}{R_{load}} + \frac{1}{jX_{load}} \tag{2.108}$$

Moreover, the admittance of dynamic aggregated IM Y_{IM} can be expressed as

$$Y_{IM} = \frac{1}{\frac{R_e}{s} + jX_e} \tag{2.109}$$

Finally, the total admittance of system loads is presented by

$$Y = \frac{1}{Z_{\mu G}} = Y_{load} + Y_{IM} = (R_0\, V^{2-\alpha})^{-1} + (jX_0\, V^{2-\beta})^{-1} + \left(\frac{R_e}{s} + jX_e\right)^{-1} \tag{2.110}$$

The derivative of Y with respect to time dY/dt is calculated by

$$\frac{dY}{dt} = \frac{dY_{load}(V)}{dV} \times \frac{dV}{dt} + \frac{dY_{IM}(s)}{ds} \times \frac{ds}{dt} \tag{2.111}$$

Combining (2.110) and (2.111) results in

$$\begin{aligned}\frac{dY}{dt} =& -\left[(2-\alpha)R_0^{-1}V^{\alpha-3} - j(2-\beta)X_0^{-1}V^{\beta-3}\right] \times \frac{dV}{dt} \\ &+ \left[\frac{R_e\left(R_e^{\,2} - X_e^{\,2}s^2\right) - 2jR_e^2X_e s}{\left(R_e^{\,2} - X_e^{\,2}s^2\right)^2 + 4R_e^2X_e^2s^2}\right] \times \frac{ds}{dt}\end{aligned} \tag{2.112}$$

The imaginary part of dY/dt is derived as

$$\text{Imag}\left\{\frac{dY}{dt}\right\} = (2-\beta)X_0^{-1}V^{\beta-3} \times \frac{dV}{dt} - \frac{2R_e^2X_es}{\left(R_e^{\ 2}-X_e^{\ 2}s^2\right)^2+4R_e^2X_e^2s^2} \times \frac{ds}{dt} \tag{2.113}$$

Assuming $\sigma = (\beta-2)X_0^{-1}$ and $\lambda = \frac{2R_e^2X_es}{\left(R_e^{\ 2}-X_e^{\ 2}s^2\right)^2+4R_e^2X_e^2s^2}$, (2.113) can be rewritten as

$$\text{Imag}\left\{\frac{dY}{dt}\right\} + \sigma V^{\beta-3} \times \frac{dV}{dt} = -\lambda \times \frac{ds}{dt} \tag{2.114}$$

where λ is a positive coefficient and σ is a constant parameter. During motor reacceleration, $ds/dt < 0$. Hence, $-\lambda \times ds/dt$, and thus the left side of (2.114) is positive during motor reacceleration. Thus, the following reacceleration detection index (*RDI*) is suggested whose positive value identifies the motor reacceleration condition.

$$\text{RDI} = \text{Imag}\left\{\frac{dY}{dt}\right\} + \sigma V^{\beta-3} \times \frac{dV}{dt} \tag{2.115}$$

Therefore, if the IM does not reaccelerate within a specified recovery time, the instability of IM is identified and then the unstable motor can be disconnected to preserve the stability of the rest of the SGBMG.

2.5.3 Using Proper Control Strategies

As discussed before, SSSGs are particularly susceptible to transient instability, primarily due to their low rotational inertia and the relatively long FCT typically observed in low-voltage distribution networks. These factors limit the generator's ability to maintain synchronism during and after disturbances. One of the primary solutions to this issue involves modifying the control system of the SSSG, specifically the governor and AVR controllers, and ensuring their parameters are properly tuned for optimal performance. A modified version of the SSSG control system is proposed by [12], which is explained in this section. This supplementary control strategy is intended to enhance the generator's dynamic performance, reduce rotor angle fluctuations, and strengthen the overall transient stability.

The swing equation of the SSSG is employed to illustrate the fundamental concept of the proposed control scheme, as follows [2]:

$$P_m - P_e = 2H\frac{d\omega}{dt} \tag{2.116}$$

This equation can be rewritten as

$$\begin{aligned} P_m &= 2H\frac{d\omega}{dt} + P_e \\ P_e &= E_GV_t/X_T \sin\delta \end{aligned} \tag{2.117}$$

where V_t is the SSSG terminal voltage. Considering $k = E_G/V_t$, the following equation is obtained:

$$P_m = 2Hd\omega/dt + kV_t^2/X_T \sin\delta \tag{2.118}$$

During a short-circuit fault, considering the worst condition ($V_t = 0$), (2.118) can be simplified as

$$P_m/2H = d\omega/dt \tag{2.119}$$

Based on (2.119), the SSSG rotor speed increases during a short-circuit fault where its rate of change is equal to $P_m/2H$. The amount of P_e decreases significantly during the faults due to the system voltage drop. Thus, under the constant power control mode, the amount of P_{er}, u, and P_m would be increased, which leads to more increment of SSSG rotor speed. Therefore, the steady-state power control loop should be blocked during short-circuit faults to prevent increasing of SSSG mechanical power and its instability. It should be noted that the reference values of ω and $d\omega/dt$ are considered equal to ω_{ref} and zero, respectively. Therefore, as shown in Fig. 2.40, the input of PI controller (λ) is proposed to be

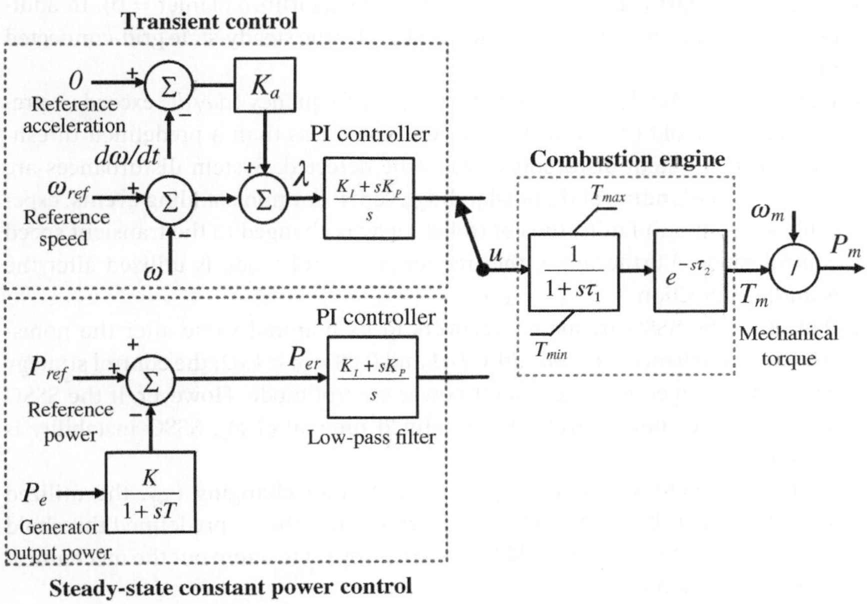

Fig. 2.40 The proposed governor controller.

determined by

$$\lambda = [\omega_{ref} - \omega] + K_a \left[\left(\frac{d\omega}{dt} \right)_{ref} - \frac{d\omega}{dt} \right] = [\omega_{ref} - \omega] - K_a \frac{d\omega}{dt} \tag{2.120}$$

where $(d\omega/dt)_{ref}$ is the reference value of rotor acceleration, which is equal to zero.

Based on (2.120), when $\omega > \omega_{ref}$ or $d\omega/dt > (d\omega/dt)_{ref}$, the PI controller input λ would be negative, and the amount of u and P_m decrease. Consequently, the amount of $d\omega/dt$ decreases based on (2.119). This can effectively manage the SSSG rotor speed increment during short-circuit faults and help improve its transient stability. It should be noted that SSSG speed may oscillate during short-circuit faults or after the fault clearance. Thus, the amount of $d\omega/dt$ may become negative for some duration. In such a condition, to prevent increasing of λ and P_m, the parameter K_a in (2.120) is considered equal to zero for $d\omega/dt \leq 0$. In addition, for smooth transition between the transient speed control and steady-state control modes, the PI controllers would be initialized by the latest amount of u. In other words, the initial value of the PI controller after transition should be equal to the latest amount of u before the transition. The mentioned point is managed by the controller flowchart as described in the next section.

The flowchart of the proposed control strategy is shown in Fig. 2.41, and the control procedure is described as follows:

- **Part 1:** This part is employed to initialize the algorithm (Timer = 0). In addition, the power control strategy is utilized under the steady-state grid-connected mode.
- **Part 2:** When the derivative of SSSG angular frequency ($d\omega/dt$) exceeds a predefined threshold ($Th1$), or its voltage become less than a predefined threshold ($Th2$), a system disturbance would be detected. System disturbances are divided into islanding and nonislanding events. For nonislanding events, especially short-circuit faults, the control strategy is changed to the transient speed control mode. Furthermore, the frequency control mode is utilized after the islanding detection.
- **Part 3:** If the SSSG frequency recovers to its nominal value after the nonislanding disturbance, i.e., $|d\omega/dt| < Th3$ and $0.98 < \omega < 1.02$, the control strategy would be changed to the constant power control mode. However, if the SSSG frequency does not recover after a defined interval ($Th4$), SSSG instability is detected.
- **Part 4:** To decrease the settling time of P_e after changing P_{ref}, the utilized low-pass filter is bypassed. When P_{er} becomes less than a predefined threshold ($Th5$), the low-pass filter should be employed again to damp out the mechanical torque oscillations.

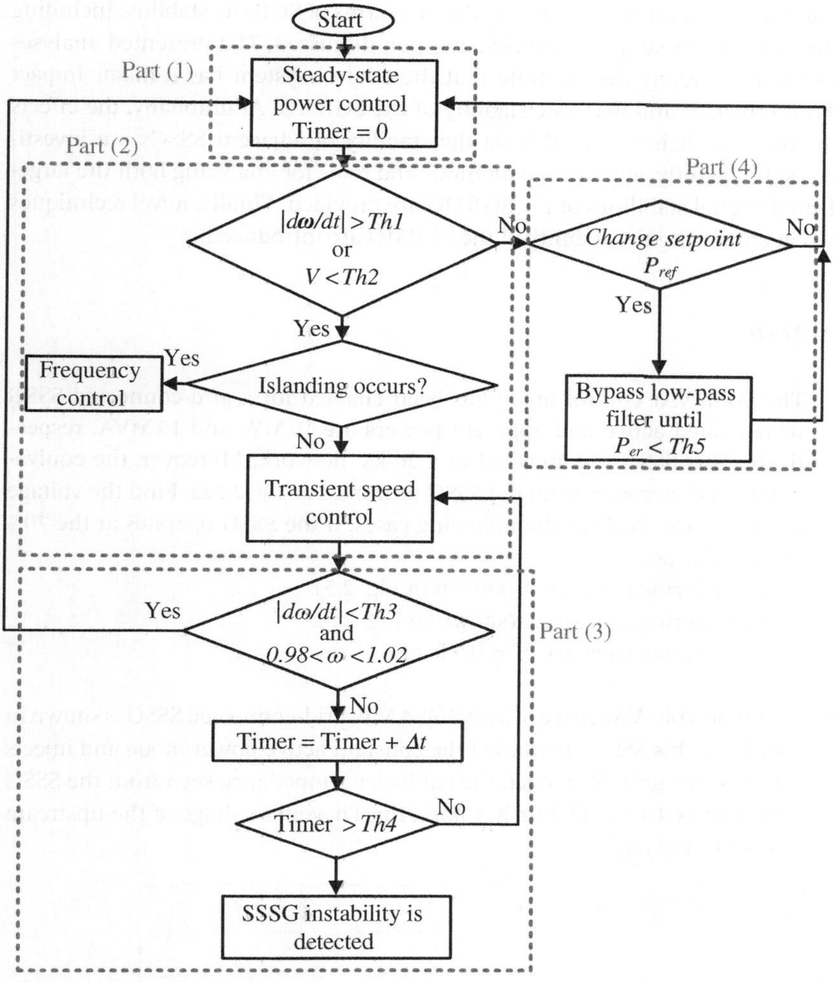

Fig. 2.41 Flowchart of the proposed control strategy.

2.6 Summary

The control system of the SGBMG significantly influences both its small-signal and large-signal stabilities. Therefore, selecting an appropriate control system configuration and properly tuning its parameters are crucial for the stable operation of the SGBMG, as well as for supporting the upstream grid. To address this, the common configurations of the SGBMG control system are first presented in this chapter, in accordance with the IEEE 1547 standard, under both the steady-state and abnormal operating conditions. Various forms of SGBMG stability, including control and power supply stabilities, are then discussed. The presented analyses and examples clearly demonstrate that the control system has a major impact on both transient and dynamic stability of the SGBMG. Additionally, the effects of the dynamic behavior of IMs on the stability of adjacent SSSGs are investigated. Subsequently, appropriate methods and tools for analyzing both the large- and small-signal stabilities of the SGBMG are provided. Finally, novel techniques aimed at enhancing the stability of the SGBMG are introduced.

Problems

2.1 The Watt–VAr control mode has been enabled for a grid-connected SSSG whose rated active and apparent powers are 10 MW and 15 MVA, respectively. This SSSG is connected to a 20-kV network. Moreover, the equivalent impedance seen from the SSSG terminals is $2+j2.3\ \Omega$. Find the voltage change at the PCC for the following cases, if the SSSG operates at the 70% rated active power.

a) Considering category A (shown in Fig. 2.5).
b) Considering category B (shown in Fig. 2.5).
c) Considering category $pf = 0.9$ lead.

2.2 A sample Volt–VAr curve of a 5 MVA/4 MW grid-connected SSSG is shown in Fig. 2.42. This SSSG operates at the constant active power mode and injects 3 MW to the grid. Moreover, the equivalent impedance seen from the SSSG terminals is $4+j4.5\ \Omega$. Furthermore, the Thevenin voltage of the upstream system is 0.95 pu.

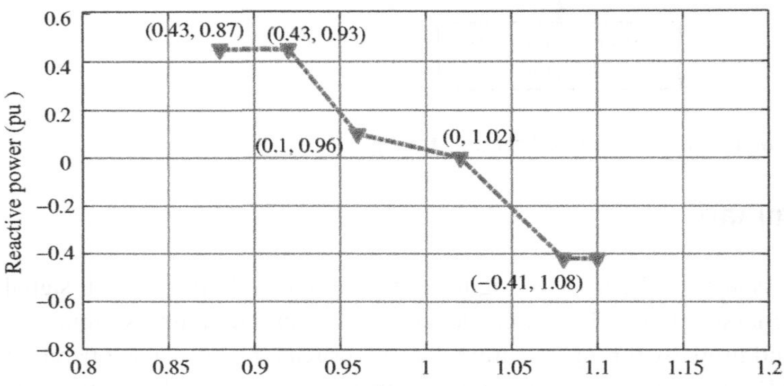

Fig. 2.42 Volt–VAr curve of a typical SSSG (Problem 2).

a) Find the voltage at the PCC.
b) Calculate the output reactive power of the SSSG.
c) Resolve this problem if the Thevenin voltage of the upstream system is 1.05 pu.

2.3 A two-phase short-circuit fault occurs in a grid-connected SGBMG, which contains SSSGs and IMs, and is cleared after 400 ms. Determine the IM speed after the fault clearance. The parameters of the aggregated IM model are $J = 1$ s, $R'_r = 0.15\ \Omega$, and $\omega_s = 314$ rad/s. The nominal terminal voltage is 6 kV, and the load torque is expressed as $T_L = 490$ N.m. The nominal slip is considered 4%.

2.4 In the previous problem, calculate the aggregated IM recovery time. It is assumed the voltage reaches to 0.95 pu after the fault clearance.

2.5 A grid-connected SSSG is depicted in Fig. 2.30 whose transient reactance is 0.2 pu and its nominal power is 9 MVA. Furthermore, its transformer reactance is 0.12 pu, and its rated power is 10 MVA. Moreover, the three-phase short-circuit level of the 20 kV upstream system is 10 kA, and its resistance is neglected. Find the CCT of this SSSG for a three-phase fault at the PCC. The SSSG delivers 8 MVA (0.95 lag) before the fault occurrence.

References

1 M. Farrokhabadi *et al.*, "Microgrid stability definitions, analysis, and examples," *IEEE Trans. Power Syst.*, vol. 35, no. 1, pp. 13–29, 2020.
2 P. S. Kundur, N. J. Balu, and M. G. Lauby, *Power System Dynamics and Stability*. CRC Press, Boca Raton: Taylor & Francis Group, 2017.
3 P. Kundur *et al.*, "Definition and classification of power system stability," *IEEE Trans. Power Syst.*, vol. 19, no. 2, pp. 1387–1401, 2004.
4 C. W. Taylor, "Concepts of undervoltage load shedding for voltage stability," *IEEE Trans. Power Deliv.*, vol. 7, no. 2, pp. 480–488, 1992.
5 G. Dellile, B. Francois, and G. Malarange, "Dynamic frequency control support by energy storage to reduce the impact of wind and solar generation on isolated power system's inertia," *IEEE Trans. Sustain. Energy*, vol. 3, no. 4, pp. 931–939, 2012.
6 "IEEE standard for interconnection and interoperability of distributed energy resources with associated electric power systems interfaces," IEEE Std 1547-2018 (Revision of IEEE Std 1547-2003), 2018.

7 "Technical guidelines for interconnection of distributed generations to Iran power grid," Iran, Tavanir, 2020. Available: https://wamp.tavanir.org.ir/rule/index.php (accessed 2020).

8 "Technical connection rules for medium-voltage," VDE-AR-N 4110, 2008.

9 C. Y. Lee and M. Tuegeh, "Virtual visualization of generator operation condition through generator capability curve," *Energies*, MDPI, vol. 14, no. 1, 2021.

10 "IEEE application guide for IEEE std 1547-2018, IEEE standard for interconnection and interoperability of distributed energy resources with associated electric power systems interfaces," IEEE Std 1547.2-2023 (Revision of IEEE Std 1547.2-2008), pp. 1–291, 2024.

11 A. Hussain, C. -H. Kim, and A. Mehdi, "A comprehensive review of intelligent islanding schemes and feature selection techniques for distributed generation system," *IEEE Access*, vol. 9, pp. 146603–146624, 2021.

12 R. Bekhradian, M. Sanaye-Pasand, M. Davarpanah, and Y. Kor, "Governor controller improvement to enhance SSSG performance in presence of electrical arc furnace," *IET Gener. Transm. Distrib.*, vol. 16, no. 18, pp. 3645–3660, 2022.

13 A. Mahdavian, A. A. Ghadimi, and M. Bayat, "Microgrid small-signal stability analysis considering dynamic load model," *IET Renew. Power Gener.*, vol. 15, no. 13, pp. 2799–2813, 2021.

14 N. Hatziargyriou *et al.*, "Definition and classification of power system stability – Revisited & extended," *IEEE Trans. Power Syst.*, vol. 36, no. 4, pp. 3271–3281, 2021.

15 X. Liang, H. Chai, and J. Ravishankar, "Analytical methods of voltage stability in renewable dominated power systems: A review," *Electricity*, MDPI, vol. 3, no. 1, pp. 75–107, 2022.

16 T. Van Cutsem and C. D. Vournas, *Voltage Stability of Electric Power Systems.* Boston: Kluwer Academic Publishers, 1998.

17 M. Eremia and M. Shahidepour, *Handbook of Electrical Power System Dynamics: Modeling, Stability, and Control.* New Jersey: John Wiley & Sons, Inc., IEEE Press, 2013.

18 H. Bai and V. Ajjarapu, "A novel load shedding strategy for mitigating fault-induced delayed voltage recovery," *IEEE Trans. Power Syst.*, vol. 26, no. 1, pp. 294–304, 2011.

19 "Fault-Induced Delayed Voltage Recovery," Technical Reference Paper, NERC, 2009.

20 E. Hajipour *et al.*, "An improved aggregated model of residential air conditioners for FIDVR studies," *IEEE Trans. Power Syst.*, vol. 35, no. 2, pp. 909–919, 2019.

21 R. Bekhradian, M. Davarpanah, and M. Sanaye-Pasand, "Current-based blocking scheme to stabilize distribution network relays against FIDVR," *Int. J. Electr. Power Energy Syst.*, vol. 132, p. 107205, 2021.

22 P. Krause, O. Wasynczuk, and S. D. Sudhoff, *Analysis of Electric Machinery and Drive Systems*, vol. 75. John Wiley & Sons, 2013.

23 IEEE Std 421.5-2005: "IEEE recommended practice for excitation system models for power system stability studies," 2006.

24 J. V. Milanovic, M. T. Aung, and S. C. Vegunta, "The influence of induction motors on voltage sag propagation—Part I: Accounting for the change in sag characteristics," *IEEE Trans. Power Deliv.*, vol. 23, no. 2, pp. 1063–1071, 2008.

25 R. Bekhradian, M. Sanaye-Pasand, and M. Davarpanah, "Innovative load shedding scheme to restore synchronous generator-based microgrids during FIDVR," *IEEE Trans. Smart Grid*, vol. 14, no. 1, pp. 388–399, 2023.

26 H. Saadat, *Power System Analysis*, 3rd ed. New York: McGraw-Hill, 1999.

27 R. Razzaghi, M. Davarpanah, and M. Sanaye-Pasand, "A novel protective scheme to protect small-scale synchronous generators against transient instability," *IEEE Trans. Ind. Electron.*, vol. 60, no. 4, pp. 1659–1667, 2013.

28 H. Khalil, *Nonlinear Systems*. Hoboken, NJ: Prentice-Hall, 2002.

29 S. H. Zak, *Systems and Control*. New York, NY: Oxford University Press, 2003.

30 S. Teimourzadeh, F. Aminifar, M. Davarpanah, and M. Shahidehpour, "Adaptive protection for preserving microgrid security," *IEEE Trans. Smart Grid*, vol. 10, no. 1, pp. 592–600, 2019.

31 H. W. Dommel, "Digital computer solution of electromagnetic transients in single and multiphase networks," *IEEE Trans. Power App. Syst.*, vol. PAS-88, no. 4, pp. 388–399, 1969.

32 W. Ren, M. Steurer, and T. Baldwin, "Improve the stability and the accuracy of power hardware-in-the-loop simulation by selecting appropriate interface algorithms," *IEEE Trans. Ind. Appl.*, vol. 44, no. 4, pp. 1286–1294, 2008.

33 M. Maniatopoulos, D. Lagos, P. Kotsampopoulos, and N. Hatziargyriou, "Combined control and power hardware in-the-loop simulation for testing smart grid control algorithms," *IET Gener. Transm. Distrib.*, vol. 11, no. 12, pp. 3009–3018, 2017.

34 M. Adelpour, M. Hamzeh, and K. Sheshyekani, "Comprehensive small-signal stability analysis of islanded synchronous generator-based microgrids," *Sustain. Energy Grids Netw.*, vol. 26, 2021.

35 M. Abedini, M. Davarpanah, M. Sanaye-Pasand, S. M. Hashemi, and R. Iravani, "Generator out-of-step prediction based on faster-than-real-time analysis: Concepts and applications," *IEEE Trans. Power Syst.*, vol. 33, no. 4, pp. 4563–4573, 2018.

36 M. Abedini, M. Sanaye-Pasand, M. Davarpanah, H. Lesani, and M. Shahidehpour, "Predictive auto-reclosure approach to enhance transient stability of grid-connected DGs," *IET Gener. Transm. Distrib.*, vol. 13, no. 14, pp. 3011–3019, 2019.

3

Stability of Inverter-Based Microgrids: Fundamentals, Classification, and Assessment

3.1 Introduction

The penetration of inverter-based distributed generations (IBDGs) into modern microgrids (MGs) has recently increased to attain the benefits of renewable and green energies such as solar and wind. As such, to increase the reliability of MGs and decrease the cost of operation, IBDGs can be utilized, especially in parallel with small-scale synchronous generators (SSSGs). Moreover, to cover the essential loads for which power interruption is not allowable, energy storage systems (ESSs) may be utilized along with the inverter-based resources. In essence, the configuration, type, and technologies of distributed generations (DGs) and also the MG topology are specified based on the financial–technical analysis. The MGs containing only IBDGs are commonly known as inverter-based MGs (IBMGs). Furthermore, the MGs that include both IBDGs and SSSGs are called hybrid MGs (HMGs).

IBMGs and HMGs may have different dynamics and transient responses compared to those of the synchronous generator-based MGs (SGBMGs), which affects the MG control and protection systems. These differences are mainly related to the converter's dynamic behavior and its associated controllers during small and large disturbances. Although some types of system stabilities may occur similarly for IBMGs and SGBMGs, their dynamic responses, especially the voltage and frequency variations, may be completely different. For instance, the transient stability of SGBMGs is considerably dependent on the inherent mechanical response of SSSGs. However, for IBMGs, the transient stability phenomenon mainly occurs due to their control system instability. Therefore, these stability types are investigated for IBMGs in this chapter. Accordingly, at first, the structure and configuration of various types of IBDG control systems are presented. Then, the transient and small signal stabilities of IBDGs are investigated. These investigations could help system engineers to modify and revise conventional protection schemes to make them compatible with the modern MGs containing IBDGs.

3.2 Control System of IBMGs

Similar to SGBMGs, various control strategies have been proposed to control the electrical parameters of IBMGs, i.e., their active/reactive power, voltage, and frequency. Meanwhile, different control schemes may be adopted to maximize the energy output of IBDGs and minimize their energy losses. Therefore, the main goals of IBMG control systems could be summarized as (i) regulating and maximizing the output active power, (ii) regulating the output reactive power, (iii) supporting the IBMG voltage and frequency in the grid-connected operation mode, and (iv) regulating the voltage and frequency in the islanded operation mode. In this regard, recently, two major control strategies have been developed for IBDGs, which are commonly referred to as (i) grid-following (GFL) and (ii) grid-forming (GFM) strategies, which are explained as follows [1].

3.2.1 Grid-Following Strategy

GFL strategy is commonly utilized when the frequency of the network is adjusted by other resources such as upstream network or other DGs connected to the IBMG. In fact, in this condition, the voltage phase angle at the point of common coupling (PCC) is dominated by other resources, and the GFL-IBDG should "follow" the grid voltage through injecting the proper amount of active/reactive current. To do so, an appropriate phase-locked loop (PLL) is utilized to determine the frequency and angle of the system voltage at the PCC and synchronize the IBDG with the power network. If the PLL fails to track the system frequency and grid voltage, the IBDG loses its stability and synchronism. Figure 3.1 illustrates a simple representation of a GFL-IBDG, which is connected to the upstream grid. Since the GFL-IBDG follows the grid through injecting the proper amount of current, it can be modeled by a controlled current source and a parallel large impedance, as shown in Fig. 3.1 [1].

The GFL-IBDGs commonly operate in the constant active and reactive power control mode. However, they can also support the upstream network in both the

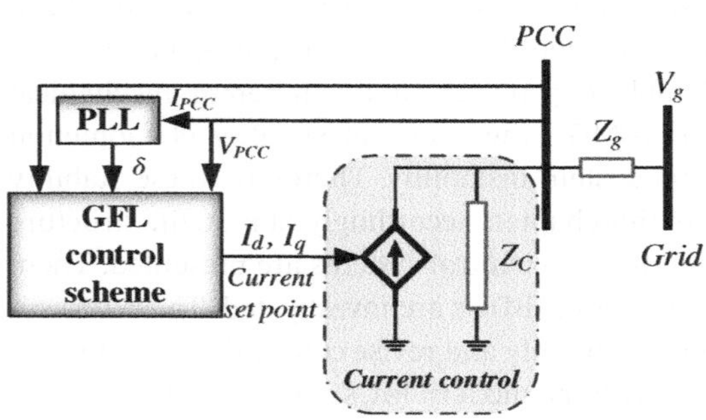

Fig. 3.1 Simple representation of a GFL-IBDG.

Table 3.1 Various control modes of grid-supporting inverters. Data from [2].

State	Control loop	Control mode
Normal (steady-state)	Voltage/reactive power	Fixed reactive power
		Fixed power factor
		Droop
		Volt–VAr
		Watt–VAr
		Constant voltage
	Frequency/active power	Fixed active power
		Droop
		Volt–Watt
Abnormal	Voltage disturbances ride-through	Dynamic voltage support
	Frequency disturbances ride-through	Fast frequency response

steady-state and transient operations, as presented in Table 3.1. Therefore, they are also referred to as "grid-supporting IBDGs." The explanations of each control loop and control mode of Table 3.1 are similar to those of the SSSGs, which are thoroughly explained in Chapter 2. The detailed control loops of a common type GFL-IBDG are depicted in Fig. 3.2, which contain the loops described in the following subsections.

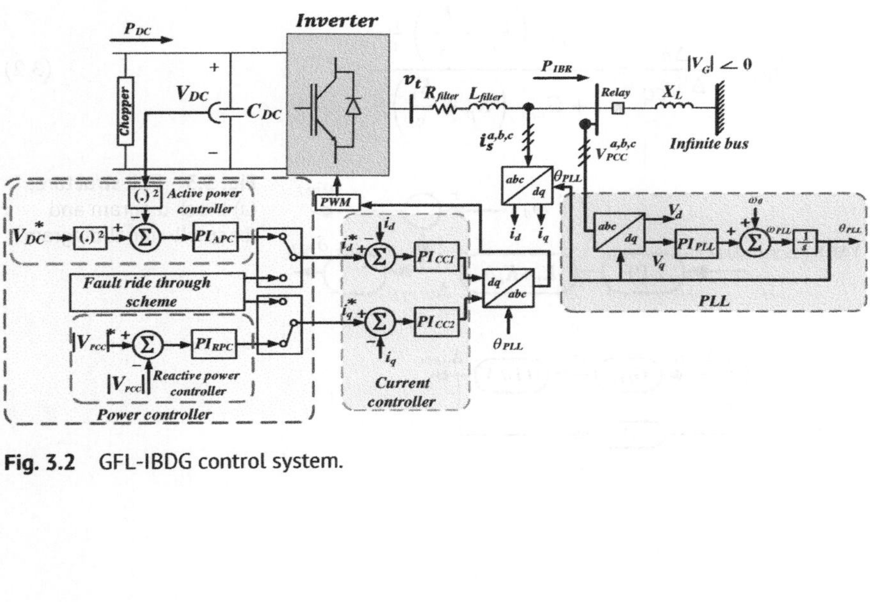

Fig. 3.2 GFL-IBDG control system.

3.2.1.1 Phase-Locked Loop

In GFL-IBDGs, usually the familiar decoupled $d - q$ reference frame control schemes are utilized. In this regard and to synchronize the GFL-IBDG with the upstream grid, it is necessary to determine the frequency and angle of the system voltage at the PCC. To do so, the commonly used PLL, shown in Fig. 3.2, is exploited. In this widely utilized PLL, the q-axis voltage is set at zero, and thus, the d-axis voltage is aligned with the PCC voltage phasor [3]. This approach provides independent control of active and reactive powers. In other words, the active power is controlled by the d-axis current, i.e., i_d, while the reactive power is regulated by the q-axis current, i.e., i_q [4].

One of the main reasons of instability of GFL-IBDGs and their outages is the loss of synchronism of the PLL, as reported by NERC [5]. The stability of GFL-IBDGs in weak distribution systems containing high Thevenin impedances is adversely affected by the PLL, especially during severe disturbances such as short-circuit faults where the voltage magnitude at the PCC severely drops.

The PLL input signal, i.e., V^q voltage at PCC, can be expressed as

$$V^q = V \sin(\delta_{PCC} - \delta_{PLL}) \tag{3.1}$$

where δ_{PCC} is the actual amount of voltage angle at PCC and δ_{PLL} is the angle estimated by the PLL. Considering this equation, the block diagram of a common type of PLL is illustrated in Fig. 3.3(a). In this figure, the LPF block represents a low-pass filter, for which the transfer function is G_{LPF}. In view of small-signal variations, the term $\sin(\delta_{PCC} - \delta_{PLL})$ can be considered equal to $(\delta_{PCC} - \delta_{PLL})$ and the voltage magnitude at PCC is assumed to be 1 pu. According to these assumptions, the small-signal model of PLL is depicted in Fig. 3.3(b). Considering Fig. 3.3(b), the closed-loop transfer function of PLL is given by [3, 4]

$$G_{PLL} = \frac{\Delta\delta}{\Delta\delta_{PCC}} = \frac{G_{LPF}\left(K_p + \frac{K_i}{s}\right)\frac{1}{s}}{1 + G_{LPF}\left(K_p + \frac{K_i}{s}\right)\frac{1}{s}} \tag{3.2}$$

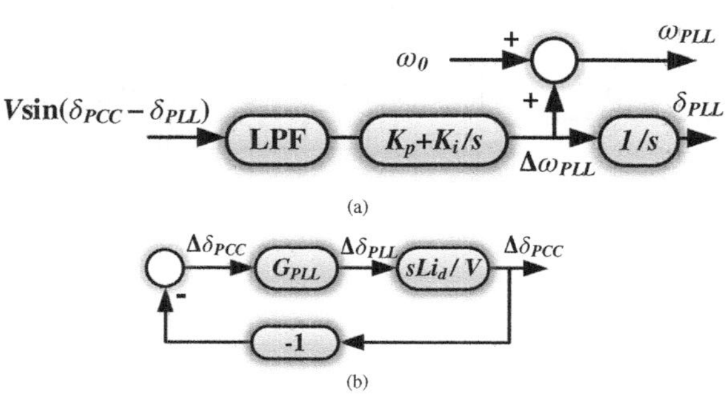

Fig. 3.3 PLL structure: (a) block diagram and (b) small-signal diagram.

The controller parameters of PLL, i.e., K_p and K_i, should be adjusted to provide an appropriate bandwidth and robust dynamic response for the PLL. Meanwhile, it should be considered that the PLL dynamics may have a significant impact on the PCC voltage angle considering the system. Thevenin impedance and thus the stability of G_{PLL} are not sufficient to guarantee the system stability. As such, the small-signal and transient stabilities of the PLL are discussed in this chapter.

3.2.1.2 Power Control Loop

During the MG continuous operation, in which the system voltage and frequency are within the permissible ranges, the output active and reactive powers are regulated based on the GFL-IBDG control mode. In fact, the power control loop adjusts (i) output active power (P_{IBDG}) and (ii) reactive power (Q_{IBDG}) at their setpoints by determining the appropriate values of the d- and q-axis current components, i.e., i_d^* and i_q^* [6]. To do so, suitable proportional-integral (PI) controllers are utilized, as depicted in Fig. 3.2, i.e., PI_{APC} and PI_{RPC}, as the PI controllers for active and reactive powers, respectively [6, 7]. Considering the ideal operation of PLL, the amounts of i_d^* and i_q^* are calculated as [7, 8]

$$i_d^* \approx P_{IBDG}^*/V^d,\ i_q^* \approx Q_{IBDG}^*/V^d \tag{3.3}$$

where V^d is the d-axis voltage component, which is determined using the $d-q$ reference frame transformation and the PCC voltage angle δ_{PCC}. Meanwhile, P_{IBDG}^* and Q_{IBDG}^* are the setpoints of IBDG active and reactive powers, respectively. Furthermore, the reference current phasor generated by the power control loop could be expressed in the abc-frame as follows:

$$\begin{aligned} I_{PCC}^* &= \left|I_{PCC}^*\right| \angle\theta_I = \frac{P_{IBDG}^* - jQ_{IBDG}^*}{(|V_{PCC}|\angle\delta_{PCC})} = \frac{\left|S_{IBDG}^*\right| \angle -\varphi}{(|V_{PCC}|\angle\delta_{PCC})} \\ &= \frac{\left|S_{IBDG}^*\right|}{|V_{PCC}|}\angle(\delta_{PCC} - \varphi) \end{aligned} \tag{3.4}$$

where $S_{IBDG}^* = \sqrt{\left(P_{IBDG}^*\right)^2 + \left(Q_{IBDG}^*\right)^2}$ is the reference apparent power of the IBDG and $\varphi = tg^{-1}\left(Q_{IBDG}^*/P_{IBDG}^*\right)$ is the power factor angle. This equation is valid only for the ideal PLL in which the PLL output is equal to the PCC voltage angle, i.e., $\delta_{PLL} = \delta_{PCC}$. However, for the nonideal PLL, the frequency (ω_{PLL}) and angle (δ_{PLL}) estimated by the PLL may not be exactly equal to the grid angular frequency (ω_G) and angle (δ_{PCC}) of V_{PCC}, especially during voltage disturbances. The difference between the PLL estimation and actual values results in the oscillations of abc-frame signals and power swing. The dynamic behavior of PLL influences the power swing characteristics, and it depends on the PI controller parameters and its bandwidth. Considering the PLL dynamic behavior, the GFL-IBDG-injected active power P_{IBDG} would not be equal to its reference value P^*_{IBDG}. In this condition, the amount of P_{IBDG} can be derived using the following equation:

$$P_{IBDG} = P^*_{IBDG} \cos(\delta_{PLL} - \delta_{PCC} - \varphi)/\cos(\varphi) \tag{3.5}$$

Therefore, based on this equation, the PLL error leads to deviation in P_{IBDG} from its reference value P^*_{IBDG}.

In conventional GFL-IBDGs, the grid-supporting functions were not implemented, and the power setpoints were considered fixed. However, in the modern GFL-IBDGs, grid-supports are provided by suitably changing the power setpoints instead of using the inflexible values. In other words, the power references are changed to support IBMG voltage and frequency and provide the control loops presented in Table 3.1. The detailed explanations for determining the proper setpoints are presented in Chapter 2 of this book. Considering the contents of Chapter 2 and the provisions of IEEE-1547 standard, the power setpoints, i.e., P^*_{IBDG} and Q^*_{IBDG}, could be determined as [9]

$$P_i^{ref} = \mathcal{F}\left[P_i^{set} + (\omega^{nom} - \omega_i)/m_{p_i}\right]_{P_{min}}^{P_{max}} \tag{3.6}$$

$$Q_i^{ref} = \mathcal{F}\left[Q_i^{nom} + \left(V_i^{set} - V_i\right)/m_{q_i}\right]_{q_{min}}^{q_{max}} \tag{3.7}$$

where P_i^{set}, Q_i^{nom} are the nominal steady-state amounts of output active and reactive powers of GFL-IBDG, respectively. P_{max}/P_{min} and q_{max}/q_{min} are the maximum/minimum allowable amount of output active and reactive powers, correspondingly. Furthermore, ω^{nom} and V_i^{set} are the nominal values of frequency and voltage in the IBMG, respectively. Parameters m_{pi} and m_{qi} represent droop coefficients for the frequency/Watt and Volt/VAr control loops, respectively. Meanwhile, $F[x]$ is a saturation function, which is defined as [9]

$$\mathcal{F}[x] = \begin{cases} x^{min}, & \text{if } x \le x^{min} \\ x, & \text{if } x^{max} \le x \le x^{min} \\ x^{max}, & \text{if } x \ge x^{max} \end{cases} \tag{3.8}$$

Equations (3.6) and (3.7) are adopted to implement frequency/Watt and Volt/VAr control loops, respectively. For other control loops, such as Volt/Watt or Watt/VAr, a similar function could be developed based on the explanations of Chapter 2.

A typical saturation function widely utilized in the GFL-IBDGs is illustrated in Fig. 3.4. As shown, this function saturates at the predefined maximum/minimum possible amounts of active/reactive powers of GFL-IBDG and prevents infeasible amount of power setpoints. The nominal steady-state amount of output active power, i.e., P_i^{set}, could be adjusted lower than the maximum available power P_{max} to provide some capacity for frequency regulation.

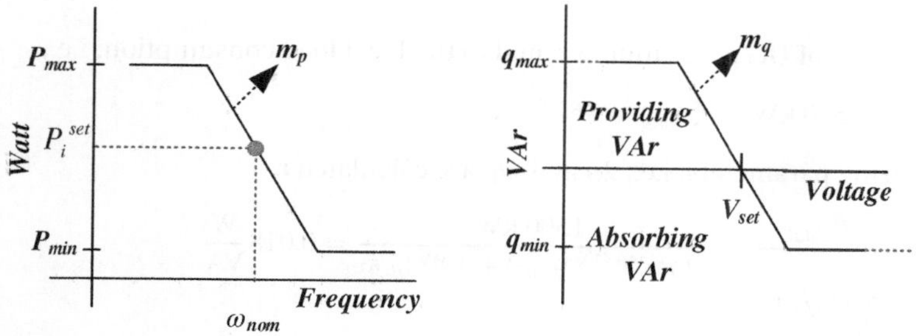

Fig. 3.4 Saturation function for frequency/Watt and Volt/VAr control loops. [9] / with permission of IEEE.

Using frequency/Watt and Volt/VAr droops could enable equal active and reactive power sharing between the IBDGs. Equal power sharing means that the powers of IBDGs and other resources located in the MG are dispatched proportional to their per-unit droop coefficients and rated power. For analytical expressions, power-sharing parameters are defined as [9]

$$\lambda p = \frac{\sum_{i=1}^{n} P_i}{\sum_{i=1}^{n} (S_i/m_{pi})}, \quad \lambda q = \frac{\sum_{i=1}^{n} Q_i}{\sum_{i=1}^{n} (S_i/m_{qi})} \tag{3.9}$$

where λp and λq are power-sharing parameters for the active and reactive powers, respectively. Meanwhile, P_i, Q_i, and S_i are the active, reactive, and rated powers of the ith DG located in the MG, respectively. Furthermore, m_{pi} and m_{qi} are the droop coefficients of frequency/Watt and Volt/VAr control loops, respectively. For equal power sharing, the active and reactive power should be dispatched as [9]

$$P_{iE} = \lambda p \frac{S_i}{m_{pi}}, \quad Q_{iE} = \lambda q \frac{S_i}{m_{qi}} \tag{3.10}$$

If the active/reactive outputs of DGs are diverged from P_{iE}/Q_{iE}, the equal power sharing is violated, which may cause high circulating of VAr among the DGs.

Example 3.1

Two identical 500-kVA IBDGs and one 1-MVA SSSG are located in a grid-connected MG. A local load (1500 kW and 562 kVAr) is supplied by these DGs, and the power exchange with the upstream grid is zero. Find the amounts of active/reactive powers of these DGs, considering equal power sharing. For IBDGs, the droop coefficients of frequency/Watt and Volt/VAr control loops are 2% and 5%, respectively, and for SSSG, they are 3% and 2%, respectively.

Solution

The total amount of DG generation is equal to the local load consumption, i.e.,

$$P_{Total} = 1500\,\text{kW}, \quad Q_{Total} = 562\,\text{kVAr}$$

Power-sharing parameters, i.e., λp and λq, are calculated as

$$\lambda p = \frac{P_{Total}}{\sum_{i=1}^{3}(S_i/m_{pi})} = \frac{1500\,\text{kW}}{2\times(^{500\,\text{kVA}}/_{0.02}) + {}^{1\,\text{MVA}}/_{0.03}} = 0.018\frac{\text{W}}{\text{VA}}$$

$$\lambda q = \frac{Q_{Total}}{\sum_{i=1}^{3}(S_i/m_{qi})} = \frac{562\,\text{kVAr}}{2\times(^{500\,\text{kVA}}/_{0.05}) + {}^{1\,\text{MVA}}/_{0.02}} = 0.008\,\frac{\text{VAr}}{\text{VA}}$$

Consequently, the active and reactive powers of each DG are determined as

$$\text{For IBDGs:}\;\begin{cases} P_{IBDG} = 0.018\,\frac{\text{W}}{\text{VA}} \times \frac{500\,\text{kVA}}{0.02} = 450\,\text{kW} \\ Q_{IBDG} = 0.008\,\frac{\text{VAr}}{\text{VA}} \times \frac{500\,\text{kVA}}{0.05} = 80\,\text{kVAr} \end{cases}$$

$$For\,SSSG:\;\begin{cases} P_{SSSG} = 0.018\,\frac{W}{VA} \times \frac{1000\,kVA}{0.03} = 600\,kW \\ Q_{SSSG} = 0.008\,\frac{VAr}{VA} \times \frac{1000\,kVA}{0.02} = 400\,kVAr \end{cases}$$

In modern GFL-IBDGs, which are exploited using renewable energy resources, such as wind and solar energies, when the primary energy resources are not available, the capacity of IBDG can be utilized to exchange reactive power (both injection and absorption) with the upstream network. As an example, for modern photovoltaic (PV) IBDGs, reactive power exchange capability at night, when the solar energy is not available, are provided, which is known as the "Q at night" function. This function could significantly help the system voltage recovery and its stability, especially after voltage disturbances.

After fault occurrences (such as short-circuit faults), the IBDG fault ride-through function will be enabled based on the requirements of the IEEE-1547 standard, which are presented in Chapter 2. In such a condition, the GFL-IBDG should inject active/reactive currents for a duration depending on the severity of the fault. Moreover, a chopper circuit is exploited to prevent a severe voltage rise in the DC link during the short-circuit fault.

The aforementioned active power control loop, shown in Fig. 3.2, can be implemented only if an ideal DC voltage source is adopted, which provides the DC-bus voltage. However, in common applications like PV systems, the DC side is not connected to an ideal voltage source, and as a result, the DC-bus voltage is not inherently fixed and shall be actively regulated. In such a condition, a capacitor is located at the DC link whose voltage should be controlled. Therefore, in this scenario, the output power of IBDG is adjusted by regulating the DC-link voltage. Since the voltage of the DC-bus is adjusted at a predefined value, the net exchange power of the DC-bus capacitor is held at about zero. Thus, the output active power of IBDG becomes equal to the injected power to the DC link.

For analytical investigations, the dynamic behavior of the DC-link voltage can be analytically expressed by the power balance equation as follows [10]:

$$P_{DC} - P_{IBDG} - P_{Loss} = P_{Cap} \tag{3.11}$$

where P_{DC} is the injected power to the DC link from the PV or wind resources, P_{IBDG} is the active power injected by the inverter to the grid, P_{Cap} is the power absorbed by the DC-link capacitor, and P_{Loss} is the inverter and its associated filter losses. It is worth mentioning that the dynamics of the source connected to the DC link can be modeled using the P_{DC} signal. Furthermore, P_{Cap} can be calculated as

$$P_{Cap} = d\left(CV_{DC}^2/2\right)/dt \tag{3.12}$$

where C_{DC} is the capacitance of the DC-link capacitor. Combining (3.11) and (3.12) and neglecting P_{loss} result in

$$dV_{DC}^2/dt = 2(P_{DC} - P_{IBDG})/C_{DC} \tag{3.13}$$

According to (3.13), if V_{DC} is adjusted at a constant value, dV_{DC}^2/dt and P_{Cap} become relatively small, and thus, P_{IBDG} will be equal to P_{DC}. The block diagram of the IBDG active power control in which the DC-link voltage is regulated instead of the output active power is depicted in Fig. 3.5. The dynamic behavior of the power control loop in the Laplace domain can be expressed by the following equation:

$$i_d^* = \left(V_{DC}^2 - \left[V_{DC}^*\right]^2\right) \times PI_{APC}(s) \tag{3.14}$$

where $PI_{APC}(s)$ is the transfer function of PI active power controller, and parameter i_d^* is the current control loop reference value, whose per-unit value is equal to the per-unit value of the reference power P_{IBDG}^*. Due to the wide bandwidth of the current control loop, its dynamics could be neglected, and thus, it is assumed that the injected current is equal to the reference current, i.e., $i_d \approx i_d^*$. Therefore, considering the ideal operation of PLL, the output active power (P_{IBDG}) can also be considered equal to its reference value $\left(P_{IBDG}^*\right)$, i.e., $P_{IBDG} \approx P_{IBDG}^*$. Consequently, (3.14) can be rewritten in the time-domain as [10]

$$dP_{IBDG}/dt \approx dP_{IBDG}^*/dt = K_I^{APC}\left(V_{DC}^2 - \left[V_{DC}^*\right]^2\right) + K_P^{APC} \times dV_{DC}^2/dt \tag{3.15}$$

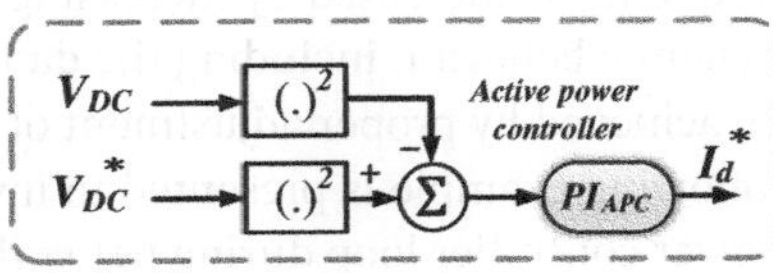

Fig. 3.5 Active power control through adjusting the DC-link voltage.

where K_P^{APC} and K_I^{APC} are the proportional and integral coefficients of the PI_{APC} controller, respectively. By substituting dV_{DC}^2/dt from (3.13) into (3.15), the following equation is obtained:

$$dP_{IBDG}/dt = K_I^{APC}\left(V_{DC}^2 - \left[V_{DC}^*\right]^2\right) + K_P^{APC}[2(P_{DC} - P_{IBDG})/C_{DC}] \tag{3.16}$$

State-space equations are utilized to investigate the dynamic behavior of P_{IBDG} and V_{DC} after the occurrence of a disturbance on the AC side. The system state variables are considered as V_{DC}^2 and P_{IBDG}. Considering (3.13) and (3.16), the system state-space equation can be written as

$$\dot{\mathbf{X}} = \mathbf{A}\cdot\mathbf{X} + \mathbf{B}\cdot\mathbf{U} \tag{3.17}$$

where

$$\mathbf{X} = \left(V_{DC}^2 \;\; P_{IBDG}\right)^T, \qquad \mathbf{U} = \left(P_{DC} \;\; \left[V_{DC}^*\right]^2\right)^T$$
$$\mathbf{A} = \begin{pmatrix} 0 & \dfrac{-2}{C_{DC}} \\ K_I^{APC} & \dfrac{-2K_P^{APC}}{C_{DC}} \end{pmatrix}, \quad \mathbf{B} = \begin{pmatrix} \dfrac{2}{C_{DC}} & 0 \\ \dfrac{2K_P^{APC}}{C_{DC}} & -K_I^{APC} \end{pmatrix} \tag{3.18}$$

Based on the developed state-space Equations (3.17) and (3.18), the characteristic equation of the system CE is derived as

$$CE(s) = s^2 + 2K_P^{APC}/(C_{DC}s) + 2K_I^{APC}/C_{DC} \tag{3.19}$$

Using (3.19), the frequency (ω_{d_IBDG}), damping coefficient (α_{IBDG}), and the damping ratio (ζ_{IBDG}) of oscillations can be calculated by

$$\omega_{d\_IBDG} = \sqrt{\left(2K_I^{APC}C_{DC} - \left(K_P^{APC}\right)^2\right)/C_{DC}^2}, \quad \alpha_{IBDG} = K_P^{APC}/C_{DC}$$
$$\zeta_{IBDG} = K_P^{APC}\sqrt{1/\left(2C_{DC}K_I^{APC}\right)} \tag{3.20}$$

Increasing the capacity of DC-link capacitor reduces the oscillation frequency. Moreover, the proportional coefficient of the active power controller, i.e., K_P^{APC}, and its integral coefficient, i.e., K_I^{APC}, correspond to the damping and synchronizing factors in SSSGs. In other words, the damping factor increases by increasing K_P^{APC}, which results in the decrement of the oscillation frequency. On the other hand, increasing K_I^{APC} increases the synchronizing factor and reduces the probability of unstable power oscillations. Meanwhile, the oscillation frequency is increased by increasing K_I^{APC}. Based on these findings, the required dynamic behavior, including the damping factor and oscillation frequency, could be achieved by proper adjustment of the active power controller parameters. The following example is presented to investigate the dynamic response of the active power controller loop during network disturbances.

Example 3.2
The single-line diagram of a typical hybrid grid-connected MG containing a GFL-IBDG and an SSSG is shown in Fig. 3.6. The parameters of these resources are presented in Table 3.2. A solid short-circuit fault occurs at the terminal of IBDG at $t = 30$ s. Investigate the impacts of power controller parameters and DC-link capacitor value on the dynamic behavior of the GFL-IBDG using simulation studies and compare the obtained results with the analytical equations. Neglect the impact of the PLL.

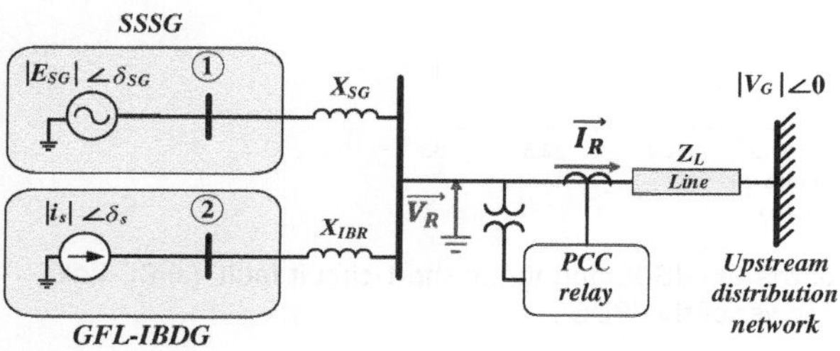

Fig. 3.6 Hybrid MG containing an SSSG and a GFL-IBDG.

Table 3.2 System parameters.

	Resistance	Reactance	Length
Line	0.261 Ω/km	0.292 Ω/km	5 km
Generator	Inertia constant (H)	Rated power	Control mode
SSSG	0.6 s	2 MVA	$P-Q$
GFL-IBDG	N/A	10 MVA	$P-Q$

Solution
In this example, the PLL control loop is considered ideal to focus solely on the effect of power control loop. In this regard, it is assumed that the angle estimated by the PLL is precisely equal to the V_{PCC} angle, and the frequency estimated by the PLL is exactly equal to the V_{PCC} frequency.

The DC-link voltage (V_{DC}) and the output active power of the GFL-IBDG (P_{IBDG}) are depicted in Fig. 3.7. As shown, after the fault clearance, the oscillations of these signals are damped out fast. The frequency spectrum of P_{IBDG} presented in Fig. 3.8 indicates that the oscillation frequency is approximately 12 Hz. Moreover, these oscillations are damped out within 0.7 s after the fault clearance. To analyze the impacts of the power controller parameters of the GFL-IBDG and also the DC-link capacitor value on the IBDG dynamic response,

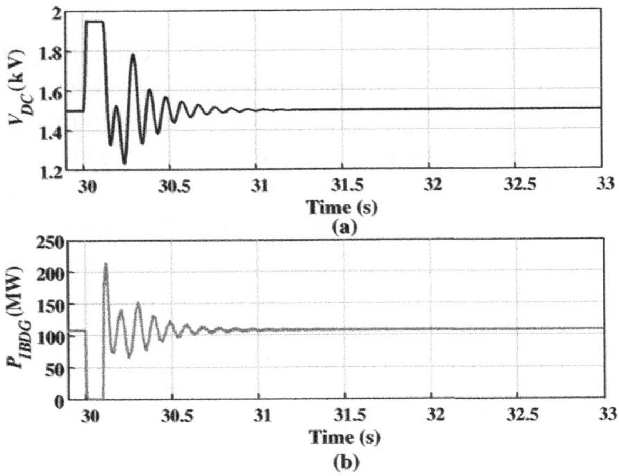

Fig. 3.7 Dynamic behavior of the GFL-IBDG during the short-circuit fault: (a) DC-link voltage and (b) output active power of the IBDG.

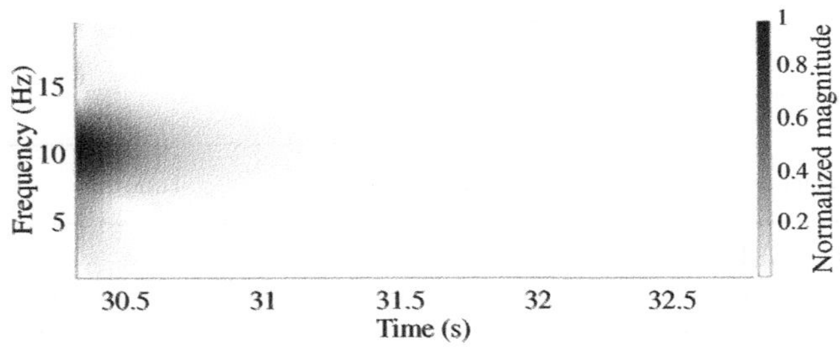

Fig. 3.8 Frequency content of P_{IBDG}.

at first, it is assumed that the controller integral gain and DC-link capacitor (K_I^{APC} and C_{DC}) are constant, and the output active power of the IBR for different values of the proportional controller parameter is illustrated in Fig. 3.9(a). As shown, with an increase in the K_P^{APC} parameter, the power swing frequency decreases, and its damping becomes faster. In the next case, the proportional gain K_P^{APC} and DC-link capacitor values are considered constant, and the integral gain has been changed. The GFL-IBDG output power is depicted in Fig. 3.9(b) for this case. As shown, with an increase in the K_I^{APC} value, the oscillation frequency of the IBDG dynamic response is increased. Meanwhile, the active power of the GFL-IBDG becomes more oscillatory, and the damping ratio decreases. Then, the proportional and integral controller coefficients are assumed constant, and the value of the DC-link capacitor has been increased. In this case, the active power of the GFL-IBDG is presented in Fig. 3.9(c). As shown, the increment of

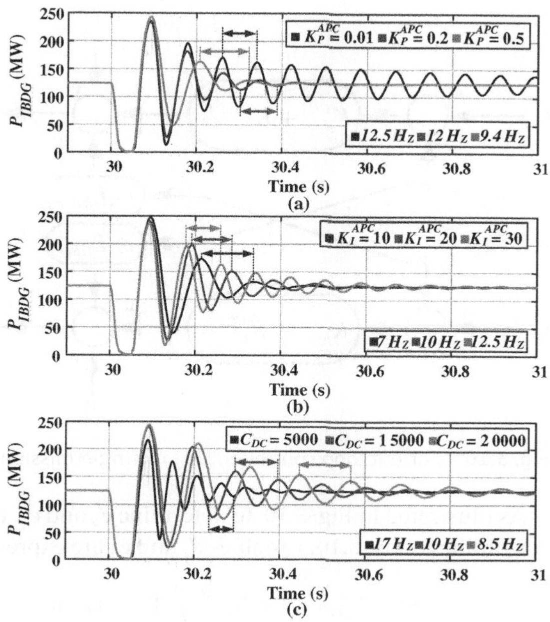

Fig. 3.9 Impact of the power controller parameters and DC-link capacitor value on the IBDG dynamic response: (a) changes in proportional gain, (b) changes in integral gain, and (c) changes in DC-link capacitor.

the DC-link capacitor decreases the frequency of the power swing and reduces the damping ratio of the power swing. In this condition, the oscillations are damped out more slowly. It should be emphasized that the presented simulation-based results appropriately ascertain the validity of (3.20).

3.2.1.3 Current Control Loop

This loop regulates the current injected into the grid at the reference values, i.e., i_d^* and i_q^*, determined by the power control loop. In this regard, the suitable voltages are generated through appropriate switching strategy to properly regulate the inverter-injected $d - q$ reference currents. The relation between the grid voltage, i.e., v_d and v_q, and the inverter voltage, i.e., e_d and e_q, can be expressed using the KVL rule as follows [11]:

$$\begin{aligned} e_d^{ref} &= e_d' - \omega_s L_{filter} i_q + v_d \\ e_q^{ref} &= e_q' + \omega_s L_{filter} i_d + v_q \end{aligned} \tag{3.21}$$

where ω_s is the angular frequency of the upstream grid, and L_{filter} is the reactance utilized at the inverter terminal, as shown in Fig. 3.10. Furthermore, e_d' and e_q' are

$$\begin{aligned} e_d' &= R_{filter} i_d + L_{filter} \frac{di_d}{dt} \\ e_q' &= R_{filter} i_q + L_{filter} \frac{di_q}{dt} \end{aligned} \tag{3.22}$$

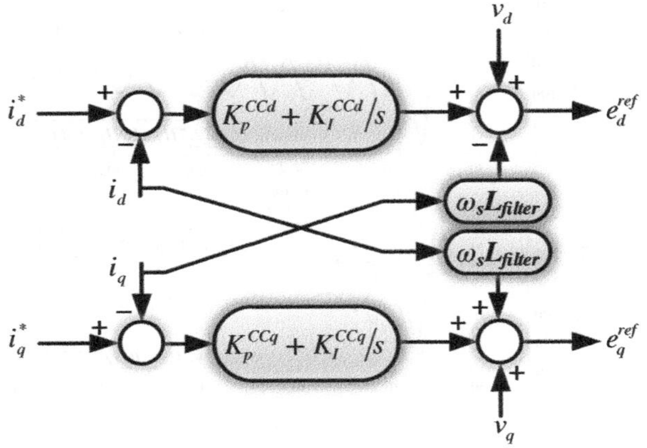

Fig. 3.10 Current loop controller. [11] / with permission of Elsevier.

As illustrated in Fig. 3.10, to determine e'_d and e'_q, two PI controllers are utilized. According to Fig. 3.10, variables e'_d and e'_q are expressed as

$$e'_d = K_p^{CCd}\left(i_d^* - i_d\right) + K_I^{CCd}\int\left(i_d^* - i_d\right)dt$$
$$e'_q = K_p^{CCq}\left(i_q^* - i_q\right) + K_I^{CCq}\int\left(i_q^* - i_q\right)dt \tag{3.23}$$

Combining (3.20) and (3.22) results in

$$e_d^{ref} = K_p^{CCd}\left(i_d^* - i_d\right) + K_I^{CCd}\int\left(i_d^* - i_d\right)dt - \omega_s L_{filter} i_q + v_d$$
$$e_q^{ref} = K_p^{CCq}\left(i_q^* - i_q\right) + K_I^{CCq}\int\left(i_q^* - i_q\right)dt + \omega_s L_{filter} i_d + v_q \tag{3.24}$$

Equations (3.22) and (3.24) can be expressed in the Laplace domain as

$$e'_d = (R_{filter} + sL_{filter})i_d$$
$$e'_q = (R_{filter} + sL_{filter})i_q \tag{3.25}$$

$$e_d^{ref} = \left(K_p^{CCd} + \frac{K_I^{CCd}}{s}\right)\left(i_d^* - i_d\right) - \omega_s L_{filter} i_q + v_d$$
$$e_q^{ref} = \left(K_p^{CCq} + \frac{K_I^{CCq}}{s}\right)\left(i_q^* - i_q\right) + \omega_s L_{filter} i_d + v_q \tag{3.26}$$

Similarly, taking Laplace transform of (3.23) provides

$$e'_d = \left(K_p^{CCd} + \frac{K_I^{CCd}}{s}\right)\left(i_d^* - i_d\right)$$

$$e_q' = \left(K_p^{CCq} + \frac{K_I^{CCq}}{s}\right)\left(i_q^* - i_q\right) \tag{3.27}$$

Therefore, by combining (3.25) and (3.27), the closed-loop transfer function of the current control loop is given by

$$\begin{aligned}\frac{i_d}{i_d^*} &= \frac{K_I^{CCd} + sK_p^{CCd}}{L_{filter}s^2 + \left(R_{filter} + K_p^{CCd}\right)s + K_I^{CCd}}\\ \frac{i_q}{i_q^*} &= \frac{K_I^{CCq} + sK_p^{CCq}}{L_{filter}s^2 + \left(R_{filter} + K_p^{CCq}\right)s + K_I^{CCq}}\end{aligned} \tag{3.28}$$

Meanwhile, the open-loop transfer function of current control loop would be obtained as follows:

$$\begin{aligned}\frac{i_d}{i_d^* - i_d} &= \left(K_p^{CCd} + \frac{K_I^{CCd}}{s}\right)\frac{1}{R_{filter} + sL_{filter}}\\ \frac{i_q}{i_q^* - i_q} &= \left(K_p^{CCq} + \frac{K_I^{CCq}}{s}\right)\frac{1}{R_{filter} + sL_{filter}}\end{aligned} \tag{3.29}$$

It should be noted that the very small delays related to the switching are neglected in these equations. To achieve a fast and robust dynamic response of the current controller loop, various methods could be utilized. In this regard, as a common practice, the pole-zero placement approach is adopted where a zero is located at the dominant pole. Therefore, K_p^{CCdq} and K_I^{CCdq} should be selected as [11]

$$K_p^{CCdq} = \frac{L_{filter}}{\tau}, \quad K_p^{CCdq} = \frac{R_{filter}}{\tau} \tag{3.30}$$

where τ is a constant parameter and could be considered as the time constant of the closed-loop transfer function of current controller. Finally, the second-order closed-loop transfer function is expressed as

$$\frac{i_d}{i_d^*} = \frac{K_I^{CCd} + sK_p^{CCd}}{L_{filter}s^2 + \left(R_{filter} + K_p^{CCd}\right)s + K_I^{CCd}} = \frac{1}{1 + \tau s} \tag{3.31}$$

The time constant of the current controller is commonly set at 5–10 times the switching period. Furthermore, this time constant is normally considered 5–10% of the power controller time constant. As a result, the amount of τ is adjusted about 2–5 ms [11].

3.2.2 Grid-Forming Strategy

GFM strategy is initially developed for islanded IBMGs and those connected to a weak power network. Such a strategy could retain and improve IBMG stability

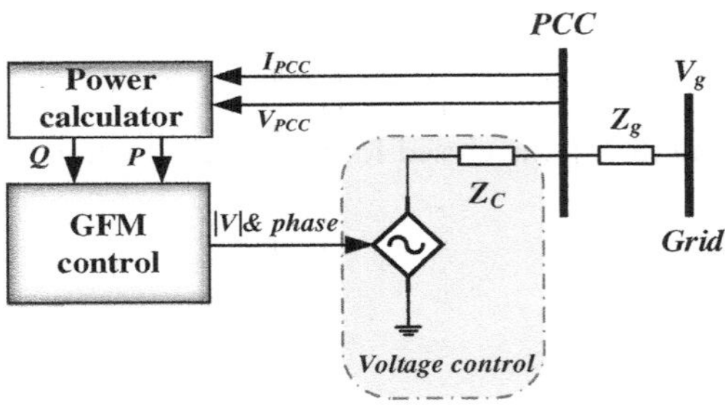

Fig. 3.11 Simple representation of a GFM-IBDG. Adapted from [1].

compared to the GFL scheme performance. Unlike GFL-IBDGs, in which PLL is employed to follow the angle and frequency of the grid, the phase angle of GFM-IBDG terminal voltage is determined by the active power controller. In fact, GFM-IBDGs could dictate the frequency and voltage angle. In this regard, various control strategies are proposed by different research works. However, the most common scheme is the virtual synchronous machine (VSM) scheme, which can be adopted in GFM-IBDGs to emulate the dynamic behavior of SGs and their mechanical inertia. Assuming that the primary energy source is available and the converter current does not reach the limits, GFM-IBDG maintains its output terminal voltage, i.e., its frequency and magnitude, during disturbances and even if the upstream system is disconnected. This feature enables GFM-IBDG to "form" a grid through regulation of its frequency and voltage. Therefore, unlike GFL-IBDGs, which are modeled through a current source, GFM-IBDGs are modeled using a voltage source and a series impedance, as depicted in Fig. 3.11.

The VSM scheme implemented in GFM-IBDGs allows these resources to support the grid through the proper frequency response. Furthermore, these resources can adaptively change their active/reactive powers for instant responding to the transient disturbances such as voltage phase angle deviation. Although GFM-IBDGs could support and enhance the IBMG stability during severe disturbances, they may still lose their stability in some scenarios. The general control scheme of a GFM-IBDG is illustrated in Fig. 3.12, whose loops are described as follows.

3.2.2.1 Power Controller

The power controller regulates the output active and reactive powers of the GFM-IBDG by proper adjustment of the output voltage magnitude and phase angle. In fact, this loop generates the voltage phasor reference, including its magnitude and phase angle. As mentioned, various strategies have been developed for the power controlling in GFM-IBDGs in recent years. Four common strategies of

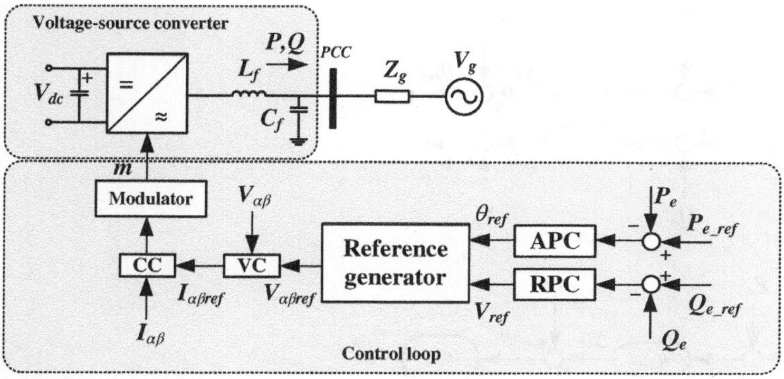

Fig. 3.12 Control scheme of a GFM-IBDG. Adapted from [12].

power regulation loop are depicted in Fig. 3.13 [12]. The first and second control strategies, shown in Fig. 3.13(a) and (b), respectively, are very similar to each other, and both provide a first-order power control. In the third control strategy, a low-pass filter (LPF) has been added compared to the second one. The last one is the common VSM control scheme, which emulates the swing equation and mechanical inertia of the SG. Both the third and last control strategies are second-order power controllers and equivalent to each other.

So far, various VSM controllers have been developed whose structures are modified to provide better dynamic response and enhance the IBMG stability. For instance, a power system stabilizer can be added in the reactive power controller to improve the system stability. Although different schemes have been proposed, the kernel of these methods are still similar and established based on the second-order swing equation of synchronous machines as depicted in Fig. 3.13. The phase angle of GFM-IBDG is generated by the virtual swing equation as [13]

$$J\frac{d^2\theta_{ref}}{dt^2} = P_{ref} - P_{IBDG} - D_p(\omega_g - \omega) \tag{3.32}$$

where J is the virtual inertia constant, D_p is the virtual damping coefficient, P_{ref} is the active power setpoint, and ω_g is the angular frequency setpoint. Therefore, both inertial and droop responses could be adopted through appropriate tuning of J and D_p. Meanwhile, the following differential equation is exploited to generate voltage magnitude (V_{ref}) of the GFM-IBDG terminal [13]:

$$\tau\frac{dV_{ref}}{dt} = Q_{ref} - Q_{IBDG} - D_q(V_g - V_{ref}) \tag{3.33}$$

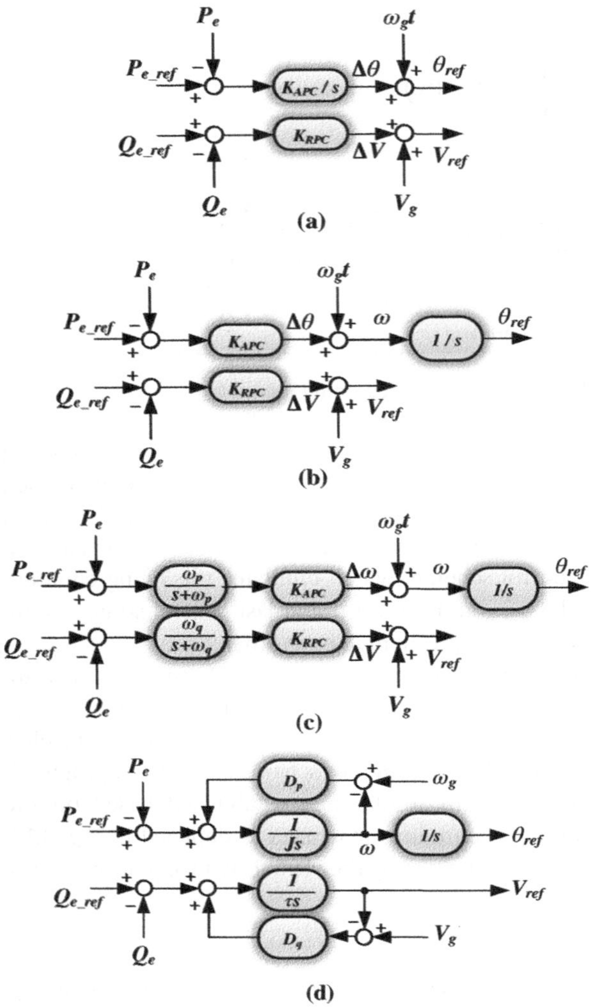

Fig. 3.13 Various control schemes of GFM-IBDG: (a) type I first-order controller, (b) type II first-order controller, (c) second-order controller, and (d) VSM controller. Adapted from [12].

where τ and D_q are constant parameters, Q_{ref} is the reactive power setpoint, and V_g is the voltage magnitude setpoint. The droop Volt–VAr control could be adjusted through D_q. Similar to the GFL-IBDG, droop control loops enable IBDGs to share power and prevent VAr circulation between parallel IBDGs. In this regard, the droop characteristics of GFM-IBDGs and GFL-IBDGs should be properly coordinated with each other. A typical droop characteristic of a GFM-IBDG is illustrated in Fig. 3.14.

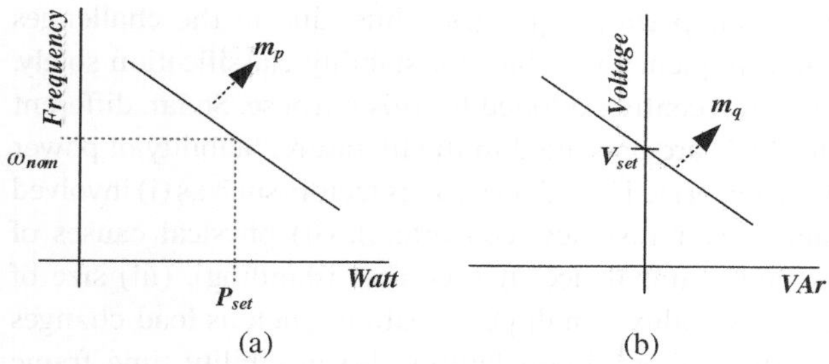

Fig. 3.14 Typical droop characteristic of a GFM-IBDG: (a) frequency-active power droop and (b) voltage reactive power droop. [9] / with permission of IEEE.

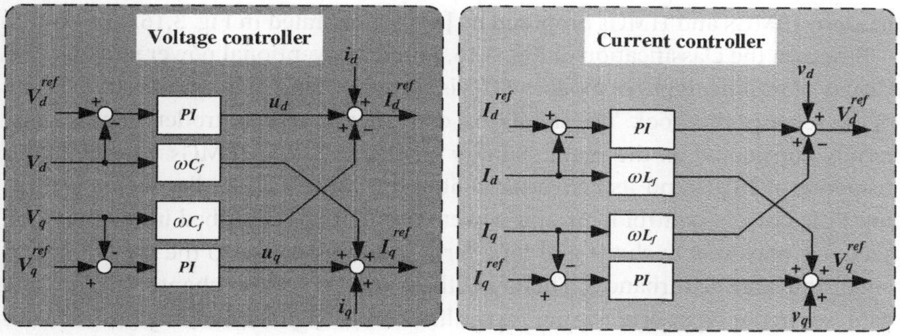

Fig. 3.15 Voltage and current controllers of the GFM-IBDG. Adapted from [12].

3.2.2.2 Voltage and Current Controllers

After the determination of the voltage phasor references through the power control loops, these signals are injected into the voltage control loop, whose structure is depicted in Fig. 3.15. As shown, the reference current signals, i.e., I_d^{ref} and I_q^{ref}, are generated by the voltage controller through PI controllers. Feed-forward signals are utilized to improve the dynamic response and decrease the settling time. These reference current signals will be injected into the current control loop. Then, similar to the current control loop of GFL-IBDGs, proper reference voltages for PWM-switching are generated.

3.3 Stability of IBMGs

For conventional power systems, system instability is divided into different categories mainly based on the significant alterations of one of its key variables, such as voltage and frequency. Unlike conventional power systems, the instability of MGs drives all system variables to fluctuate considerably, and they might even increase/decrease beyond their permitted ranges. Thus, due to the challenges of utilizing the voltage and frequency variables for stability classification solely, physical components are also recently exploited for this purpose. So far, different stability classifications for MGs are presented in the literature. Stability of power systems and MGs is newly categorized based on various factors such as (i) involved equipment in the instability (for instance, converters), (ii) physical causes of instability (for instance, power imbalance in MG after islanding), (iii) size of the disturbance leading to instability (small perturbations such as load changes or large perturbations such as short-circuit faults), (iv) instability time frame (for instance, the timespan could be varied from few seconds, i.e., short term, to several minutes, i.e., long term), and (v) unstable variable (for instance, frequency, voltage) [14]. Based on these factors, one of the useful stability classifications for modern IBMGs and HMGs proposed by [14] is illustrated in Fig. 3.16.

Based on the classification of Fig. 3.16, unlike conventional power systems, the variety of stability types of modern MGs is very wide, and their investigation is out of the scope of this book. The stability of converter and its controller loops is commonly considered as the main stability challenge in the IBMGs. The dynamic response of IBDG and its converter containing various controller loops, including PLL, current control loop, and power control loop (described in the previous section), may lead to short- and long-term instabilities due to the occurrence of small and large disturbances. Therefore, the main focus of this chapter is to investigate two major types of converter instabilities: large-signal instability (also known as transient stability) and small-signal instability, which may threaten the stable operation of the whole IBMG and even the adjacent SSSGs.

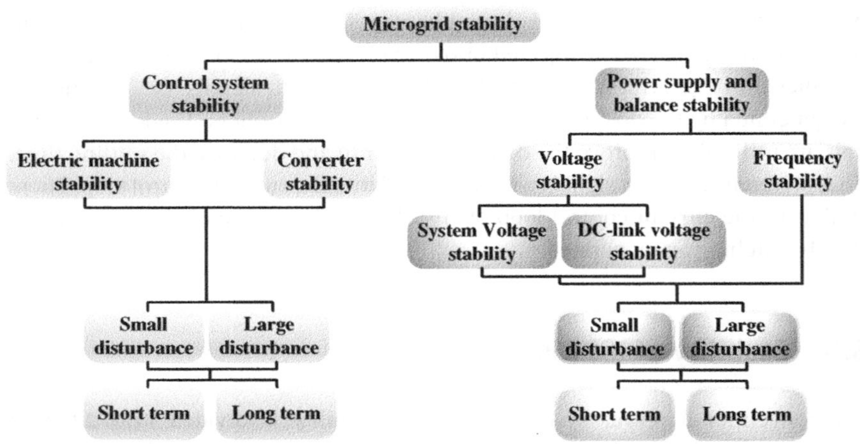

Fig. 3.16 Classification of the modern MG's stability. Adapted from [14].

3.3.1 Transient Stability

Transient instability for both GFL-IBDGs and GFM-IBDGs may occur during large-signal disturbances due to the instability of controller loops responsible for grid synchronization. For GFL-IBDGs, transient instability takes place due to the dynamic response of its PLL, while for GFM-IBDGs, this issue occurs because of the active power controller behavior. It is worth mentioning that the transient stability of IBDGs could be somehow considered equivalent to the rotor angle stability in SSSGs. In the following sections, transient instability for both types of these IBDGs is investigated.

3.3.1.1 GFL-IBDGs

For a stable operation, the PLL of GFL-IBDG should be able to follow the phase angle of the PCC by regulating the q-axis voltage (V_q) at zero. In fact, during transient instability, the PLL would not be able to control V_q to zero. Considering Fig. 3.17, using the KVL rule, the q-axis component of PCC voltage (V_{PCC}) can be expressed as [12, 15, 16]

$$V_q = |V_g| \sin(\delta_g - \delta_{PLL}) + |I_{PCC}||Z_{Th}| \sin(\theta_I + \varphi_Z) \tag{3.34}$$

where δ_g and δ_{PLL} are the phase angle of the upstream system voltage and the output angle of the PLL, respectively. The PLL could only control the $\sin(\delta_g - \delta_{PLL})$ term, and other variables are not controllable by the PLL. Thus, for weak grids, i.e., large amounts of Z_{Th}, the second-term at the right side of (3.34), i.e., $|I_{PCC}||Z_{Th}| \sin(\theta_I + \varphi_Z)$ is large, which behaves as positive feedback and endangers the PLL stability. Furthermore, during severe voltage drops for which the amount of $|V_g|$ is small, the first term at the right side of (3.34) decreases, which may lead to the PLL instability. Considering $V_q = 0$ in (3.34), for stable operation of the PLL, the following condition should be fulfilled:

$$|\sin(\delta_g - \delta_{PLL})| = \frac{|I_{PCC}||Z_{Th}|| \sin(\theta_I + \varphi_Z)|}{|V_g|} \tag{3.35}$$

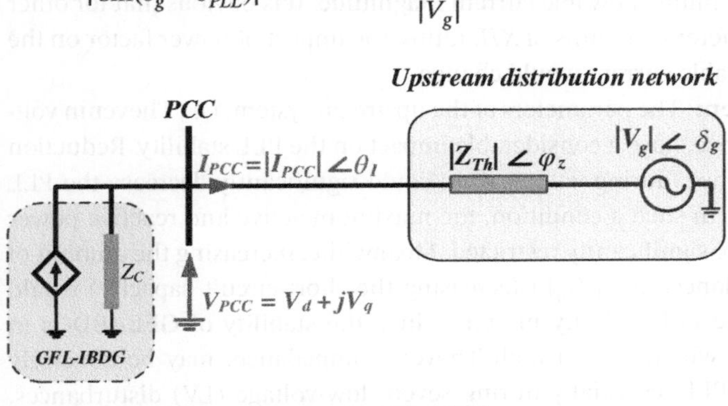

Fig. 3.17 A typical grid-connected GFL-IBDG.

Considering this equation and since $|\sin(\delta_g - \delta_{PLL})| < 1$, the following inequality should be satisfied for PLL stable operation:

$$|I_{PCC}| < \frac{|V_g|}{|Z_{Th}||\sin(\theta_I + \varphi_Z)|} \tag{3.36}$$

Thus, to prevent PLL instability at each operation point of the system, the magnitude of I_{PCC} should be limited based on (3.36). In other words, besides thermal limitations of IBDG switches, stability constraints should also be considered for determining the maximum allowable current magnitude. The following factors have significant impacts on the PLL stability [12, 15–18]:

- **X/R ratio:** Considering the active power injection in which θ_I is zero, the maximum amount of $|I_{PCC}|$ is limited by the upstream system reactance (X_{Th}), where $X_{Th} = |Z_{Th}||\sin(\varphi_Z)|$. In other words, the maximum allowable active current injection is restricted by the system reactance. As the system reactance, X/R ratio and φ_Z decrease, and the maximum active current injection increases. On the other hand, for reactive power injection, the amount of θ_I is about $-90°$. In this condition, the maximum reactive power injection is limited by the resistive part of the impedance, i.e., $|Z_{Th}||\cos(\varphi_Z)|$. Therefore, during reactive power injection intervals, such as the low-voltage ride-through (LVRT) period, an increment in the resistance of the upstream impedance significantly decreases the reactive current injection capability of GFL-IBDG.
- **Power factor:** The impact of the GFL-IBDG power factor on the PLL stability depends on the upstream system impedance and the amount of φ_Z. For a typical distribution system whose X/R ratio is about 1 (i.e., φ_Z is 45°), decreasing θ_I from zero (full active current) to $-45°$ (equal injection of active and reactive currents) considerably increases the maximum allowable current magnitude. On the other hand, increasing θ_I from zero to $+45°$ (reactive current absorption) reduces the maximum allowable current magnitude. It is obvious that for other systems with different amounts of X/R ratios, the impact of power factor on the maximum allowable current would change.
- **Upstream system:** The parameters of the upstream system, i.e., Thevenin voltage and impedance, have a considerable impact on the PLL stability. Reduction of $|V_g|$, for instance, during voltage dips, could significantly decrease the PLL stability margin. In such a condition, the maximum active and reactive power exchanges will be significantly restricted. Meanwhile, increasing the amount of upstream impedance, i.e., $|Z_{Th}|$ (decreasing the short-circuit capacity) would also decrease the PLL stability margin. Thus, the stability of GFL-IBDGs in weak networks, which have a high Thevenin impedance, may be adversely affected by the PLL, especially during severe low-voltage (LV) disturbances, such as short-circuit faults.

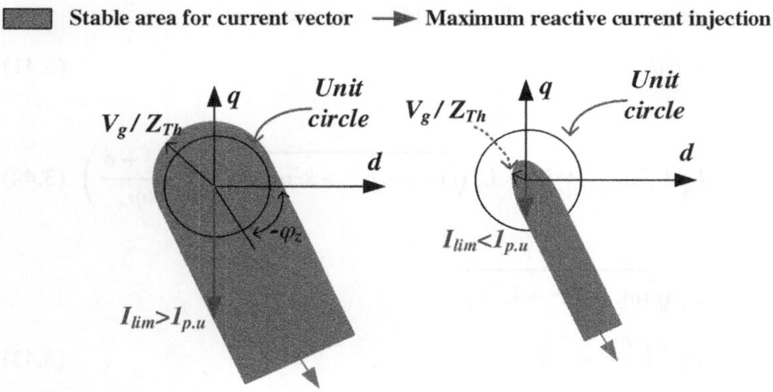

Fig. 3.18 Stable operation region of GFL-IBDG. [12] / with permission of IEEE.

For better presentation of the influence of the aforementioned factors, the stable operation region of the GFL-IBDG is depicted in Fig. 3.18 for various working conditions [12]. It is worth mentioning that (3.36) provides the necessary condition for PLL stability and not the sufficient one. In fact, even if this condition is fulfilled, transient instability and PLL loss of synchronism may still occur because of insufficient damping of the PLL. For a better explanation and to investigate the transient stability of the GFL-IBDGs, the nonlinear second-order differential equation of the PLL could be adopted. Considering Fig. 3.18, the PLL output can be expressed by [18–21]

$$\delta_{PLL} = \int \left[\omega_n + \left(K_p + K_i \int V_q \right) \right] \tag{3.37}$$

where ω_n is the upstream grid angular frequency. Meanwhile, K_p and K_i are the proportional and integral gains of the PLL PI controller, respectively. Combining (3.34) and (3.37) results in

$$\delta_{PLL} = \int \left\{ \omega_n + \left(K_p + K_i \int (|V_g| \sin(\delta_g - \delta_{PLL}) + |I_{PCC}||Z_{Th}| \sin(\theta_I + \varphi_Z)) \right) \right\} \tag{3.38}$$

This can be rewritten as

$$\delta = \int \left\{ \left(K_p + K_i \int (|I_{PCC}||Z_{Th}| \sin(\theta_I + \varphi_Z) - V_g \sin(\delta)) \right) \right\} \tag{3.39}$$

where

$$\delta = \delta_{PLL} - \delta_g \tag{3.40}$$

Furthermore, for the distribution system, it is assumed that the R/X ratio would be equal to k at the nominal angular frequency ω_n. Moreover, the equivalent reactance of the upstream system, X_{Th}, could be expressed as

$$X_{Th} = (\omega_n + \dot{\delta})L_g \tag{3.41}$$

Therefore,

$$Z_{Th} = k\omega_n L_g + j(\omega_n + \dot{\delta})L_g = L_g\sqrt{(\omega_n + \dot{\delta})^2 + k^2\omega_n^2}\angle tg^{-1}\left(\frac{\omega_n + \dot{\delta}}{k\omega_n}\right) \tag{3.42}$$

Consequently,

$$\begin{aligned} |Z_{Th}(\dot{\delta})| &= L_g\sqrt{(\omega_n + \dot{\delta})^2 + k^2\omega_n^2} \\ \varphi_Z(\dot{\delta}) &= tg^{-1}\left(\frac{\omega_n + \dot{\delta}}{k\omega_n}\right) \end{aligned} \tag{3.43}$$

Equation (3.39) presents a second-order nonlinear equation, which can be rearranged as

$$\ddot{\delta} + D(\delta)\dot{\delta} + F(\delta, \dot{\delta}) = 0 \tag{3.44}$$

where function $D(\delta)$ is a nonlinear function of δ and $F(\delta, \dot{\delta})$ is a nonlinear function of both δ and $\dot{\delta}$.

For $X/R \gg 1$, the amount of k is negligible, and Z_{Th} can be simplified as

$$Z_{Th} = L_g(\omega_n + \dot{\delta})\angle\frac{\pi}{2} \tag{3.45}$$

Consequently, based on (3.38), (3.40), and (3.45), the second-derivative equation of δ is obtained as

$$(1 - K_p I_g L_g)\ddot{\delta} = K_i\left(I_g\omega_n L_g + I_g\dot{\delta}L_g - |V_g|\sin\delta - \frac{K_p|V_g|\cos\delta}{K_i}\dot{\delta}\right) \tag{3.46}$$

where $I_g = |I_{PCC}|\sin(\theta_I + \pi/2) = |I_{PCC}|\cos(\theta_I)$.

To provide a deeper physical insight into the PLL transient stability, this nonlinear equation can be rearranged as follows to make it similar to conventional power-angle swing equation of synchronous machines [12]:

$$\underbrace{\frac{1 - K_p I_g L_g}{K_i}}_{\text{Equivalent moment of inertia}}\ddot{\delta} = \underbrace{|Z_{Th}|I_g}_{\substack{v_{zq} \\ \text{Equivalent mechanical power}}} - \underbrace{|V_g|\sin(\delta)}_{\substack{v_{gq} \\ \text{Equivalent electrical power}}} \underbrace{- \dot{\delta}\frac{K_p}{K_i}|V_g|\cos(\delta)}_{\text{Damping}} \tag{3.47}$$

where v_{zq} and v_{gq} are the q-axis components of the voltage across the line impedance and the grid voltage, respectively. Similar to the power-angle swing equation of SGs, this equation presents a voltage-angle swing equation. Based on this equation, the equivalent virtual inertia of the GFL-IBDG increases by reducing the PLL integral gain (K_i) or increasing the PLL proportional gain (K_p). Meanwhile, the damping factor could be adjusted by proper selection of the PLL PI controller parameters. Furthermore, compared to the power-angle curve of the SGs, v_{zq} and v_{gq} could be interpreted as the equivalent virtual mechanical and electrical powers, respectively.

For the stability assessment, similar to the SSSGs, the equal-area criterion (EAC) can be adopted using the voltage-angle curve. If the equivalent damping factor given in (3.47) is neglected, the voltage-angle curve could be illustrated as Fig. 3.19. As shown, the following two conditions may occur, which result in transient instability [12, 22]:

- **First swing instability:** During severe disturbances, the PLL output angle increases beyond the maximum critical angle. Before the fault occurrence, as shown in Fig. 3.19, the PLL works at δ_i point on the "Before fault" curve. At the instant of fault occurrence, δ_i jumps on the "After fault" curve. Since v_{zq} becomes greater than V_g sin (δ) after the fault occurrence, the amount of δ increases based on (3.47), neglecting the damping factor. Based on EAC, the maximum critical point is π-δ_a, and if δ becomes greater than π-δ_a, the PLL loses its synchronism at the first swing without oscillations. In this situation, the condition stated in (3.36) is violated.
- **Multi-swing instability:** In this case, the PLL output angle does not cross the maximum critical angle at the first swing. After clearing the disturbance, the PLL output angle reduces before reaching the maximum critical angle. However, based on the EAC shown in Fig. 3.19, since K_{acc} is greater than K_{max}, the PLL output angle increases again, and the PLL becomes unstable. This is referred to as multi-swing instability.

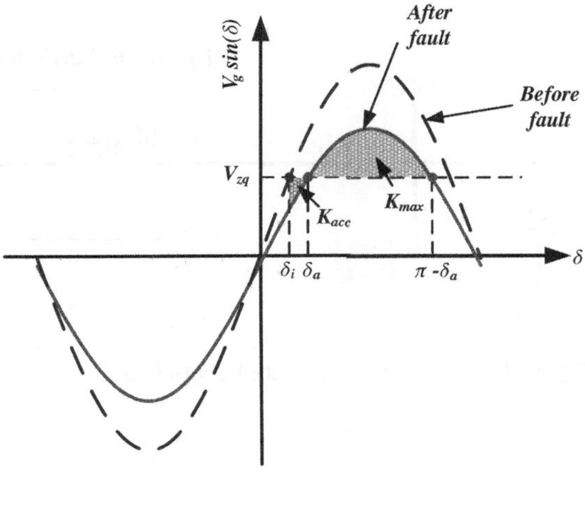

Fig. 3.19 Equivalent voltage-angle curve of GFL-IBDG. [12] / with permission of IEEE.

The damping factor of GFL-IBDGs is determined by the parameters of PLL PI controller. Thus, unlike synchronous machines, considering zero damping factor for GFL-IBDGs may not be realistic and using the EAC, neglecting the damping factor, leads to a conservative assessment. To address this issue and consider the damping factor, other stability analysis techniques mentioned in Chapter 2 might be adopted for the GFL-IBDGs, including transient simulation-based studies and Lyapunov methods.

3.3.1.2 GFM-IBDGs

Considering Fig. 3.20, the GFM-IBDG can be modeled as a voltage source along with a series impedance, which is similar to synchronous machines, and thus, the output active (P) and reactive (Q) powers of GFM-IBDG are determined by [18]

$$\begin{aligned} P &= \frac{3}{2}\frac{E\,V\sin\delta}{X_{Th}} \\ Q &= \frac{3}{2}\frac{V^2 - E\,V\cos\delta}{X_{Th}} \end{aligned} \tag{3.48}$$

where X_{Th} is the upstream grid equivalent reactance. To simplify the analysis, these equations are derived assuming $X/R \gg 1$. For smaller X/R ratios, similar power equations can be derived, as discussed in Chapter 2.

The $P-\delta$ curves of GFM-IBDGs are similar to those of the well-known curves of synchronous machines as depicted in Fig. 3.21. The equilibrium points located at the left side of this curve, i.e., $\delta < \pi/2$, are stable; while for $\delta > \pi/2$, these points are unstable. For stable operation of the GFM-IBDG, P_{ref} should be smaller than P_{max}. Similar to SSSGs, and considering Fig. 3.21, δ_1 represents the equilibrium operating point, while δ_u denotes the critical angle. If the power angle exceeds δ_u, the GFM-IBDG may lose its synchronism. A detailed analysis is provided as follows.

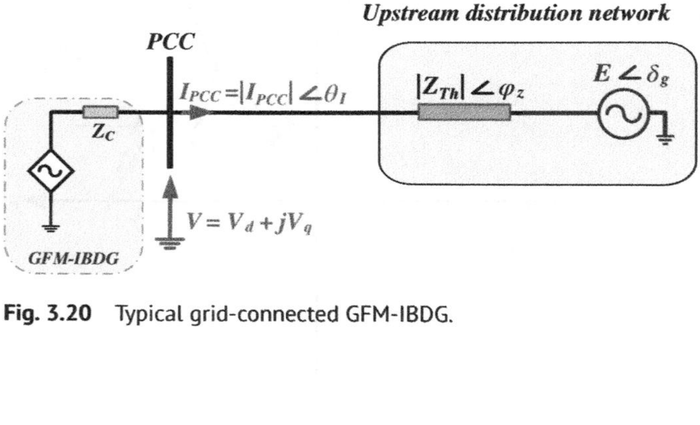

Fig. 3.20 Typical grid-connected GFM-IBDG.

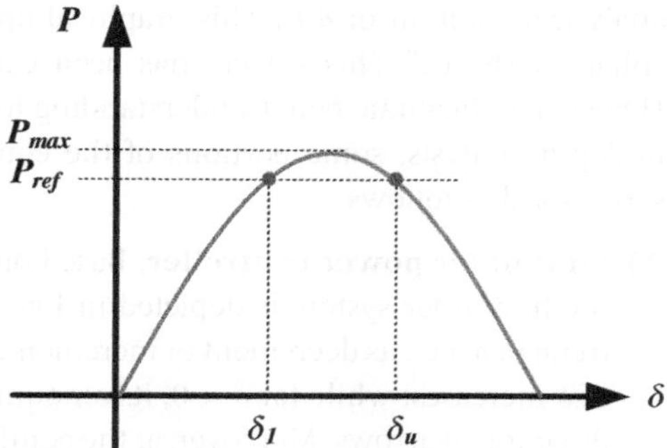

Fig. 3.21 $P - \delta$ curve of GFM-IBDGs. [12] / with permission of IEEE.

Assuming the second-order GFM-IBDG, shown in Fig. 3.13, the phase angle of GFM-IBDG terminal (with respect to the infinite bus), i.e., δ, can be expressed as [12, 18, 23]

$$\delta = \frac{K_p}{s}\frac{\omega_p}{s+\omega_p}(P_{ref} - P) \tag{3.49}$$

Furthermore, for the first-order system, δ can be expressed as

$$\delta = \frac{K_{APC}}{s}(P_{ref} - P) \tag{3.50}$$

For a second-order system, combining (3.48) and (3.49) results in [12, 18, 23]

$$\ddot{\delta} = -\omega_p\dot{\delta} + \omega_p K_p\left(P_{ref} - \frac{3}{2}\frac{E\,V}{X_{Th}}\sin\delta\right) \tag{3.51}$$

In addition, for the first-order system,

$$\dot{\delta} = K_{APC}\left(P_{ref} - \frac{3}{2}\frac{E\,V}{X_{Th}}\sin\delta\right) \tag{3.52}$$

Considering these equations, the GFM-IBDG may lose its synchronism with the system during severe disturbances. However, the loss of synchronism of GFM-IBDGs differs from that of the GFL-IBDGs. Meanwhile, the loss of synchronism occurring in the first- and second-order power controllers of GFM-IBDGs is not the same, as described as follows. Since providing the analytical closed-form solution of the differential equations stated earlier is not straightforward, these solutions may not provide a physical insight into the transient stability of GFM-IBDGs. Thus, instead of solving the abovementioned equations, a graphical approach is adopted here to provide a more intuitive and simpler assessment for transient stability. To do so, the $\dot{\delta} - \delta$ curve is exploited instead of the analytical solution of $\delta(t)$. This graphical approach is commonly referred to as "phase-portraits." This subject has been examined in various research works. However, to facilitate better understanding for the reader and to allow for a more in-depth analysis, some portions of the contents presented in [23, 24] will be scrutinized as follows:

A) **First-order power controller**: Based on (3.52), the phase-portrait curve for the first-order system is depicted in Fig. 3.22, which is sinusoidal. Variation trend of δ, i.e., its decrement or increment, is dictated by $\dot{\delta}$. If $\dot{\delta} > 0$, the amount of δ increases; while for $\dot{\delta} < 0$, its amount decreases, which are shown by the direction of arrows. Moreover, at the equilibrium points, $\dot{\delta}$ is equal to zero. The arrows entering toward the stable equilibrium point (SEP), are named "sink," while those exiting from the unstable equilibrium point (UEP) are named "source." If the equilibrium point exists, i.e., the point whose $\dot{\delta}$ be equal to zero, the operation point will monotonically approach the adjacent SEP for any scenario. In fact, if the equilibrium point exists, the IBDG always remains stable, regardless of the duration of disturbance or fault. Furthermore, the phase angle approaches monotonically to its final value without any overshoot or oscillations. Thus, this dynamic response is completely different from that of SSSGs or GFL-IBDGs. If the equilibrium point does not exist, the amount of δ increases/decreases and does not approach any value. In such a condition, the amount of $\dot{\delta}$ is always positive/negative. Thus, the frequency of IBDG increases/decreases, and it finally loses its synchronism with the system.

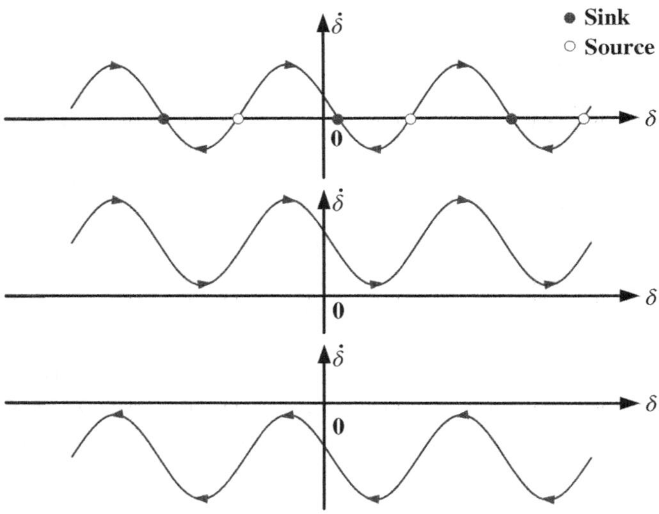

Fig. 3.22 Phase-portrait curves for the GFM-IBDG first-order system. [23] / with permission of IEEE.

For better physical insight, the following conceptual examples are presented:

Example 3.3 It is assumed that a typical GFM-IBDG using a first-order power controller is connected to a strong distribution network (Fig. 3.23) through a double-circuit line. One of the line circuits is tripped out and becomes out-of-service due to the maintenance operation. The phase-portrait curve for this condition is depicted in Fig. 3.24(a). As shown, due to the opening of one of the circuits, the amount of X_{Th} increases, and thus, the maximum active power decreases. Thus, after the disturbance occurrence, the operating point jumps from "a" to "b." As depicted in Fig. 3.24(a), an equilibrium point still exists after the disturbance occurrence. Consequently, after a while, the operating point approaches to the nearest SEP point, i.e., point "c," without any overshoot and oscillation. Meanwhile, according to (3.52), the travel time between the points "b" and "c" is dependent on the proportional gain of the power controller, i.e., K_{APC}. It should be noted that point "d" is a UEP. Furthermore, the PLL angle is depicted in Fig. 3.24(b). As shown, the first-order GFM-IBDG could provide a suitable "overdamped response" without any overshoot, and the second derivative of δ remains negative until δ reaches δ_c.

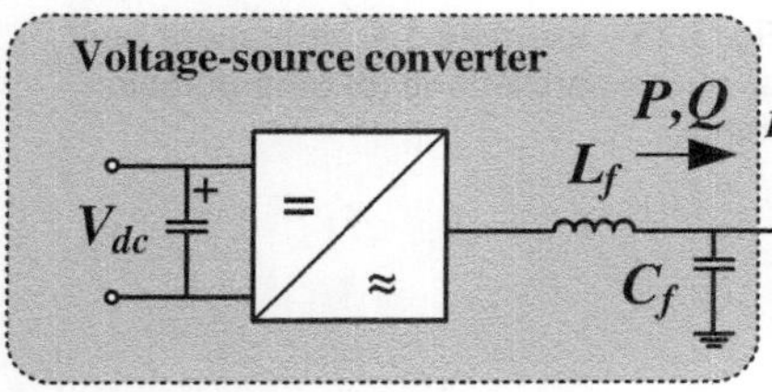

Fig. 3.23 A grid-connected IBDG.

Example 3.4 Considering the distribution system presented in Fig. 3.23, a severe three-phase short-circuit fault takes place at one of the circuits of the line. In this example, two cases are investigated: (i) permanent fault and (ii) clearing the fault by opening the faulty circuit by the protection system. The phase-portrait of the first case is illustrated in Fig. 3.25(a). After the fault occurrence, the maximum active power immediately reduces considerably, and the operating point jumps from point "a" to point "b" due to the change of phase-portrait curve. It should be noted that this new curve does not have any equilibrium point and the amount of $\dot{\delta}$ is always positive. Thus, during the fault event, the amount of δ increases, and if the fault is permanent, the GFM-IBDG loses its synchronism with the system. In this case, the time-domain signal of the PLL angle is depicted in Fig. 3.25(b). As observed, the second-derivative of δ is negative for $\delta < \pi/2$, and it becomes positive for $\delta > \pi/2$.

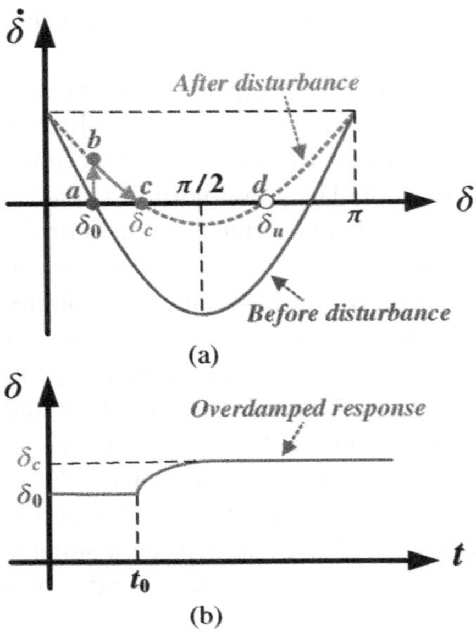

Fig. 3.24 Example 3.3: (a) phase-portrait and (b) time-domain δ signal.

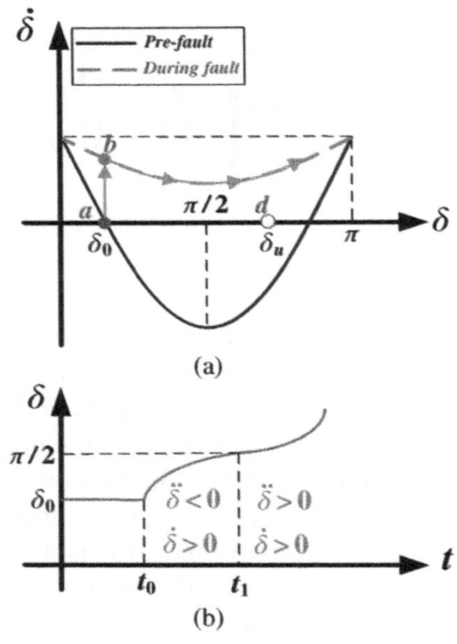

Fig. 3.25 Example 3.4, permanent fault: (a) phase-portrait and (b) time-domain δ signal.

If the fault is cleared, the following conditions may occur:

- If the amount of δ at the fault clearing instant (t_{FCT}) be less than δ_u (i.e., left side of point *d*), considering Fig. 3.26(a), the portrait curve changes and the operation point jumps to the new curve. In this region, $\dot{\delta}$ is negative and the amount of δ decreases. Thus, the operating point moves to point "c." This point is a new SEP in which the amount of $\dot{\delta}$ is equal to zero. The time-domain of δ signal is depicted in Fig. 3.26(b). As shown, for $t < t_{FCT}$, the second derivative of δ is negative when $\delta < \pi/2$, and positive when $\delta > \pi/2$. Conversely, after the fault clearance ($t > t_{FCT}$), the second derivative of δ becomes positive for $\delta < \pi/2$ and negative for $\delta > \pi/2$.
- If the amount of δ at the fault clearing instant be greater than δ_u (i.e., right side of point *d*), the portrait curve changes and the operation point jumps to the new curve, as shown in Fig. 3.27(a). In this region, $\dot{\delta}$ is positive and the amount of δ increases. Thus, δ increases to π and the GFM-IBDG loses its synchronism. Afterward, without any action, the amount of δ approaches to point "c" and the GFM-IBDG will automatically re-synchronize with the system. The time-domain signal of δ is illustrated in Fig. 3.27(b), which indicates the IBDG automatic re-synchronization.

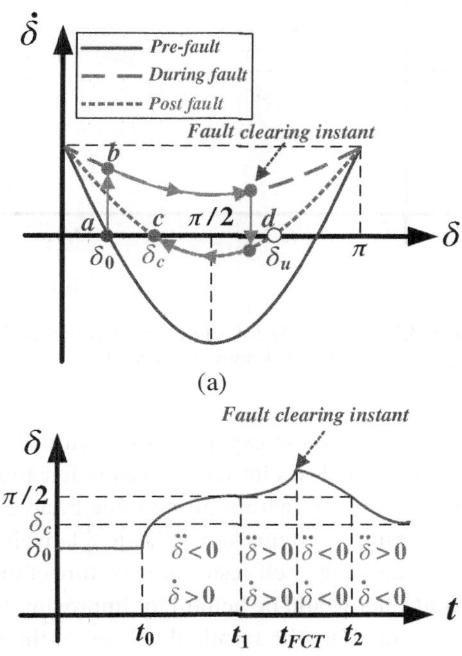

Fig. 3.26 Example 3.4, fault clearance before δ_u: (a) phase-portrait and (b) time-domain δ signal.

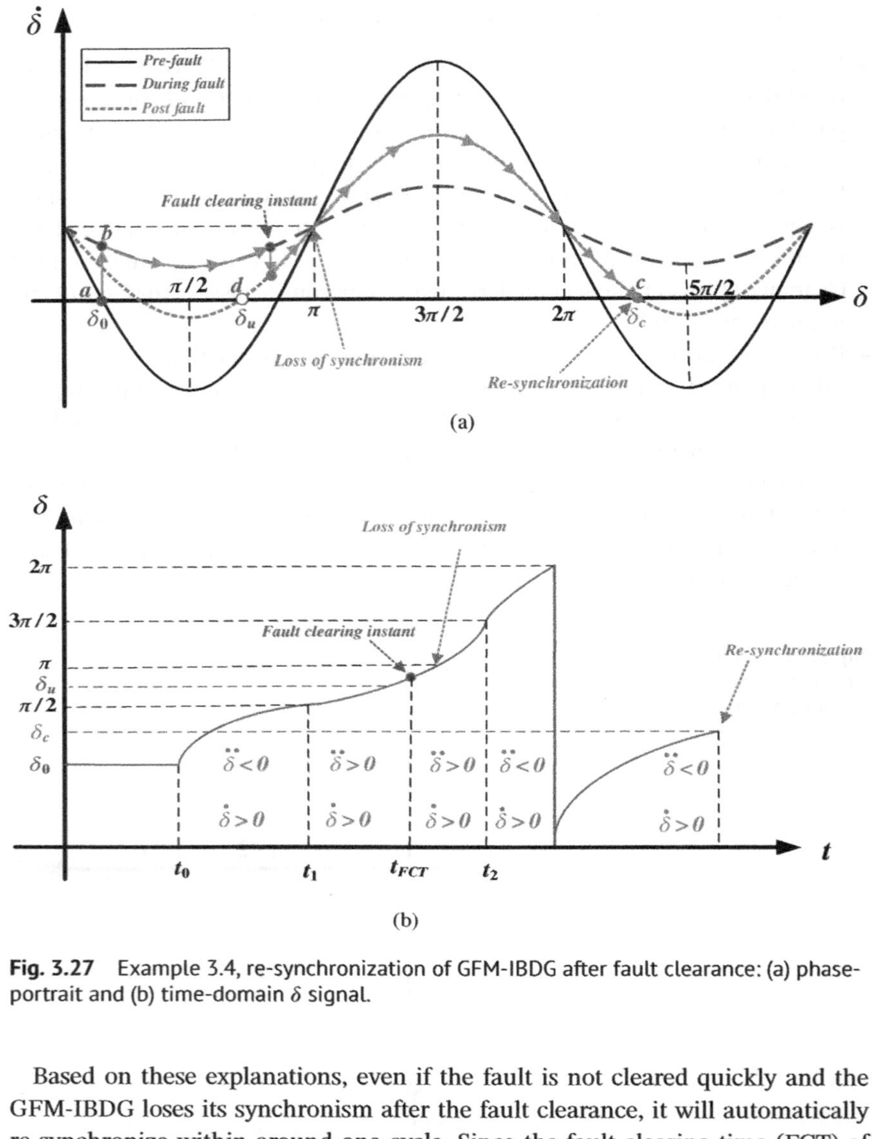

Fig. 3.27 Example 3.4, re-synchronization of GFM-IBDG after fault clearance: (a) phase-portrait and (b) time-domain δ signal.

Based on these explanations, even if the fault is not cleared quickly and the GFM-IBDG loses its synchronism after the fault clearance, it will automatically re-synchronize within around one cycle. Since the fault clearing time (FCT) of distribution systems and MGs are relatively long due to using overcurrent protection relays, this self-restoration feature of the GFM-IBDG with a first-order power controller would be helpful for improving MG transient stability.

Meanwhile, the terminal voltage of the GFM-IBDG may reduce considerably during distribution system voltage drops. The amount of terminal voltage change

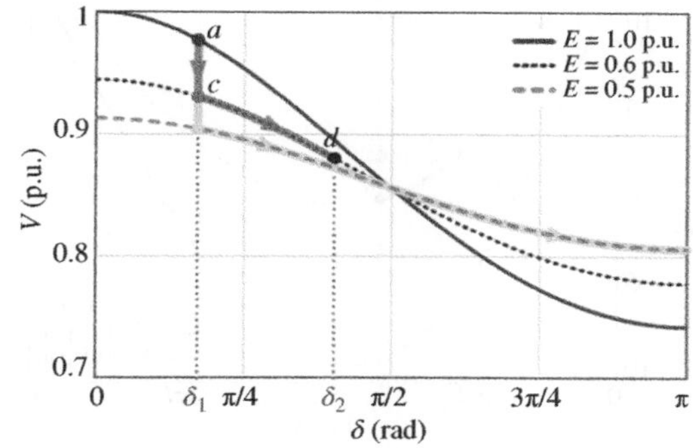

Fig. 3.28 $V-\delta$ curve of a first-order GFM-IBDG for various amounts of grid voltage magnitudes. [24] / with permission of IEEE.

depends on the voltage reactive power control loop. Thus, to analyze the dynamic response of IBDG terminal voltage, the equation of reactive power control loop should be scrutinized:

$$V = V_0 + K_q(Q_0 - Q) \tag{3.53}$$

where Q and Q_0 are the output reactive power and its reference value. Combining (3.48) and (3.53) results in [24]

$$V = V_0 + K_q\left(Q_0 - \frac{3}{2}\frac{V^2 - EV\cos\delta}{X_{Th}}\right) \tag{3.54}$$

Solving this equation provides the amount of V as a function of δ:

$$V = \frac{1.5K_qE\cos\delta - X_{Th} + \sqrt{(X_{Th} - 1.5K_qE\cos\delta)^2 + 6K_qX_g(V_0 + K_qQ_0)}}{3K_q} \tag{3.55}$$

Based on this equation, the $V-\delta$ curve of a grid-connected GFM-IBDG is depicted in Fig. 3.28 for various amounts of grid voltage magnitudes, i.e., for $E=1$, 0.6, and 0.5 pu. Meanwhile, the phase-portrait curve for this case is shown in Fig. 3.29. When the grid voltage drops, the output reactive power increases instantaneously based on (3.54) and the operating point voltage jumps from (a) to (c), as shown in Fig. 3.28. Then, as δ increases, the amount of output reactive power increases and the terminal voltage reduces due to the $Q-V$ characteristic. For the $E=0.6$ pu case, the reduction of terminal voltage stops at "d." However, in the $E=0.5$ pu case where no equilibrium point exists in the phase-portrait curve, the voltage drop continues. The amount of output active power P is a function of IBDG terminal voltage. Therefore, the $Q-V$ control loop could have great impact on the output active power and transient stability of GFM-IBDG.

B) **Second-order (VSM) power controller:** First-order LPFs can be added to the first-order power controller to diminish fluctuations of measured signals,

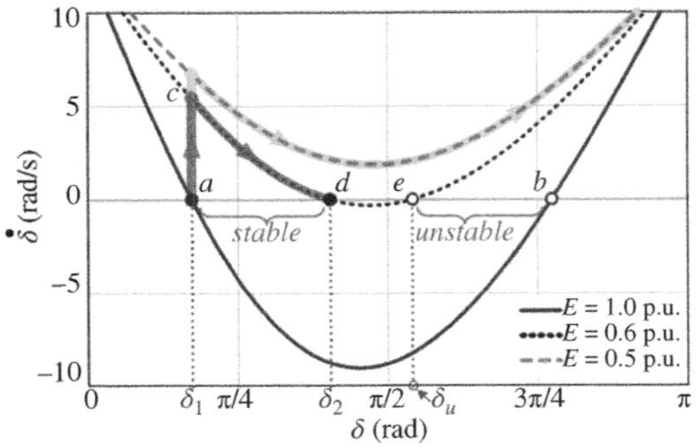

Fig. 3.29 Phase-portrait curve of the first-order GFM-IBDG for various amounts of grid voltage magnitudes. [24] / with permission of IEEE.

as shown in Fig. 3.13(c). This strategy is known as second-order power controller, in which the angular frequency and voltage could be expressed as [24]

$$\omega = \omega_0 + K_p \frac{\omega_p}{s + \omega_p}(P_0 - P) \tag{3.56}$$

$$V = V_0 + K_q \frac{\omega_q}{s + \omega_q}(Q_0 - Q) \tag{3.57}$$

These equations could also be utilized to express the VSM-based control loops, shown in Fig. 3.13(d). In fact, the models containing the first-order LPF and VSM are identical. To prove this, the equations of VSM active/reactive power controller are written as [24]

$$\omega = \frac{1}{Js}[D_p(\omega_0 - \omega) + P_0 - P] \tag{3.58}$$

$$V = \frac{1}{\tau s}[D_q(V_0 - V) + Q_0 - Q] \tag{3.59}$$

These equations can be simplified as

$$\omega = \frac{D_p}{Js + D_p}\omega_0 + \frac{1}{Js + D_p}(P_0 - P) \tag{3.60}$$

$$V = \frac{D_q}{\tau s + D_q}V_0 + \frac{1}{\tau s + D_q}(Q_0 - Q) \tag{3.61}$$

Comparing (3.56) and (3.57) with (3.60) and (3.61), it can be found that these equations are equivalent. In fact, the terms J, D_p, τ, D_q in VSM power controller correspond to $1/(\omega_p K_p)$, $1/K_p$, $1/(\omega_q K_q)$, $D_q = 1/K_q$ in the second-order power controller, respectively. Therefore, the transient and dynamic responses of both second-order and VSM power controllers are identical.

For transient stability analysis, at first, it is assumed that only the active power controller is second order, while the reactive power controller is first order. In fact, no LPF is exploited for the reactive power controller, i.e., $\omega_q = \infty$. For this case, combining (3.61) and (3.48) provides the dynamic equation of δ:

$$\ddot{\delta} = -\omega_p\dot{\delta} + \omega_p K_p \left(P_0 - \frac{3}{2}\frac{E\sin\delta}{X_{Th}} \frac{1.5K_qE\cos\delta - X_{Th} + \sqrt{\begin{matrix}(X_{Th} - 1.5K_qE\cos\delta)^2 \\ +6K_qX_{Th}(V_0 + K_qQ_0)\end{matrix}}}{3K_q} \right) \tag{3.62}$$

Based on this dynamic equation, the phase-portrait for decreasing E amount from $E = 1$ pu to $E = 0.6$ pu is depicted in Fig. 3.30. Moreover, in this figure, the phase-portrait of the first-order active power controller, i.e., $\omega_q = \infty$, is illustrated. As shown, the equilibrium points of both the first- and second-order power controllers are the same, and in both of them, the operating point travels from "a" to "d." In fact, since the amounts of V and E for the first- and second-order active power controllers are the same, the transfer powers and equilibrium points are identical considering (3.48). However, the trajectory between these equilibrium points is completely different. As shown, the first-order power controller reaches to the final equilibrium point without any power angle overshoot, while the second-order power controller includes an overshoot. The power angle overshoot is defined as the difference between the maximum power angle during the transient response of the GFM-IBDG and the steady-state equilibrium point. Moreover, if the maximum power angle reaches to the point E (i.e., δ_u) during the transient response, the GFM-IBDG will lose its synchronism with the system. In other words, even if an equilibrium point exists in the second-order power controller, the GFM-IBDG may not approach this equilibrium point due to the large overshoot. Thus, to preserve GFM-IBDG, the power angle overshoot should be kept small enough.

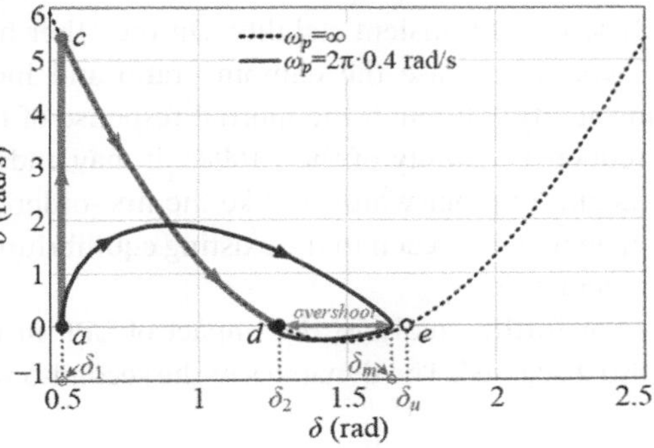

Fig. 3.30 Phase-portrait curve of second-order GFM-IBDG for decreasing E from 1 to 0.6 pu. [24] / with permission of IEEE.

For a second-order power controller, the overshoot is specified by its damping ratio. Thus, to determine the power angle overshoot, and using the Laplace transform, the active power controller loop dynamics in the s-domain can be expressed as follows, assuming $\sin\delta \approx \delta$ [24]:

$$\delta = \frac{K_p}{s}\frac{\omega_p}{s+\omega_p}(P_0 - G_P\delta) \tag{3.63}$$

where $G_P = 3EV/2X_g$.

Therefore, the active power control loop can be expressed as

$$\delta = \frac{K_p\omega_p}{s^2 + \omega_p s + K_p\omega_p G_P}P_0 \tag{3.64}$$

Thus, the damping ratio of this system could be calculated by

$$\zeta = \frac{1}{2}\sqrt{\frac{\omega_p}{K_p G_P}} \tag{3.65}$$

For various amounts of ω_p and K_p, Fig. 3.31 illustrates the phase-portrait curves if the amount of E decreases from 1 to 0.6 pu. As shown, the overshoot is the same for similar values of ω_p / K_p. Increasing K_p reduces the damping ratio and increases the power angle overshoot, which may endanger the GFM-IBDG transient stability. Furthermore, increment of ω_p increases the damping ratio and reduces the overshoot, which enhances the transient stability margin. For $K_p = 0.04$ pu and $\omega_p = 2\pi.0.3$ rad/s, the power angle overshoot is very large, and the power angle exceeds the critical angle δ_u. Beyond this point, the amount of $\dot{\delta}$ remains positive and, finally, the GFM-IBDG loses its synchronism with the system. The damping ratio could be rewritten as follows, considering the VSM control strategy:

$$\zeta = \frac{D_p}{2\sqrt{JG_P}} \tag{3.66}$$

Based on this equation, increasing the damping factor, i.e., D_p of the VSM, increases the damping ratio and decreases the overshoot. This would enhance the GFM-IBDG transient stability. On the other hand, increasing the virtual inertia J would decrease the damping ratio and increase the possibility of transient instability. Although the inertial response of the GFM-IBDG could improve the frequency stability of the IBMGs, it may endanger the transient stability of the system. In other words, unlike the first-order active power controller, the power angle may not reach to the existing equilibrium point due to the increment of the system order.

For further analysis, the impact of LPF in the reactive power control loop is also discussed. The dynamics of the second-order reactive power controller in the

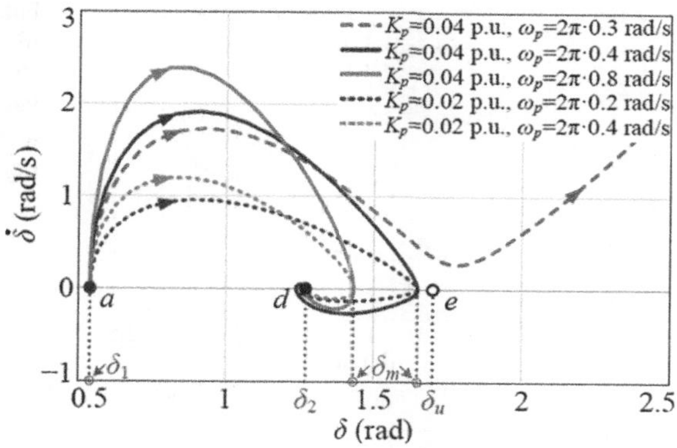

Fig. 3.31 Phase-portrait curve of second-order GFM-IBDG for decreasing E from 1 to 0.6 pu, and different amounts of ω_p and K_p. [24] / with permission of IEEE.

s-domain can be expressed as follows [24]:

$$V = V_0 + K_q \frac{\omega_q}{s + \omega_q} \left(Q_0 - \frac{3}{2} \frac{V^2 - EV \cos \delta}{X_{Th}} \right) \tag{3.67}$$

Moreover, in the time-domain, this equation is rewritten as

$$\dot{V} = -\omega_q V + \omega_q V_0 + \omega_q K_q \left(Q_0 - \frac{3}{2} \frac{V^2 - EV \cos \delta}{X_{Th}} \right) \tag{3.68}$$

Based on (3.67) and (3.68), the phase-portrait and $V - \delta$ curves for various amounts of ω_p are depicted in Fig. 3.32(a) and (b), respectively. In this study, it is assumed that E decreases from 1 to 0.6 pu, and K_p and ω_p are considered 0.04 pu and 2π.0.3 rad/s, respectively. As shown, for a first-order reactive power controller, i.e., $\omega_q = \infty$, the GFM-IBDG becomes unstable, while a second-order reactive power controller prevents transient instability. As illustrated in Fig. 3.32, after decreasing E, for a first-order reactive power controller, the IBDG voltage jumps from point "a" to point "f." However, a second-order reactive power controller decreases the voltage drop and helps increase the output active power. Increment of ω_q could improve the terminal voltage of GFM-IBDG and help the GFM-IBDG retain its stability.

As a conclusion, the second-order active and reactive power controllers have opposite impacts on the GFM-IBDG transient stability. Using the second-order active power controller degrades the transient stability, while exploiting the second-order reactive power controller could improve this stability.

3.3.1.3 Energy Function Models for Transient Stability Assessment

As explained in Chapter 2 of this book, one of the major techniques for the evaluation of the MG transient stability is the Lyapunov direct method. To obtain the proper Lyapunov function, the energy function of the DGs should be selected.

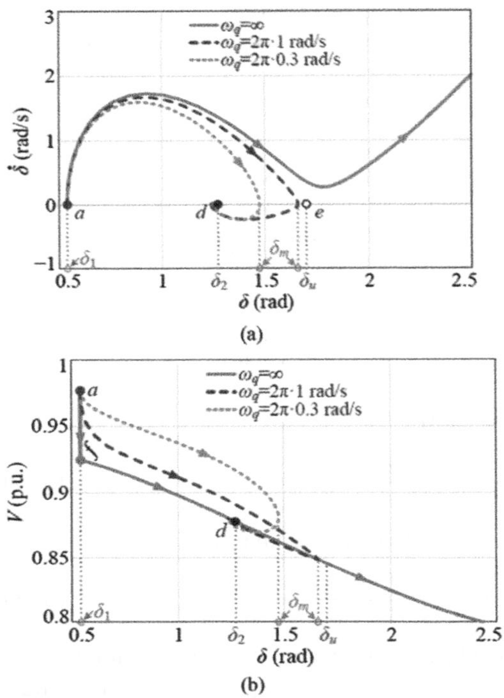

Fig. 3.32 Second-order GFM-IBDG response for decreasing E from 1 to 0.6 pu for various values of ω_q: (a) phase-portrait and (b) $V-\delta$ curve. [24] / with permission of IEEE.

In this regard, recently, various energy functions including different features are adopted in the research studies. In this section, the common energy functions for IBDGs are presented. Before discussing these functions for analyzing the transient stability of the GFL-IBDGs/GFM-IBDGs, the traditional energy function of an SG is briefly explained. Although this subject has been investigated in various studies, for improved clarity and reader comprehension, the explanations provided in [18] are employed in the following discussion.

3.3.1.3.1 Synchronous Generators The swing equation of an SG, also known as the motion equation, is commonly utilized to adopt a proper energy function for the transient stability assessment. The motion equation can be expressed as [18]

$$H\ddot{\delta} = P_m - \frac{3V_{PCC}V_g}{2X_g}\sin\delta - D\dot{\delta} \tag{3.69}$$

where δ is the SG rotor angle, H is the SG inertia, and D is the damping coefficient. Moreover, V_{PCC} is the PCC voltage, V_g is the Thevenin voltage of the upstream grid, and X_g is the equivalent reactance seen from the PCC. The SG energy function is obtained as

$$V(\delta,\dot{\delta}) = -P_m\delta - \frac{3V_{PCC}V_g}{2X_g}\cos\delta + \frac{1}{2}H\dot{\delta}^2 + E_0 \tag{3.70}$$

where E_0 is constant. Moreover, the derivative of the SG energy function is presented by the following equation by combining (3.69) and (3.70):

$$\begin{aligned}\dot{V}(\delta,\dot{\delta}) &= \frac{\partial V(\delta,\dot{\delta})}{\partial \delta'}\dot{\delta} + \frac{\partial V(\delta,\dot{\delta})}{\partial \dot{\delta}}\ddot{\delta} \\ &= \left(-P_m + \frac{3V_{PCC}V_g}{2X_g}\sin\delta + H\ddot{\delta}\right)\dot{\delta} = -D\dot{\delta}^2\end{aligned} \tag{3.71}$$

This equation will be exploited to establish the proper energy functions for GFL-IBDGs/GFM-IBDGs as presented as follows.

3.3.1.3.2 *GFL-IBDG* By comparing the power-angle equation of GFL-IBDG (3.47) and SG motion Equation (3.69), it is found that these equations are similar and each parameter of the GFL-IBDG power-angle equation can be considered equivalent to the parameters of the SG swing equation presented as follows [18]:

$$\begin{cases} P_{2ref} = K_i I_g \omega_n L_g \\ P_2 = K_i V_g \\ H_2 = (1 - K_p I_g L_g) \\ D_2(\delta) = -K_i I_g L_g + K_p V_g \cos\delta \end{cases} \tag{3.72}$$

Thus, the power-angle equation of GFL-IBDG could be written in the form of the SG swing equation as

$$H_2\ddot{\delta} = P_{2ref} - P_2\sin\delta - D_2(\delta)\dot{\delta} \tag{3.73}$$

Thus, similar to the SG energy function, i.e., (3.70), the energy function of the GFL-IBDG can be written as

$$V(\delta,\dot{\delta}) = -P_{2ref}\delta - P_2\cos\delta + \frac{1}{2}H_2\dot{\delta}^2 + E_0 \tag{3.74}$$

Moreover, the derivative of the energy function is given by [18]

$$\dot{V}(\delta,\dot{\delta}) = -P_{2\,\mathrm{ref}}\dot{\delta} + P_2\sin\delta\cdot\dot{\delta} + H_2\dot{\delta}\ddot{\delta} = -D_2(\delta)\dot{\delta}^2 \tag{3.75}$$

Unlike SGs in which the amount of damping coefficient, i.e., D, is constant and positive, the equivalent damping coefficient of GFL-IBDG, i.e., D_2, varies with changing the amount of δ. In fact, its value may be positive or negative as shown in Fig. 3.33. Meanwhile, the equivalent negative damping coefficient in GFL-IBDGs may reduce the transient stability margin.

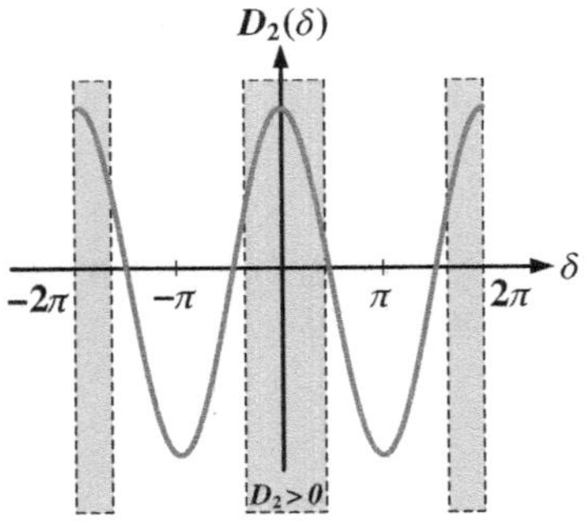

Fig. 3.33 Damping factor of a GFL-IBDG.

Moreover, similar to the SGs, the equivalent potential and kinetic energy of the GFL-IBDG are given by

$$E_p = -P_{2ref}\delta - P_2 \cos\delta + E_0$$
$$E_k = \frac{1}{2}H_2\dot{\delta}^2 \tag{3.76}$$

3.3.1.3.3 GFM-IBDG The power-angle equation of a second-order power controller GFM-IBDG and the SG equation motion are similar, considering the following analogy [18]:

$$\begin{cases} P_{1ref} = P_{ref} \\ P_1 = 3V_{PCC}V_g/2X_g \\ H_1 = 1/(\omega_p \cdot K_P) \\ D_1 = 1/K_P \end{cases} \tag{3.77}$$

Consequently, the power-angle equation of GFM-IBDG can be presented in the form of the SG swing equation as

$$H_1\ddot{\delta} = P_{1ref} - P_1 \sin\delta - D_1\dot{\delta} \tag{3.78}$$

Similar to the SGs, the energy function of the GFM-IBDG would be developed as

$$V(\delta,\dot{\delta}) = -P_{1ref}\delta - P_1\cos\delta + \frac{1}{2}H_1\dot{\delta}^2 + E_0 \tag{3.79}$$

Moreover, combining (3.78) and (3.79), the derivative of energy function is obtained by

$$\dot{V}(\delta,\dot{\delta}) = -P_{1ref}\dot{\delta} + P_1\sin\delta\cdot\dot{\delta} + H_1\dot{\delta}\ddot{\delta} = -D_1\dot{\delta}^2 \tag{3.80}$$

The equivalent damping coefficient of the GFM-IBDG depends only on the GFM-IBDG proportional gain, and thus, unlike GFL-IBDGs, its value is always positive. In addition, the equivalent potential and kinetic energies are, respectively, expressed by

$$E_p = -P_{1ref}\delta - P_1 \cos\delta + E_0 \tag{3.81}$$

$$E_k = \frac{1}{2}H_1\dot{\delta}^2 \tag{3.82}$$

3.3.1.3.4 Evaluating Transient Stability Using Energy Function La Salle's Invariance Principle (LSIP) could be adopted for transient stability assessment of GFL-IBDGs/GFM-IBDGs using their local energy-based functions. In the Lyapunov technique, the main focus is on the stability of an equilibrium solution, while LSIP investigates the behavior of all solutions as $t \to \infty$. The fundamentals of LSIP are explained as [18]

$$\begin{aligned} \Omega_l(y) &= \{y \in R^2 | V(y) < l, l > 0, \dot{V}(y) \leq 0\} \\ R(y) &= \{y \in \Omega_l(y) | \dot{V}(y) = 0\} \end{aligned} \tag{3.83}$$

where Ω_l (y) is an equipotential surface and $R(y)$ includes points for which $\dot{V}(y) = 0$. Moreover, the largest invariant set in the R is named M. In other words, M is an SEP located in Ω_l (y). Based on LSIP, any initial state points located in region Ω_l (y) will finally converge to M when $t \to \infty$. For a better illustration of the LSIP rule, Fig. 3.34 is presented. The initial state of the system after a disturbance is denoted by y_0 (y_{10}, y_{20}) in this figure. As depicted, when $t \to \infty$, the system state converges to M. If the largest invariant set, i.e., M, be outside the Ω_l (y) region, the IBDG stability cannot be evaluated through the conditions of LSIP.

- **GFL-IBDG:** To exploit LSIP for analyzing the transient stability, at first, the proper finite Ω_l (y) should be found to fulfill the condition $\dot{V}(y) \leq 0$. However, the damping coefficient of the GFL-IBDG is not constant and may become negative depending on the value of δ. In fact, the coefficient of the system differential

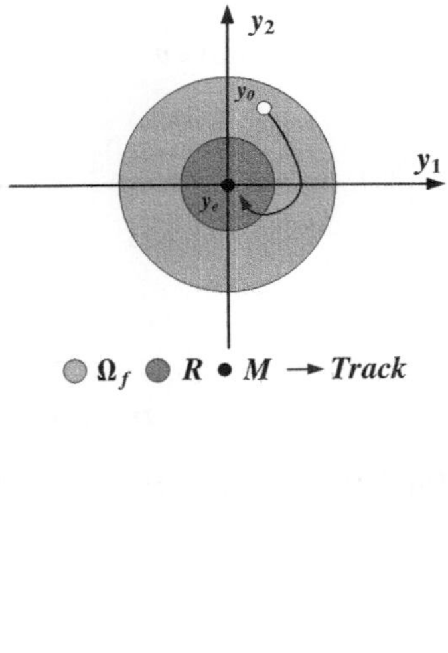

Fig. 3.34 Simple representation of the LSIP. Adapted from [18].

equation changes, which may threaten the IBDG stability. Thus, to guarantee the stability of IBDG, the $D(\delta) \geq 0$ constraints should be considered for choosing Ω_l (y). Based on (3.72), the amount of δ for which the value of D becomes zero could be given by

$$\delta_{max2} = \arccos \frac{K_i I_g L_g}{K_p V_g} \tag{3.84}$$

Moreover, based on (3.76), the system energy function at δ_{max2} is defined as

$$E_{max2} = -(P_{2ref}\delta_{max2} + P_{2PF} \cos \delta_{max2}) \tag{3.85}$$

where P_{2PF} is the maximum active power that can be delivered by GFL-IBDG after the fault clearance. As shown in Fig. 3.35, the GFL-IBDG retains its stability if the system states, i.e., $(\delta, \dot{\delta})$, enter and remain within the stability boundary, i.e., $E \leq E_{max2}$ for δ_{max2}, under the positive damping coefficient $D(\delta) \geq 0$ [18].

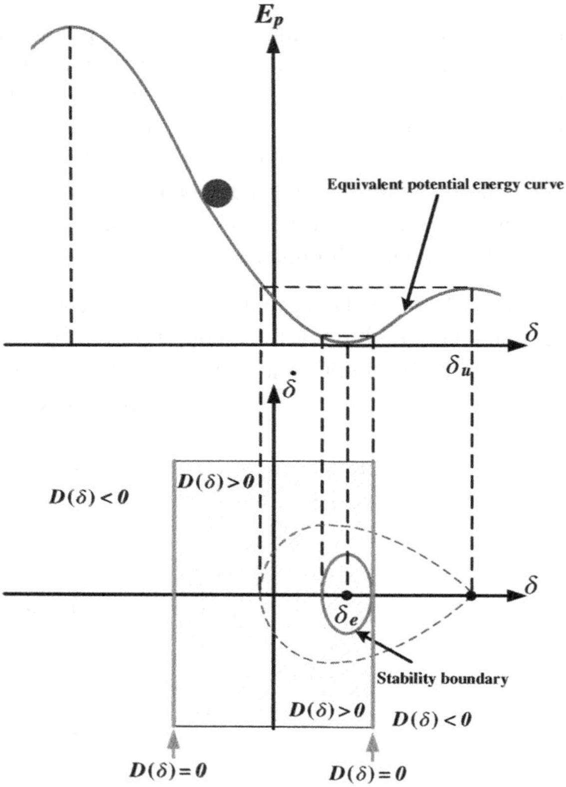

Fig. 3.35 Energy function of GFL-IBDG. Adapted from [18].

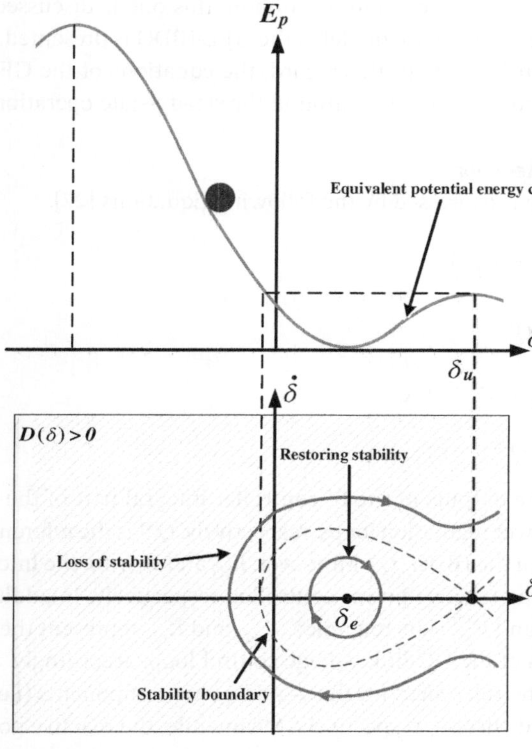

Fig. 3.36 Energy function of GFM-IBDG. Adapted from [18].

- **GFM-IBDG:** Unlike GFL-IBDGs, the GFM ones always have a constant positive damping coefficient. Thus, their proper energy function could be defined as [18]

$$E_{max1} = -(P_{1ref}\delta_U + P_{1PF}\cos\delta_U) \tag{3.86}$$

where $P_{1\,PF}$ is the maximum active power that can be delivered by GFM-IBDG after fault clearance. As shown in Fig. 3.36, the stability boundary of GFM-IBDGs can be considered the equipotential surface $E = E_{max1}$. If the system states $(\delta, \dot{\delta})$ located within the stability boundary return and remain inside this boundary, the system restores its stability, as depicted in Fig. 3.36.

3.3.2 Small-Signal Stability

To investigate the small-signal stability of IBDGs, two major approaches are developed in research works: (i) eigenvalue analysis based on small-signal model and (ii) impedance-based analysis and stability analysis using Nyquist stability criteria

[25, 26]. Since the first technique is more common, this one is discussed in this book. In this section, the state-space model of the GFL-IBDG is presented, considering small-signal perturbations. In this regard, the equations of the GFL-IBDG controllers are presented and linearized around the steady-state operation point.

3.3.2.1 Power Controller Loop

The power control loop is expressed by the following Equations [27]:

$$\begin{aligned}
\frac{dx_v}{dt} &= K_{idc}\left(V_{dc} - V_{dc}^{ref}\right) \\
\frac{dx_Q}{dt} &= K_{iQ}(Q - Q^{ref}) \\
I_d^{ref} &= K_{pdc}\left(V_{dc} - V_{dc}^{ref}\right) + x_v \\
I_q^{ref} &= K_{pQ}(Q - Q^{ref}) + x_Q
\end{aligned} \tag{3.87}$$

where x_Q and x_v are the outputs of the PI controller integral part of the DC-link voltage and reactive power regulation loops, respectively. Q^{ref} is the reference of the output reactive power of the IBDG (Q). Moreover, K_{iQ} and K_{pQ} are the integral and proportional gains of the reactive power controller, respectively. In addition, V_{dc} is the DC-link voltage and V_{dc}^{ref} is its reference. K_{pdc} and K_{idc} represent the integral and proportional gains of the DC-link voltage control loop, accordingly. Furthermore, I_d^{ref} and I_q^{ref} are the references for the d- and q-axis components (i.e., I_d and I_q) of the IBDG injected current, respectively. Meanwhile, the reactive power Q is calculated using

$$Q = V_d I_q - V_q I_d \tag{3.88}$$

where V_d and V_q are the d- and q-axis components of the PCC voltage, respectively. By combining (3.87) and (3.13), the linearized equations of power controller loops could be expressed as [28]

$$\begin{aligned}
\Delta I_d^{ref} &= -\frac{K_{pdc}s + K_{idc}}{s}\frac{1}{C_{dc}V_{dc^0}s}\Delta P = G_d(s)\Delta P \\
\Delta I_q^{ref} &= \frac{K_{PQ}s + K_{iQ}}{s}\Delta Q = G_q(s)\Delta Q
\end{aligned} \tag{3.89}$$

Moreover, the linearized equation of the GFL-IBDG power output is obtained as

$$\begin{aligned}
\Delta P &= \mathbf{c}_1{}^T \Delta \mathbf{V_{dq}} + V_{d0}\Delta I_d \\
\Delta Q &= \mathbf{c}_2{}^T \Delta \mathbf{V}_{\mathrm{dq}} - V_{d0}\Delta I_q
\end{aligned} \tag{3.90}$$

where

$$\Delta \mathbf{V_{dq}} = \left[\Delta V_d \ \Delta V_q\right]^T, \quad \mathbf{c}_1^T = \left[I_{d0} \ I_{q0}\right], \mathbf{c}_2^T = \left[-I_{q0} \ I_{d0}\right] \tag{3.91}$$

3.3.2.2 Current Controller Loop

Considering Fig. 3.10, the current controller loop can be expressed by [28]

$$\frac{dx_{id}}{dt} = K_{id}\left(I_d^{ref} - I_d\right)$$

$$\frac{dx_{iq}}{dt} = K_{iq}\left(I_q^{ref} - I_q\right)$$

$$V_{cd} = K_{pd}\left(I_d^{ref} - I_d\right) + x_{id} - x_f I_q + V_d$$

$$V_{cq} = K_{pq}\left(I_q^{ref} - I_q\right) + x_{iq} + x_f I_d + V_q \tag{3.92}$$

where x_{id} and x_{iq} are the outputs of the integral controller of the *d*- and *q*-axis current regulation loops, respectively. On the other hand, using the KVL rule over the output filter of the GFL-IBDG results in

$$x_f \frac{dI_d}{dt} = \omega_0(V_{cd} - V_d) + \omega_0 x_f I_q$$

$$x_f \frac{d1_q}{dt} = \omega_0(V_{cq} - V_q) - \omega_0 x_f I_d \tag{3.93}$$

Consequently, the linearized equations of the current controller loop are expressed as follows [28]:

$$\Delta I_d = \frac{\omega_0 K_{pd} s + \omega_0 K_{id}}{x_f s^2 + \omega_0 K_{pd} s + \omega_0 K_{id}} \Delta I_d^{ref} = G_{id}(s)\Delta I_d^{ref}$$

$$\Delta I_q = \frac{\omega_0 K_{pq} s + \omega_0 K_{iq}}{x_f s^2 + \omega_0 K_{pq} s + \omega_0 K_{iq}} \Delta I_q^{ref} = G_{iq}(s)\Delta I_q^{ref} \tag{3.94}$$

3.3.2.3 Phase-Locked Loop

The linearized model of the widely used PLL is shown in Fig. 3.37. As described earlier, the LPF is commonly adjusted to diminish the high-frequency contents above 300 Hz. For small-signal stability analysis, it is assumed that $F(s) = 1$. Considering this assumption, the PLL linearized model could be expressed by the

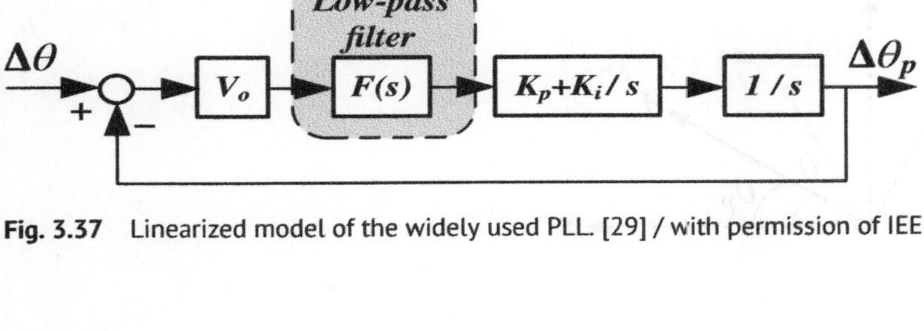

Fig. 3.37 Linearized model of the widely used PLL. [29] / with permission of IEEE.

following Equations [28, 29]:

$$\frac{dx_p}{dt} = K_i V_0(\theta - \theta_p)$$
$$\frac{d\theta_p}{dt} = K_p V_0(\theta - \theta_p) + x_p \tag{3.95}$$

where θ is the phase angle of the PCC voltage and θ_p is the PLL output angle. Moreover, x_p is the output of the PLL integrator. In addition, V_0 is a constant value, which is considered equal to the nominal voltage at PCC. Besides, K_i and K_p are the integral and proportional gains of the associated PI controller, respectively. Considering (3.95), the linearized model of the PLL is expressed using the following transfer function:

$$\Delta\theta_p = \frac{V_0 F(s)(K_p s + K_i)}{s^2 + V_0 F(s) K_p s + V_0 F(s) K_i}\Delta\theta = P_{LL}(s)\Delta\theta \tag{3.96}$$

The relation between $x-y$ and $d-q$ reference frames is depicted in Fig. 3.38. Considering this figure, $\Delta\theta$ can be calculated as [19]

$$\Delta\theta = \mathbf{c}_3^T \Delta\mathbf{V}_{\mathbf{xy}} \tag{3.97}$$

where [25]

$$\Delta\mathbf{V}_{\mathbf{xy}} = [\Delta V_x \Delta V_y]^T, \quad \mathbf{c}_3^T = \left[-\frac{V_{y0}}{V_0^2}\frac{V_{x0}}{V_0^2}\right] \tag{3.98}$$

Meanwhile, the relation between the injected current of the GFL-IBDG and its terminal voltage in the $x-y$ and $d-q$ coordinates is given by [19]

$$\Delta\mathbf{I}_{\mathbf{xy}} = K_2 \Delta\mathbf{I}_{dq} + c_5 \Delta\theta_p$$
$$\Delta\mathbf{V}_{dq} = K_1 \Delta\mathbf{V}_{\mathbf{xy}} + c_4 \Delta\theta_p \tag{3.99}$$

where

$$\mathbf{K}_1 = \begin{bmatrix} \cos\theta_{p0} & \sin\theta_{p0} \\ -\sin\theta_{p0} & \cos\theta_{p0} \end{bmatrix}, \quad \mathbf{K}_2 = \begin{bmatrix} \cos\theta_{p0} & -\sin\theta_{p0} \\ \sin\theta_{p0} & \cos\theta_{p0} \end{bmatrix} \tag{3.100}$$

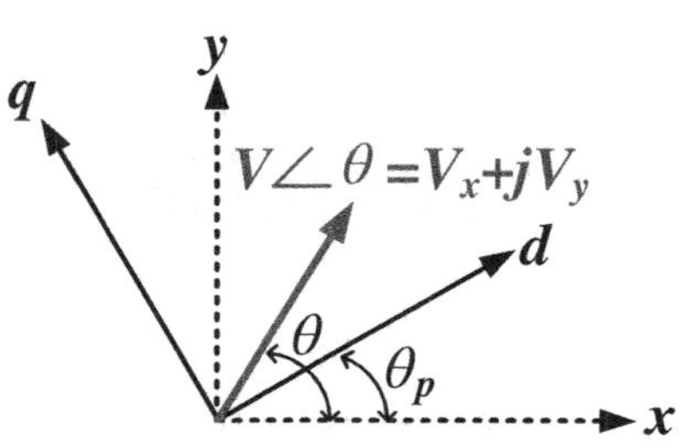

Fig. 3.38 The relation between the $x-y$ and $d-q$ reference frames.

where

$$\Delta \mathbf{I_{xy}} = \begin{bmatrix} \Delta I_x \\ \Delta I_y \end{bmatrix}; \quad c_4 = \begin{bmatrix} V_{q0} & -V_{d0} \end{bmatrix}^T; \quad c_5 = \begin{bmatrix} -I_{y0} & -I_{x0} \end{bmatrix}^T;$$
$$\Delta \mathbf{I}_{dq} = \begin{bmatrix} \Delta I_d & \Delta I_q \end{bmatrix}^T \tag{3.101}$$

3.3.2.4 Complete Linearized Model

The terminal voltage of the IBDG could be expressed based on the upstream grid voltage. Meanwhile, to reduce the complexities, it is assumed that the upstream system is modeled by an ideal voltage source and a series lumped reactance (x_L). Thus, the voltage at PCC is expressed using the KVL rule as [29]

$$\Delta \mathbf{V_{xy}} = \frac{x_L}{\omega_0} \mathbf{U}(s) \Delta \mathbf{I_{xy}} + \Delta \mathbf{V}_{\mathbf{bxy}} \tag{3.102}$$

where V_{bx} and V_{by} are the voltages of upstream grid expressed in the $x-y$ coordinate. Moreover,

$$\Delta \mathbf{V_{bxy}} = \begin{bmatrix} \Delta V_{bx} \\ \Delta V_{by} \end{bmatrix}; \quad \mathbf{U}(s) = \begin{bmatrix} s & -\omega_0 \\ \omega_0 & s \end{bmatrix} \tag{3.103}$$

Therefore, the complete linearized model of a grid-connected IBDG for small-signal assessment purposes is depicted in Fig. 3.39 [29]. This model can be exploited to investigate IBDG stability for both high- and low-frequency bands, as described in the following.

3.3.2.5 Small-Signal Stability for the High-Frequency Band

For easier analysis of IBDG small-signal stability and to simplify the block diagram shown in Fig. 3.39, the stability of IBDG is separately investigated for high- and low-frequency bands. As mentioned earlier, the bandwidth of current control loop

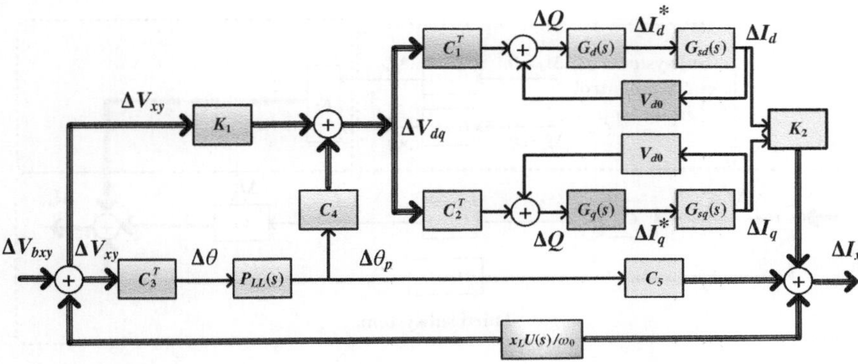

Fig. 3.39 The complete linearized model of a grid-connected GFL-IBDG. [29] / with permission of IEEE.

is generally about 10 times that of the power control loop. It means that the power control loop is slower than the current control loop by about 10 times. Therefore, for studying the high-frequency band stability, the slow dynamics and variations of power control loop could be neglected. In other words, the changes of the power controller output are assumed to be zero:

$$\Delta I_d^{ref} \approx 0, \Delta I_q^{ref} \approx 0 \tag{3.104}$$

Therefore, the block diagram of the IBDG model is simplified as depicted in Fig. 3.40. As shown, this system consists of current controller and PLL subsystems, which are decoupled from each other. The current control loop includes two subsystems whose small-signal stability in high-frequency band does not depend on the upstream system condition. In such a condition, the stability of the GFL-IBDG is mainly determined by the third subsystem containing the PLL and upstream system models, whose characteristic equation is presented by [28]

$$\left(1 - \frac{x_L K_p P_0}{\omega_0 V_0}\right) s^2 + \left(-\frac{x_L K_i P_0}{\omega_0 V_0} - \frac{x_L K_p Q_0}{V_0} + K_p V_0\right) s - \left(\frac{x_L K_i Q_0}{V_0} - K_i V_0\right) = 0 \tag{3.105}$$

where P_0 and Q_0 are the steady-state active and reactive powers of the IBDG, respectively. Using the Routh–Hurwitz rule, the GFL-IBDG is stable for the high-frequency band if and only if the following conditions are fulfilled:

$$x_L < \frac{V_0^2}{Q_0} \text{ and } x_L < \frac{\omega_0 V_0^2 K_p}{K_i P_0 + \omega_0 K_p Q_0} \tag{3.106}$$

This can be rewritten as

$$\begin{aligned} &x_L Q_0 < V_0^2 \\ &x_L K_i P_0 + x_L \omega_0 K_p Q_0 < \omega_0 V_0^2 K_p \end{aligned} \tag{3.107}$$

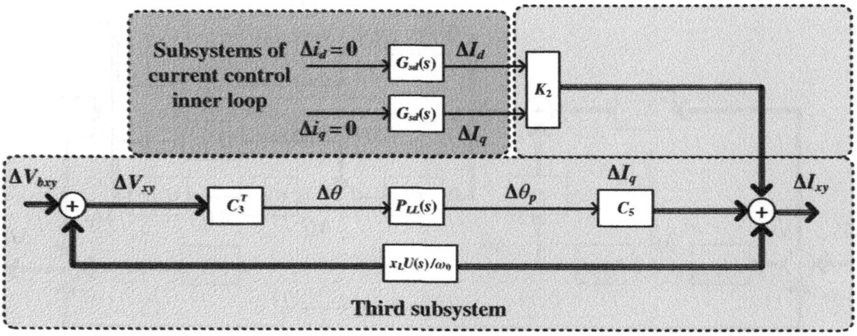

Fig. 3.40 Simplified linearized model of a grid-connected GFL-IBDG for high-frequency band. [29] / with permission of IEEE.

If the second condition is fulfilled, the first condition shall also be fulfilled due to the positive amount of $x_L K_i P_0$. Therefore, the necessary and sufficient condition for small-signal stability of IBDG in the high-frequency band could be given by [28]

$$\frac{K_i}{K_p} < \frac{\omega_0 V_0^2}{x_L P_0} - \frac{\omega_0 Q_0}{P_0} \tag{3.108}$$

Moreover, the short-circuit ratio (SCR) of the MG would be expressed as

$$\text{SCR} \approx \frac{1}{x_L} \tag{3.109}$$

where X_L is the per-unit impedance seen from the IBDG terminal. If X_L increases, the grid strength and its SCR decrease, and vice versa. For SCR < 2, the grid-connection is assumed weak, which may endanger the small-signal stability of the IBDG. Therefore, the stability criteria for the IBDG in high-frequency band can also be expressed based on the SCR value as

$$\frac{K_i}{K_p} < \frac{\omega_0 V_0^2}{x_L P_0} - \frac{\omega_0 Q_0}{P_0} \tag{3.110}$$

In this method, the stability of PLL in the high-frequency band could be simply investigated without requiring complex modal analysis. This approach will be very useful for tuning and updating the PLL parameters, considering the upstream system condition. It should be noted that the interaction between the power and current control loops is neglected, and these control loops are assumed to be decoupled as expressed earlier.

3.3.2.6 Small-Signal Stability for the Low-Frequency Band

In the low-frequency band, the dynamics of current control loop could be neglected due to its very fast response. In other words, the current control loop responds immediately after receiving the power controller command and thus $I_{dref} = I_d$ and $I_{qref} = I_q$. Thus, the transfer function of the d- and q-axis current controllers, i.e., $G_{id}(s)$ and $G_{iq}(s)$, are assumed to be equal to 1. For the low-frequency band, the dynamic behavior of the upstream impedance is neglected ($s \approx 0$), i.e., $U(s) = U(0)$. For this case, the necessary and sufficient condition for the small-signal stability would be expressed as follows using the Routh–Hurwitz criterion [28]:

$$s_0 > \frac{x_L P_0}{V_0} \text{ and } 1 > \frac{x_L P_0}{V_0} \tag{3.111}$$

where s_0 is the solution of the following equation:

$$b_5 s^5 + b_4 s^4 + b_3 s^3 + b_2 s^2 + b_1 s + b_0 > 0 \tag{3.112}$$

where b_i parameters ($i = 1, 2, 3, 4, 5$) are calculated using the PLL and power controller parameters. A weaker upstream system and the lower value of SCR lead to increment of x_L. Thus, based on (3.111), the stability margin decreases by increment of x_L. Moreover, as expressed by (3.111), increment of the IBDG loading (P_0) may endanger the IBDG stability.

The PLL bandwidth is normally set at the high-frequency band, and thus, the PLL could track the angle variations very quickly. Therefore, to derive the closed-form analytical expression for the low-frequency band, the PLL dynamic response can be neglected, and its transfer function is considered equal to 1. Considering this assumption, the characteristic equation is given by

$$s^3 + a_2 s^2 + a_1 s + a_0 = 0 \tag{3.113}$$

where

$$a_2 = \frac{K_{pdc}V_0(1 + K_{pQ}V_0) + C_{dc}V_{dc0}K_{iQ}V_0 - K_{pQ}K_{pdc}\left(\dfrac{x_L P_0}{V_0}\right)^2}{C_{dc}V_{dc0}(1 + K_{pQ}V_0)}$$

$$a_1 = \frac{\left[K_{idc}V_0 + V_0^2(K_{iQ}K_{pdc} + K_{pQ}K_{idc})\right] - (K_{iQ}K_{pdc} + K_{pQ}K_{idc})\left(\dfrac{x_L P_0}{V_0}\right)^2}{C_{dc}V_{dc0}(1 + K_{pQ}V_0)}$$

$$a_0 = \frac{K_{iQ}K_{idc}V_0^2 - K_{iQ}K_{idc}\left(\dfrac{x_L P_0}{V_0}\right)^2}{C_{dc}V_{dc0}(1 + K_{pQ}V_0)} \tag{3.114}$$

where C_{dc} is the capacitance of the DC-link capacitor and V_{dc0} is its steady-state voltage. Exploiting the Routh–Hurwitz rule, the GFL-IBDG is stable in the low-frequency band if and only if [28]

$$k_0 > \frac{x_L P_0}{V_0} \text{ and } 1 > \frac{x_L P_0}{V_0} \tag{3.115}$$

where

$$k_0 = \sqrt{\frac{K_b - \sqrt{K_b^2 - 4K_c K_{pQ}K_{pdc}(K_{iQ}K_{pdc} + K_{pQ}K_{idc})}}{2K_{pQ}K_{pdc}(K_{iQ}K_{pdc} + K_{pQ}K_{idc})}} \tag{3.116}$$

$$\begin{aligned} K_b &= K_{pdc}V_0(K_{iQ}K_{pdc} + K_{pQ}K_{idc})(1 + 2K_{pQ}V_0) + K_{pQ}K_{pdc}K_{idc}V_0 \\ &\quad - C_{dc}V_{dc0}K_{iQ}(K_{idc} - K_{pdc}K_{iQ}V_0) \\ K_c &= K_{pdc}V_0^2(1 + K_{pQ}V_0)\left(K_{idc} + K_{iQ}K_{pdc}V_0 + K_{pQ}K_{idc}V_0\right) \\ &\quad + C_{dc}V_{dc0}K_{iQ}^2K_{pdc}V_0^3 \end{aligned} \tag{3.117}$$

As a conclusion, the improper parameter selection of the GFL-IBDG power controller may lead to low-frequency oscillations and even result in the IBDG instability, especially for weak distribution systems (low amount of SCR) and heavy-loading conditions. In fact, even for a PLL with a high-frequency band, instability may occur in the low-frequency band due to the improper adjustment of the power controller.

3.4 PLL-Less Structure

The GFL strategy is widely utilized in the grid-connected inverters for both solar and wind resources. As explained in this chapter, GFL-IBDGs exploit the PLL to determine the angle of the PCC voltage and synchronize with the upstream grid. However, as discussed, the GFL-IBDG suffers from serious stability issues, especially in weak grids with low SCRs, due to the dynamic behavior of PLL. To overcome the drawbacks raised from the PLL behavior, various GFL PLL-less configurations have been recently suggested in the literature [30–33].

One of the novel types of the PLL-less strategies having a proper dynamic behavior is the power-synchronized grid-following (PSGFL) scheme. The structure of PSGFL-IBDG and its power controller is depicted in Fig. 3.41 [30, 31]. As shown, in this strategy, the terminal voltage of the inverter is exploited for power control instead of the PCC voltage. This approach improves the stability of the inverter in ultra-weak distribution systems, which is a great advantage. Moreover, the inverter does not require the voltage measurement at the connection point. Instead, the *d*- and *q*-axis components of the reference voltage signals (injected into the PWM block) are utilized by the control system for calculating the active and reactive powers. Then, a second-order LPF is exploited whose transfer function is represented by *H*. In this strategy, the rotating $d-q$ reference frame has been aligned with the injected current of the IBDG through adjusting the *q*-axis component of the injected current (i_q) at zero. In this reference, the injected current and terminal

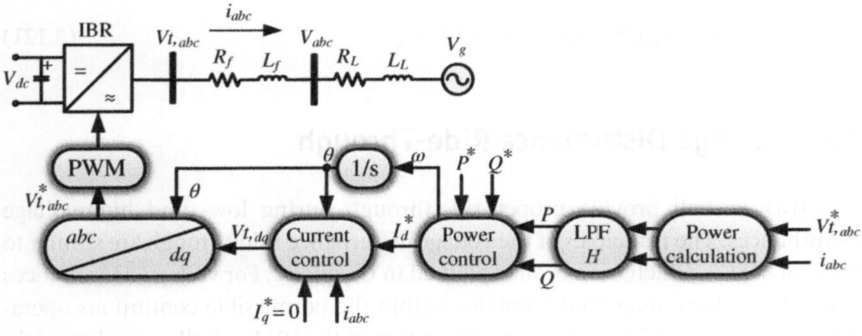

Fig. 3.41 Structure of the PLL-less GFL-IBDG [30, 31].

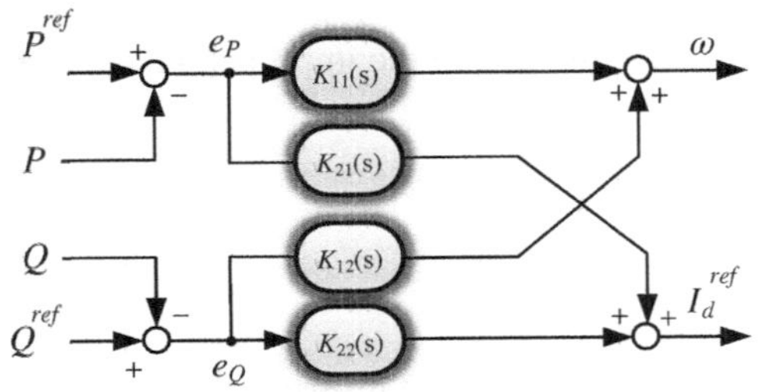

Fig. 3.42 Power controller structure of the PLL-less GFL-IBDG.

voltage could be expressed as follows [30]:

$$\begin{aligned} i_{dq} &= \widetilde{I}_{\mathrm{d}} + j0 \\ v_{t,dq} &= V_{\mathrm{t}} \cos \widetilde{\theta}_I + jV_{\mathrm{t}} \sin \widetilde{\theta}_I \end{aligned} \tag{3.118}$$

where θ_I is the angle between the inverter-injected current and its output voltage. i_{dq} represents the dq components of the IBDG current. Moreover, I_d is the d-axis component of the injected current, and V_t is the terminal voltage magnitude.

The structure of GFL-IBDG power controller utilized for grid synchronization is shown in Fig. 3.42 [30, 31]. In this strategy, the terminal voltage angle could be calculated by

$$\theta = \int (\omega_0 + \Delta\omega)dt \tag{3.119}$$

Since the grid angle is $\theta_g = \int \omega_0 dt$, the difference between the terminal voltage angle and the grid voltage angle is given by

$$\delta = \theta - \theta_g = \int \Delta\omega \cdot dt \tag{3.120}$$

Based on Fig. 3.42, $\Delta\omega$ is determined as follows using the power-synchronized controller:

$$\dot{\delta} = \Delta\omega = K_{11}(P^{ref} - P) + K_{12}(Q^{ref} - Q) \tag{3.121}$$

3.5 Voltage Disturbance Ride-Through

The IBDGs shall provide proper ride-through during low- and high-voltage disturbances. The principles of the voltage disturbance ride-through are similar to those of SSSGs, which have been explained in Chapter 2. For voltage disturbances where the voltage magnitude remains within the permissible continuous operation region of the IBDG, the protection system of the IBMG shall securely remain stable. In this case, the IBDG should remain operative and inject the available active power into the system. The procedure for determining the DG active power has been discussed in Chapter 2. If the voltage drop is large and voltage magnitude becomes lower than the permissible range, the IBDG, especially its PLL, shall ride-through the disturbance and remain in synchronism with the upstream network for a predefined duration, which is known as the mandatory operation region. The duration of the ride-through decreases for higher voltage dips. The capability of IBDG and requirements of the upstream system shall be considered for determining the duration of the ride-through for various voltage ranges.

Considering the capability of DGs for providing voltage ride-through, three categories are introduced by the IEEE 1547-2018 standard, namely category I, category II, and category III [2]. Category III affords a longer voltage disturbance compared to category II and category I, and thus, it could provide a better voltage support. Furthermore, category II and category III shall provide current exchange with the system in the mandatory operation region. For severe voltage dips (commonly below 0.5 pu) or for voltage rises (above 1.1 pu), the momentary cessation capability of the IBDG shall be enabled. In the momentary cessation region, the IBDG ceases the active power injection, while remaining connected to the system. In this region, the IBDG may exchange a limited amount of reactive power with the system. A sample voltage ride-through capability curve of an IBDG-category III is illustrated in Fig. 3.43 [2]. As shown, the mandatory operation region of the IBDG

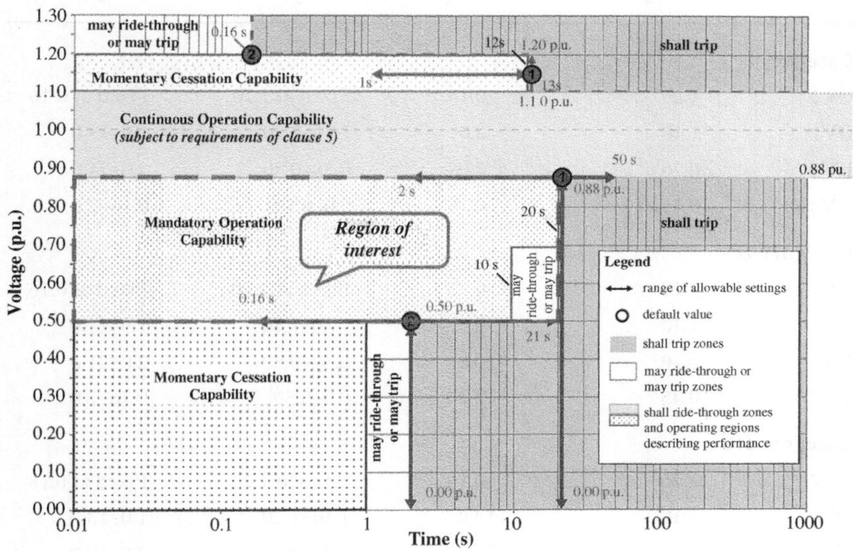

Fig. 3.43 Voltage ride-through capability curve of an IBDG-category III. Adapted from [2].

is 0.5 pu < U < 0.9 pu. For U = 0.5 pu, the ride-through duration is 0.16 s, while for 0.9 pu, it is about 2 s. Moreover, the trip region in which the voltage protection should operate is specified in this figure.

The protection system of the MG, especially overvoltage (OV) and undervoltage (UV) functions, shall be coordinated with the voltage ride-through capability of the IBDG. The recommended settings for voltage functions for the three different categories of DGs are presented in Table 3.3 [2]. Based on these settings, the overvoltage function of IBDGs shall trip for U > 1.2 pu with clearing time of 0.16 s. Moreover, for category I and category II, the UV function shall trip for U < 0.45 pu with clearing time of 0.16 s. However, for category III, which provides a longer voltage disturbance ride-through, the time delay of UV for U < 0.5 pu is recommended to be 2 s.

The IBDGs of category II and category III could provide dynamic voltage support through the injection of suitable current during voltage disturbance ride-through in the mandatory operation region. This dynamic support may

Table 3.3 Recommended settings for DG voltage functions. [2] / with permission of IEEE.

	Typical settings		Ranges of permissible settings	
Shall trip function	**Voltage (pu of nominal voltage)**	**Clearing time (s)**	**Voltage (pu of nominal voltage)**	**Clearing time (s)**
Category I				
OV2	1.20	0.16	Fixed at 1.20	Fixed at 0.16
OV1	1.10	2.0	1.10–1.20	1.0–13.0
UV1	0.70	2.0	0.58–0.80	2.0–21.0
UV2	0.45	0.16	0.0–0.50	0.16–2.0
Category II				
OV2	1.20	0.16	Fixed at 1.20	Fixed at 0.16
OV1	1.10	2.0	1.10–1.20	1.0–13.0
UV1	0.70	10.0	0.58–0.80	2.0–21.0
UV2	0.45	2.0	0.0–0.50	0.16–2.0
Category III				
OV2	1.20	0.16	Fixed at 1.20	Fixed at 0.16
OV1	1.10	13.0	1.10–1.20	1.0–13.0
UV1	0.88	21.0	0.70–0.88	2.0–50.0
UV2	0.50	2.0	0.0–0.50	0.16–21.0

help the system voltage stability, especially in the presence of the IMs. In the dynamic voltage support condition, the IBDG may exchange reactive current with the system. In this situation, if the IBDG reaches to its current limit, its active current is reduced due to the reactive current exchange. The effectiveness of dynamic voltage support based on the reactive power exchange considerably depends on the system X/R ratio, as explained in Chapter 2. In the transmission and subtransmission power systems, due to the large amount of X/R ratio, the reactive current injection is the most effective solution for voltage support. For the IBDG connected to the medium-voltage level of the distribution system, i.e., primary distribution systems, this approach may be still effective. However, for micro-scale IBDGs, which are connected to the LV distribution feeders, i.e., secondary distribution networks, a combination of active and reactive current exchange should be considered for effective voltage support.

One the main purposes of dynamic voltage support is mitigating the fault-induced delayed voltage recovery (FIDVR) phenomenon, which will be discussed in the following section. A sample of dynamic reactive current-based voltage support is shown in Fig. 3.44 [2]. As shown, when the voltage deviation becomes less than the predefined threshold, i.e., V_{Min}, the voltage support capability of the IBDG is enabled and the reactive current is injected into the system for voltage support. The magnitude of this reactive current is commonly determined based on the amount of voltage drop using the following equation:

$$\Delta i_q = -\frac{\Delta V}{k} \tag{3.122}$$

where Δi_q is the required change of IBDG-injected reactive current for voltage support and ΔV denotes the amount of voltage deviation from the nominal voltage. Moreover, k is a constant parameter, which is defined by the system

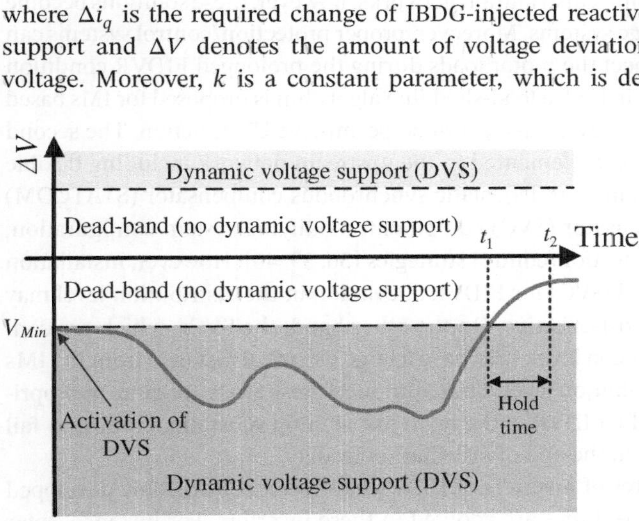

Fig. 3.44 Dynamic reactive current-based voltage support. [2] / with permission of IEEE.

operator. After the voltage recovery, when voltage deviation recovers within the permissible range, the voltage dynamic support could remain enabled for a dead-time, i.e., "Hold Time" shown in Fig. 3.44, and then, the dynamic voltage support will be disabled [2]. It should be noted that the IBDG has a ramp-rate limitation and, thus, restoring the IBDG active power input to its nominal value could not be performed instantaneously.

3.5.1 FIDVR Mitigation Using IBDGs

FIDVR is a condition in which the voltage magnitude remains low for a long duration and slowly recovers to its nominal range. This event is mainly caused by the dynamic interactions of the loads, especially IMs, after severe voltage drops such as short-circuit faults. The duration of FIDVR is typically between 3 and 20 s, which depends on various system parameters including IM characteristics, fault clearance time, magnitude of voltage sag, and etc. Stalling of IMs during the FIDVR phenomenon may result in both thermal damage and prolonged voltage recovery of IM. In addition, both SSSGs and IBDGs may become unstable during FIDVR, especially after the fault initiated islanding events [34, 35]. However, for the stable FIDVR phenomenon, DGs remain stable and IM rotor speed recovers to its nominal value.

Two major solutions are proposed in the literature to reduce the FIDVR duration and severity. The first category of solutions is mainly implemented at the customer level to enhance IM LVRT capability [36]. Customer-based solutions can be properly implemented in small distribution networks; however, these solutions become very expensive for large systems. Moreover, proper protection/control systems can be utilized to disconnect the motor loads during the prolonged FIDVR condition [36, 37]. For instance, in [38], a load-shedding algorithm is proposed for IMs based on the rate of voltage recovery using a slope-permissive UV function. The second category of solutions is implemented in the upstream network including flexible AC transmission system (FACTS), static synchronous compensator (STATCOM) and static VAr compensator (SVC) equipment, modifying system configuration, and application of protection/control strategies [36, 39, 40]. However, installation a large number of small SVCs for FIDVR alleviation in the distribution level may not be practical and cost-effective. On the other hand, the SVCs, which are connected to the transmission level, may have a large electrical distance from the IMs connected to the distribution level. Thus, although these SVCs are quite appropriate for mitigation of the FIDVR at the main distribution substation, they may fail to recover the voltage at the end of distribution feeders.

Recently, the features of inverters of IBDGs have been considerably developed and various smart capabilities are adopted in these inverters. For instance, solar inverters could be utilized to exchange reactive power with the upstream system

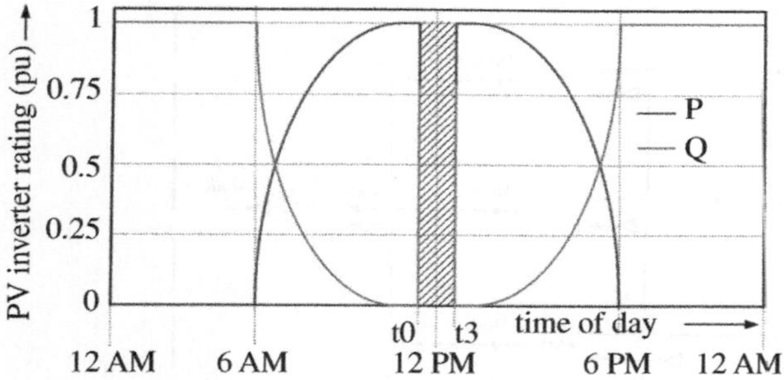

Fig. 3.45 Output power of a typical smart inverter. [40] / with permission of IEEE.

during night and help the voltage stability. The power output of a typical smart PV inverter is shown in Fig. 3.45 [36]. As shown, during night, the inverter exchanges reactive power with the system for voltage support. This dynamic voltage support of IBDGs could be effectively utilized to mitigate the FIDVR condition and helps the adjacent IM reacceleration. Since the X/R ratio in the distribution system levels is relatively low, the active power injection could also support the system voltage. To take the benefits of dynamic voltage support capability of IBDGs, the UV function of IBDG should be properly coordinated with the duration and magnitude of the FIDVR supposed to be common in a region. As a common practice, UV function is utilized as a backup function for islanding detection schemes, and the intentional time delay of this function is usually considered less than 2 s. Actual field measurements at distribution levels indicate that the voltage level could remain 0.8 pu or even lower for 5 s or longer [36]. In such a condition, the UV function prevents IBDG to support the system voltage during FIDVR. Thus, the intentional time delay of UV function should be revised to be coordinated with the FIDVR condition.

Various schemes have been proposed by research works to utilize IBDGs for mitigating the FIDVR condition and improving the system voltage stability [36, 40, 41]. It is worth mentioning that, since the solar inverter could be utilized as a STATCOM, this concept is known as PV-STATCOM. A practical configuration of a grid-connected PV-STATCOM proposed by [36] is shown in Fig. 3.46. As shown, the proposed controller is adopted to determine i_{dref} and i_{qref} for supporting the distribution system and IBMG during disturbances. For dynamic support during FIDVR, the PV-STATCOM controller includes three operation modes. In the normal operation, namely, *mode* 0, the output active power of the inverter is adjusted, considering the available active power ($P_{available}$). It should be noted that for providing frequency response, a headroom may be considered for active power. Moreover, the reactive power is adjusted by the IBDG and system operators. If a severe voltage drop takes place and the PCC voltage (V_{PCC}) becomes

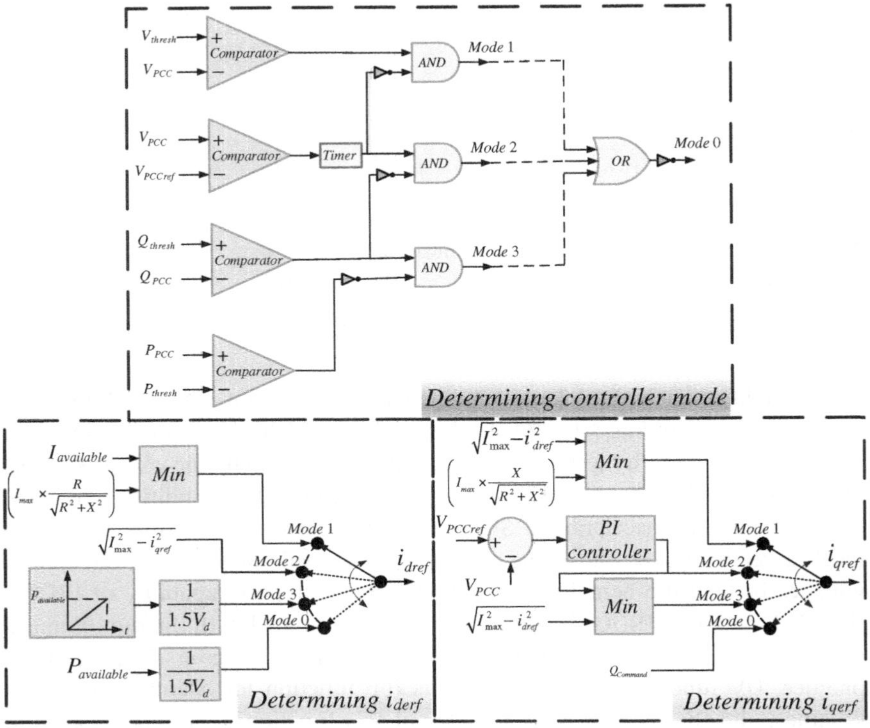

Fig. 3.46 PV-STATCOM structure. [40] / with permission of IEEE.

less than a predefined threshold (V_{thresh}), *mode* 1 will be enabled. Threshold V_{thresh} indicates the maximum voltage magnitude during LV events for which the IMs may stall and FIDVR condition occurs. This threshold is determined based on the distribution feeder specifications and operation experiences and is commonly considered between 70% and 80%. In *mode* 1, the inverter changes its output active and reactive powers to P_1 and Q_1, respectively. The amount of P_1 and Q_1 could be calculated using the system X/R ratio and the amount of voltage drop, as described in Chapter 2. If $P_1 < P_{available}$, the inverter output power will be limited to P_1, and its output reactive power is adjusted at Q_1. Moreover, if $P_1 > P_{available}$, the available active power will be generated by the inverter, and the remaining capacity will be exploited to inject reactive power to the system.

The inverter works at *mode* 1 till the short-circuit fault is cleared, and the system voltage increases above 1 pu at the IBDG location. When the system voltage increases, i.e., $V_{pcc} > V_{pccref}$, *mode* 1 will be disabled. In such a condition, reactive power should be injected by the inverter to help the IMs reaccelerate. Thus, in this regard, *mode* 2 is enabled in which the reactive power is still the first priority, similar to *mode* 1. In this mode, since the system voltage recovers, the amount of injected reactive power could be smaller than that of *mode* 1. Thus, the inverter has a larger capacity for injection of active power in this mode. *Mode* 2 operation remains enabled till the IMs recover to their nominal speed and reactive power of IBDG decreases. Thus, if $Q_{pcc} < Q_{thresh}$, *mode* 2 will be disabled. Parameter Q_{thresh} is considered 110% of the steady-state reactive power. After disabling *mode* 2, the inverter shall restore its active power output to at least the predisturbance active power, if the primary solar energy source is available. To do so, *mode* 3 is initiated in which the active power of the inverter linearly increases with a constant rate within a prespecified period t_1. After restoring the inverter output active power, *mode* 3 will be disabled and the inverter shall return to the normal operation mode, i.e., *mode* 0.

The reference currents of the inverter, i.e., i_{dref} and i_{qref}, are calculated at each mode separately. For the first mode, i.e., *mode* 1, these references could be calculated by [36]

$$i_{dref} = \min\left(I_{available}, \mathrm{I}_{max} \times \frac{R}{\sqrt{R^2+X^2}}\right)$$

$$i_{qref} = \max\left(\sqrt{\mathrm{I}_{max}^2 - i_{dref}^2}, \mathrm{I}_{max} \times \frac{X}{\sqrt{R^2+X^2}}\right) \tag{3.123}$$

where I_{max} is the inverter rated current. In the second mode, i.e., *mode* 2, a PI controller is utilized to regulate V_{pcc} at V_{pccref} by determining the proper amount of i_{qref}. Then, the remaining capacity of the inverter is exploited for the injection of i_{dref}. Moreover, for *mode* 3, the priority is injection of the available active power, and thus, the remaining capacity is utilized for reactive power injection:

$$i_{qref} = \min\left(\sqrt{\mathrm{I}_{max}^2 - i_{dref}^2}, \mathrm{I}_{VC}\right) \tag{3.124}$$

where I_{VC} is the reactive current generated by the PI controller to adjust the PCC voltage at V_{pccref}.

To evaluate the effectiveness of the PV-STATCOM in mitigating FIDVR and supporting the restoration of IMs, a hybrid distribution system, shown in Fig. 3.47, is modeled and simulated using the PSCAD/EMTDC software. The system is connected to the subtransmission level via two transformers (Tr1 and Tr2) and includes a 500-kW SSSG, a 2-MW IBDG, and some IMs and static loads. System parameters are detailed in Table 3.4. Meanwhile, the impedance of the line per unit length is considered 0.3 + j0.4 Ω/km.

A three-phase fault is applied at $t = 8$ s, and cleared after 300 ms, as indicated in Fig. 3.47. The rotor speed of the IMs, with and without using the proposed

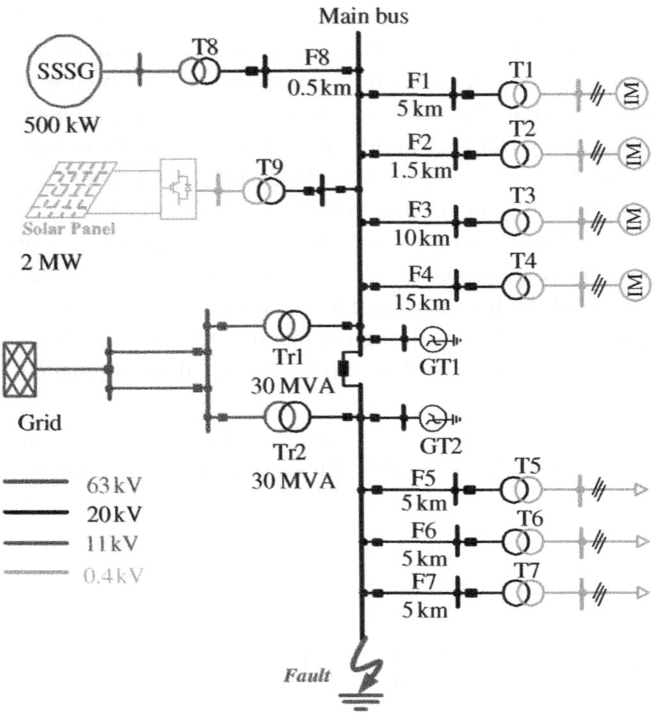

Fig. 3.47 Hybrid distribution system.

Table 3.4 Parameters of the system shown in Fig. 3.47.

Transformer	Impedance (%)	Ratio (kV)	Connection	Rated power (MVA)
Tr1, Tr2	12.5	63/20	YNd	30
T1–T7	6	20/0.4	Dyn	3
T8	6	20/0.4	Dyn	0.6
T9	6	20/0.4	Dy	2.5

Induction motor ID	Inertia (s)	Torque (pu)	Rated power (MW)
IM1	0.4	$0.3\omega + 0.1$	3
IM2	0.45	0.38	5
IM3	0.25	$0.5\omega^2 + 0.1$	2
IM4	0.15	$0.72\omega + 0.1$	1

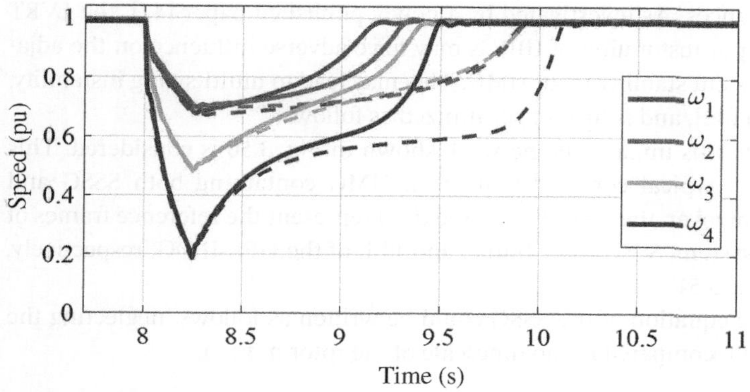

Fig. 3.48 Rotor speed of induction motors with and without using PV-STATCOM (solid line: using reactive capability of PV-STATCOM and dashed line: without using reactive capability of PV-STATCOM).

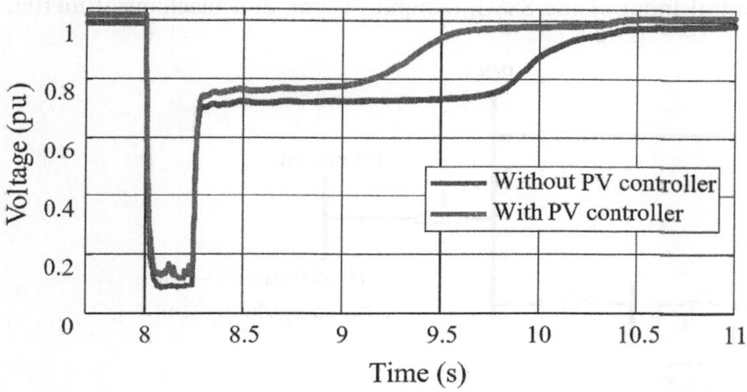

Fig. 3.49 Voltage magnitude of the main bus of the distribution system.

PV controller, is illustrated in Fig. 3.48. Results show that the PV-STATCOM significantly improves the reacceleration of the IMs and reduces the associated recovery time. Furthermore, the voltage profile of the main bus is presented in Fig. 3.49. The results demonstrate that the PV-STATCOM effectively reduces the FIDVR duration and enhances the voltage recovery of the system.

3.6 Influence of IBDG on Adjacent SSSG Transient Stability

In modern MGs, SSSGs and IBDGs may operate in parallel, and the behavior of their control loops may adversely influence the MG transient stability, especially during disturbances. As investigated by a newly published paper [42], the LVRT and power output restoration of IBDGs may have adverse influence on the adjacent SSSG transient stability in an HMG. This may lead to multi-swing instability, as discussed in [42], and is further scrutinized as follows.

To investigate this impact, the network shown in Fig. 3.50 is considered. This figure depicts a typical configuration of an HMG containing both SSSG and GFL-IBDG. Moreover, the $x-y$, d_s-q_s, and $d-q$ represent the reference frames of the upstream system, SSSG rotor frame, and PLL of the GFL-IBDG, respectively, as shown in Fig. 3.51.

The dynamic equation of the SSSG could be written as follows, neglecting the dynamics of PLL compared to the timescale of the rotor motion:

$$\begin{aligned} \dot{\delta} &= \omega_g(\omega_s - 1) = \omega_g \Delta\omega \\ T_J \Delta\dot{\omega} &= P_M - P_E - D\Delta\omega \end{aligned} \tag{3.125}$$

where parameters P_E, P_M, D, and T_J are the output active power of the SSSG, mechanical input of the SSSG, damping factor, and mechanical inertia,

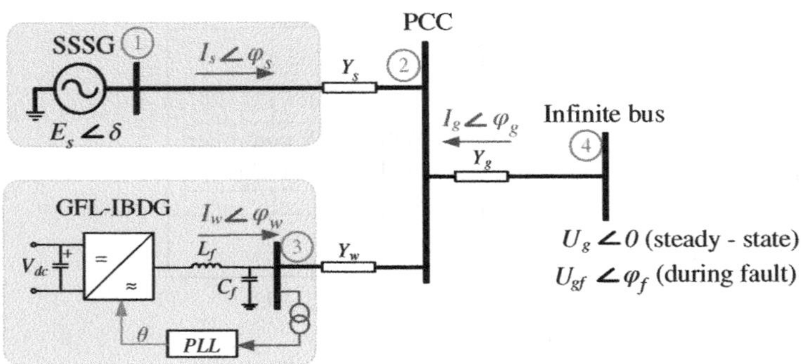

Fig. 3.50 Hybrid MG contain SSSG and GFL-IBDG [42].

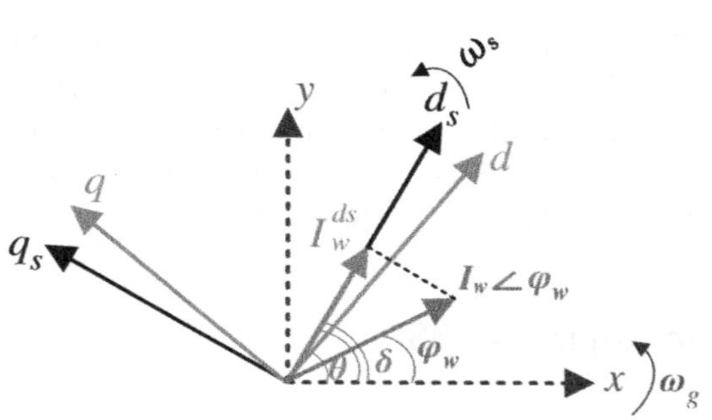

Fig. 3.51 $x-y$, d_s-q_s, and $d-q$ representing the reference frames. [42] / with permission of IEEE.

respectively. Moreover, ω_g and ω_s are the grid and SSSG angular frequencies, respectively. P_E can be calculated by

$$P_E = \mathrm{Re}(E_s\angle\delta . I_s\angle\varphi_s) \tag{3.126}$$

Consequently,

$$P_E = \underbrace{E_s U_g |Y_{sg}| \sin\delta}_{Psg} - \underbrace{\alpha E_s I_w \cos(\delta - \varphi_w)}_{Pw} \tag{3.127}$$

where $\alpha = Y_s/(Y_s + Y_g)$, $Y_{sg} = Y_s Y_g/(Y_s + Y_g)$. The term P_w could be considered as a coupling term between GFL-IBDG and SSSG. This term depends on the magnitude and angle of the injected current by GFL-IBDG, i.e., I_w. Therefore, changing the injected current by the LVRT scheme will change P_w and may have positive or negative impacts on the SSSG transient stability. Moreover, the equilibrium point of the SSSG is calculated by [42]

$$\delta_s = \arcsin\left(\frac{P_M + P_w}{E_s U_g \mid Y_{sg} \mid}\right) \tag{3.128}$$

Energy function of the SSSG is adopted as follows to investigate its transient stability:

$$V(\delta, \Delta\omega) = \frac{1}{2} T_J \omega_g \Delta\omega^2 + \int_{\delta_s}^{\delta} (P_{sg} - P_w - P_M) d\delta \tag{3.129}$$

After the fault occurrence, the grid voltage becomes $U_{gf}\angle\varphi_f$. Moreover, P_{sg} and P_w change to P_{sgf} and P_{wf} after the fault occurrence. Consequently,

$$P_{sgf} = E_s U_{gf} \mid Y_{sg} \mid \sin(\delta + \varphi_f), \quad P_{wf} = \alpha E_s I_{wf}^{ds} \tag{3.130}$$

where $I_{wf}^{ds} = I_{wf} \cos(\delta - \varphi_{wf})$ is the projection of GFL-IBDG active current (I_w^d) on the d_s axis. The $P - \delta$ curves of the SSSG before and after the fault occurrence, i.e., $P_{sg}(\delta)$ and $P_{sgf}(\delta)$, are shown in Fig. 3.52. Point A is the SEP whose power angle is equal to δ_s. Once a fault takes place, the $P - \delta$ curve changes from $P_{sg}(\delta)$ to $P_{sgf}(\delta)$ and the operation point moves from A to B. After the fault occurrence, the LVRT capability of the GFL-IBDG is enabled and the reactive current (I_w^q) increases and hence the active current (I_w^d) decreases. Thus, the projection of GFL-IBDG active current on the d_s axis, i.e., $I_{wf,}^{ds}$ decreases. Accordingly, the amount of P_{wf} becomes less than P_w $(P_{wf} < P_w)$. After the fault occurrence, if $P_{sgf}(\delta) < P_M + P_{wf}$, the SSSG will accelerate $(\Delta\omega > 0)$, and if $P_{sgf}(\delta) > P_M + P_{wf}$, the SSSG decelerates $(\Delta\omega < 0)$. Decrement of the rotor angular frequency, i.e., $\Delta\omega < 0$, during the first swing cannot guarantee the stable operation of the SSSG due to the GFL-IBDG behavior. Consequently, to evaluate the influence of the GFL-IBDG on the SSSG stability,

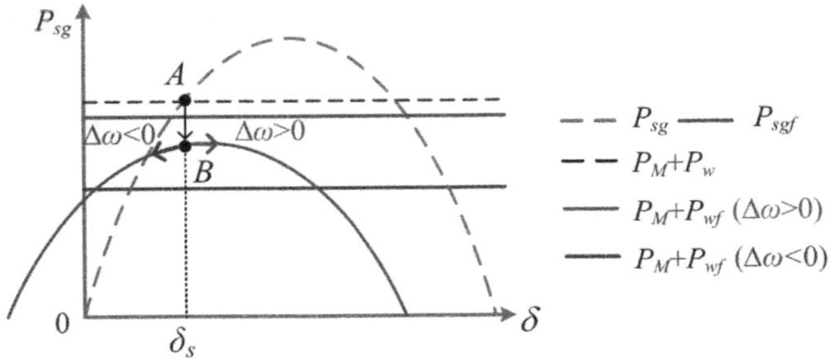

Fig. 3.52 $P-\delta$ curves of the SSSG. [42] / with permission of IEEE.

the derivative of the SSSG energy function $V(\delta, \Delta\omega)$ is exploited as follows:

$$\frac{dV(\delta,\Delta\omega)}{dt}=\frac{\partial V}{\partial \delta}\times\frac{d\delta}{dt}+\frac{\partial V}{\partial \Delta\omega}\times\frac{d\Delta\omega}{dt}=\omega_g(\Delta P_{wf}\Delta\omega-\Delta P_{sgf}\Delta\omega-D\Delta\omega^2) \tag{3.131}$$

where $\Delta P_{sgf}=P_{sgf}-P_{sg}$ and $\Delta P_{wf}=P_{wf}-P_w$.

The term $\Delta P_{wf}.\Delta\omega$ indicates the impact of the LVRT scheme of the GFL-IBDG on the energy function of the SSSG. If this term becomes positive ($\Delta P_{wf}.\Delta\omega>0$), the GFL-IBDG increases the amount of energy function, and thus, it has a negative impact on the SSSG transient stability. On the other hand, negative value of this term ($\Delta P_{wf}.\Delta\omega<0$) reduces the amount of V, and thus, it has a positive damping influence on the SSSG transient stability. Since $\Delta P_{wf}<0$, this term becomes positive for $\Delta\omega<0$ and negative for $\Delta\omega>0$. Therefore, during acceleration period of the SSSG in the first swing after the fault occurrence, i.e., $\Delta\omega>0$, the LVRT scheme has a positive damping impact and helps the SSSG to damp the accelerating energy. However, when SSSG decelerates, i.e., $\Delta\omega<0$, the LVRT scheme has a negative damping impact and increases the SSSG energy during the fault. These positive and negative impacts are depicted in Fig. 3.53(a) and (b), respectively. After the fault clearance, the LVRT should be disabled, and the active output of the GFL-IBDG should be restored.

Commonly, restoring the output active power of the GFL-IBDG to its nominal value has a ramp rate limitation, as shown in Fig. 3.54, and thus, the output active power of the GFL-IBDG gradually increases [42]. In such a condition, the term P_w after the fault clearance, namely, P_{wr}, increases gradually to reach its steady-state value and thus $P_{wf}<P_{wr}<P_w$. During the LVRT recovery stage, the derivative of the SSSG energy function can be expressed as

$$\dot{V}(\delta,\Delta\omega)=\Delta P_{wr}\omega_g\Delta\omega-D\omega_g\Delta\omega^2 \tag{3.132}$$

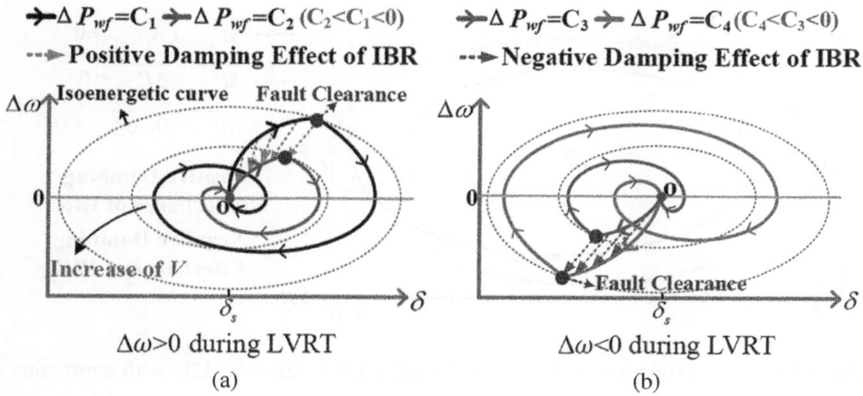

Fig. 3.53 $\Delta\omega - \delta$ trajectory of SSSG. [42] / with permission of IEEE.

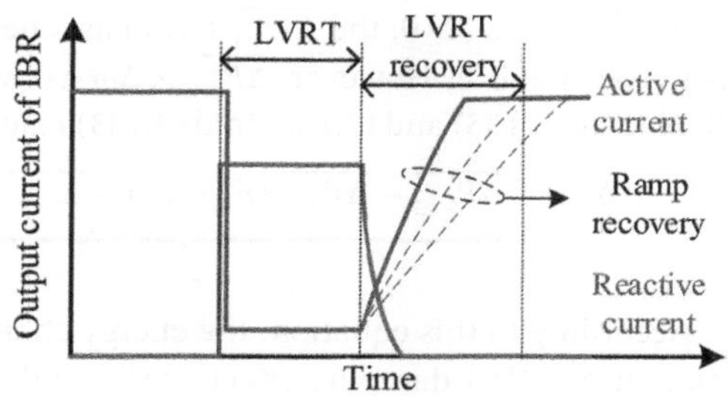

Fig. 3.54 Ramp rate of IBDG active power recovery. [42] / with permission of IEEE.

The amount of energy change during the LVRT recovery cycle can be calculated as

$$\Delta V = \int_{V_0}^{V} dV = \underbrace{\int_0^T (\Delta P_{wr}\omega_g \Delta\omega)dt}_{\Delta V_w} - \underbrace{\int_0^T (D\omega_g \Delta\omega^2)dt}_{\Delta V_D} \tag{3.133}$$

where V_0 is the initial energy at the beginning of the LVRT recovery stage and $\Delta P_{wr}(t) = P_{wr}(t) - P_w$. Thus, V can be determined by

$$V = V_0 - V_D + \omega_g \int_{\delta_A}^{\delta_B} \Delta P_w(t)d\delta + \omega_g \int_{\delta_B}^{\delta_C} \Delta P_w(t)d\delta \tag{3.134}$$

The term $\Delta P_{wr}(t)$ is negative during the restoration of the IBDG active power output. Besides, considering the "Mean Value Theorem" yields

$$\int_{\delta_A}^{\delta_B} \Delta P_w(t)d\delta = \overline{\Delta P_{wr1}}(\delta_B - \delta_A) \tag{3.135}$$

$$\int_{\delta_B}^{\delta_C} \Delta P_w(t)d\delta = \overline{\Delta P_{wr2}}(\delta_C - \delta_B) \tag{3.136}$$

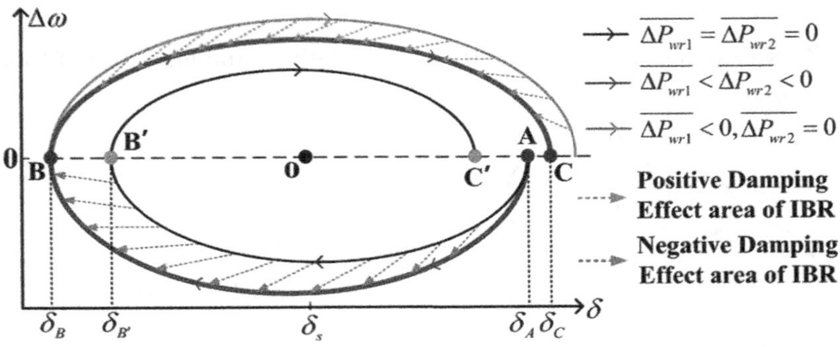

Fig. 3.55 $\Delta\omega - \delta$ trajectories of SSSG during IBDG LVRT recovery. [42] / with permission of IEEE.

where $\overline{\Delta P_{wr1}}$ and $\overline{\Delta P_{wr2}}$ are constant and also $\overline{\Delta P_{wr1}} < \overline{\Delta P_{wr2}} < 0$. Based on (3.132), since $\overline{\Delta P_{wr1}} \times \Delta\omega > 0$, the $\overline{\Delta P_{wr1}}$ term has a negative damping influence traveling between A and B. However, $\overline{\Delta P_{wr2}} \times \Delta\omega < 0$, which reduces the energy function. Replacing (3.135) and (3.136) into (3.133) results in

$$\Delta V = \underbrace{(\overline{\Delta P_{wr2}} - \overline{\Delta P_{wr1}})(\delta_A - \delta_B) + \overline{\Delta P_{wr2}}(\delta_C - \delta_A)}_{\Delta V_w} - \Delta V_D \tag{3.137}$$

According to this equation, the energy change ΔV is calculated from the summation of IBDG damping effect (ΔV_w) and SSSG mechanical damping (ΔV_D). If $\delta_A < \delta_C$, ΔV is greater than zero. Since $\Delta V_D > 0$, ΔV_w will also be positive. If $\delta_A > \delta_C$, then $\Delta V_w > 0$ because $\delta_A > \delta_C$. This reveals that the IBDG has considerable adverse impact on the transient stability when $\Delta\omega < 0$. The phase trajectory of the SSSG during IBDG LVRT recovery is shown in Fig. 3.55. As shown, negative value of $\overline{\Delta P_{wr1}}$ results in the increment of energy function during the A–B transition, which may lead to the SSSG instability. As a conclusion, the time-varying damping impact of the IBDG during angle-decreasing swings may lead to instability in the subsequent swings.

3.7 Summary

In this chapter, the configurations of various types of GFM- and GFL-IBDGs were first introduced. Subsequently, different forms of stability in IBMGs were summarized, with a particular emphasis on transient and small-signal stability due to their critical importance. The chapter provided an in-depth discussion on the transient stability of various IBDG types, highlighting their dynamic behavior. The findings from these analyses offer valuable insights for the design and tuning of protection systems. A key conclusion is that while certain types of IBDGs can maintain stability even under severe fault conditions, others may become unstable rapidly. As such, the implementation of out-of-step protection relays may be essential for some IBDGs, but not for others.

The small-signal stability of GFL-IBDGs was also examined, and an appropriate linearized model was presented. Furthermore, the chapter addressed the ride-through capabilities of inverters during system disturbances, an essential feature that supports voltage recovery and helps prevent unnecessary load shedding. However, while ride-through functionality can enhance voltage stability, it may negatively affect the transient stability of neighboring SSSGs. Therefore, enabling ride-through capabilities should be based on a comprehensive stability assessment to ensure overall system resilience.

Problems

3.1 A 3 MW PV plant that operates as a GFL-IBDG is illustrated in Fig. 3.56. The three-phase short-circuit level of the grid is assumed 10 kA, and the X/R ratio is considered 3. A solid three-phase short-circuit occurs at high-voltage terminal of the transformer. This fault takes place at $t = 30$ s and is cleared at $t = 30.5$ s. Investigate the PLL stability during fault and after fault clearance using simulation studies. The amounts of integral and proportional gains of the PLL are considered 0.1 and 24.7 pu, respectively. Repeat this example if the fault is cleared at 30.1 s and the fault impedance is 3 Ω.

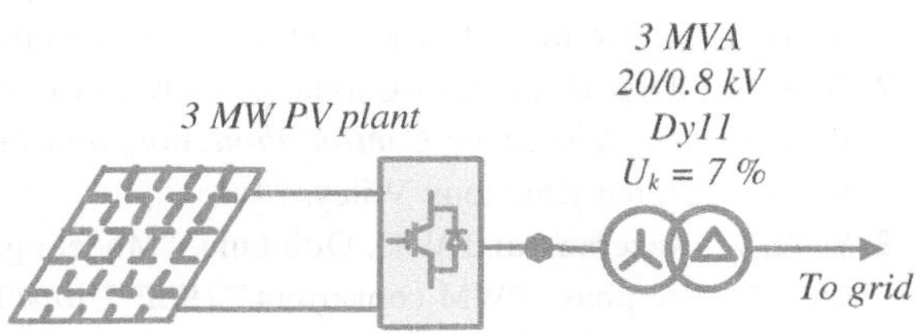

Fig. 3.56 The PV power plant of Problem 3.1.

3.2 Two IBDGs whose rated power are 600 kVA and 2 MVA, and one 1 MVA SSSG are located in a grid-connected MG. A local load (3000 kW with 0.9 lag power factor) is supplied by these DGs and the power exchange with the upstream grid is zero. Find the amounts of active/reactive powers of these DGs, considering equal power sharing. For IBDGs, the droop coefficients of frequency/Watt and Volt/VAr control loops are 1% and 6%, respectively, and for SSSG, they are 3.5% and 3%, respectively.

3.3 For the distribution system shown in Fig. 3.47:

a) Determine the maximum FCT for which the PLL of the GFL-IBDG remains stable.

b) Discuss whether enabling a PV-STATCOM controller to inject reactive power during a voltage dip can enhance transient stability.

Use appropriate transient simulation software (e.g., MATLAB/Simulink, PSCAD, or DIgSILENT PowerFactory) to perform the analyses and support your conclusions.

References

1 M. Khan, W. Wu, and L. Li, "Grid-forming control for inverter-based resources in power systems: A review on its operation, system stability, and prospective," *IET Renew. Power Gener.*, vol. 18, no. 6, pp. 887–907, 2024.

2 IEEE Std 1547-2018, "IEEE standard for interconnection and interoperability of distributed energy resources with associated electric power systems interfaces," IEEE, 2018.

3 M. Kabalan, P. Singh, and D. Niebur, "A design and optimization tool for inverter-based microgrids using large-signal nonlinear analysis," *IEEE Trans. Smart Grid*, vol. 10, no. 4, pp. 4566–4576, 2019.

4 A. Yazdani and R. Iravani, *Voltage-sourced Converters in Power Systems: Modeling, Control, and Applications*. Hoboken, New Jersey: Wiley, 2010.

5 NERC (North American Electric Reliability Corporation) report, 2023 Southwest Utah Disturbance, April 10, 2023.

6 M. Jamil, M. Rizwan, and D. P. Kothari, *Grid Integration of Solar Photovoltaic Systems*. Boca Raton, FL: CRC Press, Taylor & Francis Group, 2018.

7 O. Anaya-Lara, D. Campos-Gaona, E. Moreno-Goytia, and G. Adam, *Offshore Wind Energy Generation: Control, Protection, and Integration to Electrical Systems*. United Kingdom: Wiley, 2014.

8 Y. Ye, M. Kazerani, and V. H. Quintana, "Modeling, control and implementation of three-phase PWM converters," *IEEE Trans. Power Electron.*, vol. 18, no. 3, pp. 857–864, 2003.

9 A. Singhal, T. L. Vu, and W. Du, "Consensus control for coordinating grid-forming and grid-following inverters in microgrids," *IEEE Trans. Smart Grid*, vol. 13, no. 5, pp. 4123–4133, 2022.

10 M.-A. Nasr and A. Hooshyar, "Power swing in systems with inverter-based resources - Part I: Dynamic model development," *IEEE Trans. Power Deliv.*, vol. 39, no. 3, pp. 1889–1902, 2024.

11 N. Tleis, *Power Systems Modelling and Fault Analysis: Theory and Practice*. Academic Press, Elsevier, 2019.

12 X. Wang *et al.*, "Grid-synchronization stability of converter-based resources—an overview," *IEEE Open J. Ind. Appl.*, vol. 1, pp. 115–134, 2020.

13 D. Pan, X. Wang, F. Liu, and R. Shi, "Transient stability of voltage-source converters with grid-forming control: A design-oriented study," *IEEE J. Emerg. Sel. Top. Power Electron.*, vol. 8, no. 2, pp. 1019–1033, 2020.

14 M. Farrokhabadi *et al.*, "Microgrid stability definitions, analysis, and examples," *IEEE Trans. Power Syst.*, vol. 35, no. 1, pp. 13–29, 2019.

15 M. G. Taul, X. Wang, P. Davari, and F. Blaabjerg, "An overview of assessment methods for synchronization stability of grid-connected converters under severe symmetrical grid faults," *IEEE Trans. Power Electron.*, vol. 34, no. 10, pp. 9655–9670, 2019.

16 Ö. Göksu, R. Teodorescu, C. L. Bak, F. Iov, and P. C. Kjær, "Instability of wind turbine converters during current injection to low voltage grid faults and PLL frequency based stability solution," *IEEE Trans. Power Syst.*, vol. 29, no. 4, pp. 1683–1691, 2014.

17 D. Dong, B. Wen, D. Boroyevich, P. Mattavelli, and Y. Xue, "Analysis of phase-locked loop low-frequency stability in three-phase grid-connected power converters considering impedance interactions," *IEEE Trans. Ind. Electron.*, vol. 62, no. 1, pp. 310–321, 2015.

18 X. Fu *et al.*, "Large-signal stability of grid-forming and grid-following controls in voltage source converter: A comparative study," *IEEE Trans. Power Electron.*, vol. 36, no. 7, pp. 7832–7840, 2020.

19 M. G. Taul, X. Wang, P. Davari, and F. Blaabjerg, "An efficient reduced-order model for studying synchronization stability of grid-following converters during grid faults," in *2019 IEEE 20th Workshop on Control and Modeling for Power Electronics (COMPEL)*, pp. 1–7, 2019.

20 X. He, H. Geng, and S. Ma, "Transient stability analysis of grid-tied converters considering PLL's nonlinearity," *CPSS Trans. Power Electron. Appl.*, vol. 4, no. 1, pp. 40–49, 2019.

21 Q. Hu, J. Hu, H. Yuan, H. Tang, and Y. Li, "Synchronizing stability of DFIG-based wind turbines attached to weak ac grid," in *2014 17th International Conference on Electrical Machines and Systems (ICEMS)*, pp. 2618–2624, 2014.

22 Q. Hu, L. Fu, F. Ma, and F. Ji, "Large signal synchronizing instability of PLL-based VSC connected to weak AC grid," *IEEE Trans. Power Syst.*, vol. 34, no. 4, pp. 3220–3229, 2019.

23 H. Wu and X. Wang, "Design-oriented transient stability analysis of grid-connected converters with power synchronization control," *IEEE Trans. Ind. Electron.*, vol. 66, no. 8, pp. 6473–6482, 2018.

24 D. Pan *et al.*, "Transient stability of voltage-source converters with grid-forming control: A design-oriented study," *IEEE J. Emerg. Sel. Top. Power Electron.*, vol. 8, no. 2, pp. 1019–1033, 2019.

25 A. Etemadi, S. M. Nezamabadi, and A. Tahour, "Small signal modeling and stability analysis of a grid-following inverter with inertia emulation," *IEEE Trans. Power Electron.*, vol. 37, no. 9, pp. 10448–10457, 2022.

26 M. Mirmohammadi and S. P. Azad, "Control and stability of grid-Forming inverters: A comprehensive review," *Energies*, vol. 17, no. 13, p. 3186, 2024.

27 T. Messo, R. Luhtala, A. Aapro, and T. Roinila, "Accurate impedance model of a grid-connected inverter for small-signal stability assessment in high-impedance grids," *IEEJ J. Ind. Appl.*, vol. 8, no. 3, pp. 488–496, 2019.

28 Y. Zhou *et al.*, "Small-signal stability assessment of heterogeneous grid-following converter power systems based on grid strength analysis," *IEEE Trans. Power Syst.*, vol. 38, no. 3, pp. 2566–2579, 2023.

29 W. Du, Y. Wang, Y. Wang, H. F. Wang, and X. Xiao, "Analytical examination of oscillatory stability of a grid-connected PMSG wind farm based on the block diagram model," *IEEE Trans. Power Syst.*, vol. 36, no. 6, pp. 5670–5683, 2021.

30 H. Wu and X. Wang, "Transient stability analysis of power-synchronized PLL-less grid-following inverters," *IEEE Trans. Power Electron.*, vol. 37, no. 6, pp. 6704–6714, 2022.

31 M. Z. Mansour, M. H. Ravanji, A. Karimi, and B. Bahrani, "Linear parameter-varying control of a power-synchronized grid-following inverter," *IEEE J. Emerg. Sel. Top. Power Electron.*, vol. 10, no. 2, pp. 2547–2558, 2022.

32 N. Mohammed, M. H. Ravanji, W. Zhou, and B. Bahrani, "Enhanced frequency control for power-synchronized PLL-less grid-following inverters," *IEEE Open J. Ind. Electron. Soc.*, vol. 4, pp. 189–204, 2023.

33 N. Mohammed, W. Zhou, and B. Bahrani, "Double-synchronous reference-frame-based power-synchronized PLL-less grid-following inverters for unbalanced grid faults," *IEEE Open J. Power Electron.*, vol. 4, pp. 474–486, 2023.

34 W. Zheng, P. Crossley, B. Xu, and H. Qi, "Transient stability of a distribution subsystem during fault-initiated switching to islanded operation," *Int. J. Electr. Power Energy Syst.*, vol. 97, pp. 418–427, 2018.

35 A. H. K. Alaboudy, H. H. Zeineldin, and J. L. Kirtley, "Microgrid stability characterization subsequent to fault-triggered islanding incidents," *IEEE Trans. Power Deliv.*, vol. 27, no. 2, pp. 658–669, 2012.

36 R. K. Varma and S. Mohan, "Mitigation of fault induced delayed voltage recovery (FIDVR) by PV-STATCOM," *IEEE Trans. Power Syst.*, vol. 35, no. 6, pp. 4251–4262, 2020.

37 H. Bai and V. Ajjarapu, "A novel load shedding strategy for mitigating fault-induced delayed voltage recovery," *IEEE Trans. Power Syst.*, vol. 26, no. 1, pp. 294–304, 2011.

38 S. M. Halpin, K. A. Harley, R. A. Jones, and L. Y. Taylor, "Slope-permissive under-voltage load shed relay for delayed voltage recovery mitigation," *IEEE Trans. Power Syst.*, vol. 23, no. 3, pp. 1211–1216, 2008.

39 B. Park and M. M. Olama, "Mitigation of motor stalling and FIDVR via energy storage systems with temporal logic specifications," *IEEE Trans. Power Syst.*, vol. 36, no. 2, pp. 1164–1174, 2020.

40 W. Wang and F. D. Leon, "Quantitative evaluation of DER smart inverters for the mitigation of FIDVR in distribution systems," *IEEE Trans. Power Deliv.*, vol. 35, no. 1, pp. 420–429, 2020.

41 G. Lammert *et al.*, "Impact of fault ride-through and dynamic reactive power support of photovoltaic systems on short-term voltage stability," in *2017 IEEE Manchester PowerTech*, 2017.

42 S. Yang *et al.*, "Multi-swing transient stability of synchronous generators and IBR combined generation systems," *IEEE Trans. Power Syst.*, vol. 40, no. 1, pp. 1144–1147, 2025.

4

Microgrid Protection System Challenges

4.1 Introduction

Microgrids (MGs) should be equipped with a proper protection system that prevents maloperation during nonfaulty abnormal conditions but provides a fast and reliable reaction for faults to prevent excessive mechanical and thermal damage to the system equipment. Additionally, to maintain system stability and facilitate system restoration, the protection system should be properly coordinated with the control system and remains stable during the ride-through period as required by the system operator. In this context, various research studies and standards have proposed different protection schemes in the literature. Accordingly, this chapter first presents the widely used configurations and schemes for protecting MG equipment, including feeders, loads, transformers, small-scale synchronous generators (SSSGs), and inverter-based distributed generators (IBDGs). The issues and challenges associated with the existing protection systems, which may jeopardize the stable operation of both synchronous generator-based MGs (SGBMGs) and inverter-based MGs (IBMGs), are then discussed and analyzed through appropriate simulation studies.

One of the key challenges requiring careful consideration is the false operation of protection systems during transitions between islanded and grid-connected modes. Another challenge highlighted in several investigations is the incorrect operation of some protection functions, such as overcurrent relays (OCRs), due to dynamic system phenomena like fault-induced voltage recovery (FIDVR). The major protection issues analytically examined in this chapter include:

- Miscoordination among protection devices
- Sympathetic tripping of healthy feeders
- Reduced sensitivity and reach of protection elements
- Unsuccessful reclosing operations

Furthermore, several additional issues related to DG protection functions are examined through both analytical assessments and simulation-based studies. Since MGs often rely on load-shedding schemes to maintain voltage, frequency, and transient stability, a brief overview of the widely used load-shedding approaches for MGs is also provided. The findings of this study indicate that conventional load-shedding methods, such as frequency- and voltage-based schemes, may be unsuitable for modern MGs due to issues such as low security, false or excessive load shedding, and considerable time delays.

4.2 Protection Configuration of MGs

The MG should be equipped with a proper protection system both to preserve its required stability and to prevent unwanted mechanical and thermal damages to the equipment. The protection system of MG can be divided into three main parts: (i) collector system protective system, (ii) protection relays of DGs (i.e., SSSGs and IBDGs), and (iii) protection relay of point of common coupling (PCC) [1].

The protection relays of DGs should properly detect their internal faults and also prevent their undesirable operation under some abnormal operating conditions such as over/under-frequency (OF/UF), over/under-voltage (OV/UV), overload, and loss of synchronism. Furthermore, they should provide proper backup protection for the collector system faults. The major task of the collector system protection is to clear the short-circuit faults between the point of connection (PoC) and PCC. Moreover, it operates as the backup protection for the short-circuit faults, which may occur in power equipment, such as transformers, overhead lines, cables, and loads. The collector system structure is similar to the distribution systems, and thus, the widely used protection configurations of distribution systems are commonly exploited for the MG's collector system, as well.

The protection relay at the PCC provides a suitable backup for the faults that take place in both the upstream system and the MG feeders. Moreover, this relay should detect the unintentional islanding condition of the MG. A detailed description of each part is provided in the following subsections.

4.2.1 Collector System Protection

The collector system is categorized as a distribution grid due to its structure and voltage level. Similar to a distribution system, the MG collector system includes various feeders and equipment, such as transformers, overhead lines, cables, and capacitor banks. Therefore, the common protection configuration of distribution systems is also employed for MG collector systems, which is described as follows.

4.2.1.1 Feeder Protection

Feeders are the main parts of the collector system, which can be defined as a set of power equipment including conductors, breakers, disconnectors, and measurement devices. Some feeders are responsible for connecting the incoming overhead lines, incoming cables, DGs, and transformers to the MG busbars, which are known as "incoming feeders." In fact, incoming feeders are mainly exploited to energize the MG's busbars from the upstream system and DGs. Furthermore, some other feeders are used to supply the MG local loads by connecting an energized busbar to the downstream power equipment, such as transformers. These feeders are called "outgoing feeders." The protection scheme of medium-voltage (MV), low-voltage (LV) distribution networks and MGs whose feeders have radial configuration is commonly designed based on the unidirectional power flow [1]. However, this assumption is not valid for the ring and mesh grids, and thus, directional protection schemes such as directional overcurrent protection may be exploited. It should be noted that the implementation of the directional schemes for MGs includes some struggles due to their higher required capital expenditure, and thus, the unidirectional configuration may be preferred in some cases. In the following subsections, the widely used protection schemes of MV and LV feeders are presented.

4.2.1.1.1 Medium-Voltage Feeder To isolate the fault and faulty region in an MV feeder of an MG, various devices can be exploited, including circuit breakers (CBs), load break switches (LBSs), reclosers, and fuses [2, 3]. Furthermore, a combination of the mentioned equipment can be utilized for some applications. CBs are normally utilized along with protection relays to isolate the MG faults. However, considering the high cost of CBs, the required capital expenditure for exploiting CBs along with different protection relays may not be reasonable in some applications [2, 3]. In such a condition, LBSs, which are economical switching devices, can be utilized instead of CBs. The main purpose of using LBSs is feeder sectionalizing and interrupting/switching the load current. Indeed, LBSs cannot interrupt the fault currents whose magnitude is larger than the LBS nominal current. Therefore, to clear severe short-circuit faults, a fuse should be exploited along with LBS. If the LBS is utilized without a fuse element, the upstream CB should always operate when a severe fault occurs at the downstream region of the LBS. In other words, for severe faults whose magnitude is larger than the LBS nominal current, the associated protection relay should remain stable and must prevent issuing the trip command to the LBS. However, for the faults whose magnitude is smaller than the LBS nominal current, the fault should be cleared by LBS before fuse blowing. In this regard, the intentional delay of the protection relay should be smaller than the minimum melting time (MMT) of the fuse.

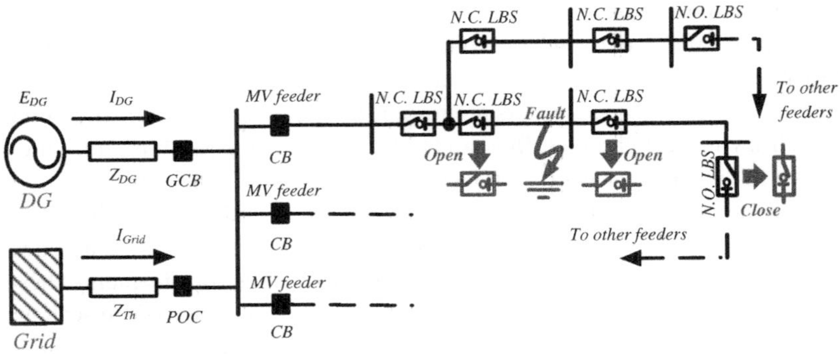

Fig. 4.1 Sample distribution system.

A sample protection configuration of a MV feeder is depicted in Fig. 4.1. In this configuration, CB is only exploited at the beginning of the main feeder and is not utilized for the laterals due to its high cost. For switching and protection of the laterals, LBSs without fuse elements are utilized. For this system, the following stages are involved to manage the "Fault" that occurs at laterals and its magnitude is larger than the LBS nominal current [2]: stage 1: CB is opened by the protection relay, stage 2: the faulty section is detected by the field crews and then isolated by operating the Normally Closed (N.C.) LBSs, stage 3: service is restored for the loads that are not located in the faulty section, and stage 4: load is transferred from the other feeders through the operations of Normally Open (N.O.) LBSs.

Based on the abovementioned explanations, various protection configurations are shown in Fig. 4.2, which can be categorized as [3]:

- ***Configuration (a):*** Using CB (or recloser) along with the protection relay
- ***Configuration (b):*** Using LBS along with the protection relay
- ***Configuration (c):*** Using LBS along with the protection relay and MV fuse element
- ***Configuration (d):*** Using MV cutout fuse.

As explained earlier, although the most suitable configuration from the technical point of view is using CBs, i.e., ***configuration (a)***, other configurations may also be exploited due to their financial benefits.

The minimum requirement of protection functions that should be included in the MV feeder protection relays are shown in Fig. 4.3. The brief explanations of the main functions are provided in the following subsections.

Phase Overcurrent Function (ANSI Code 51/50/67) The functionality of the overcurrent functions of MG feeders is mainly similar to that of similar OCRs exploited for common distribution feeders. The simple nondirectional overcurrent function

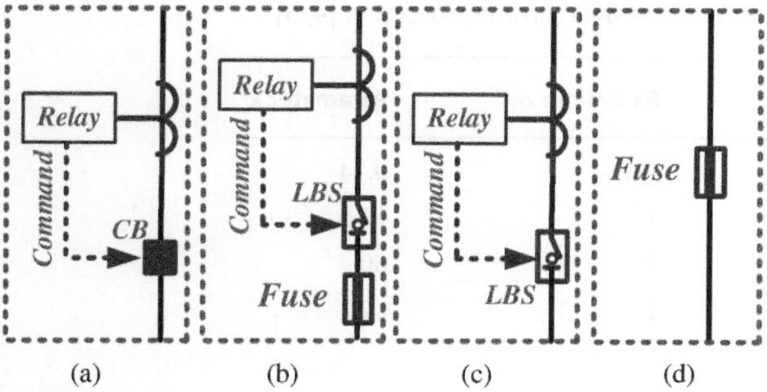

Fig. 4.2 Various protection configurations of SGBMG MV feeders.

Fig. 4.3 Minimum function requirement of MV feeder protection relays.

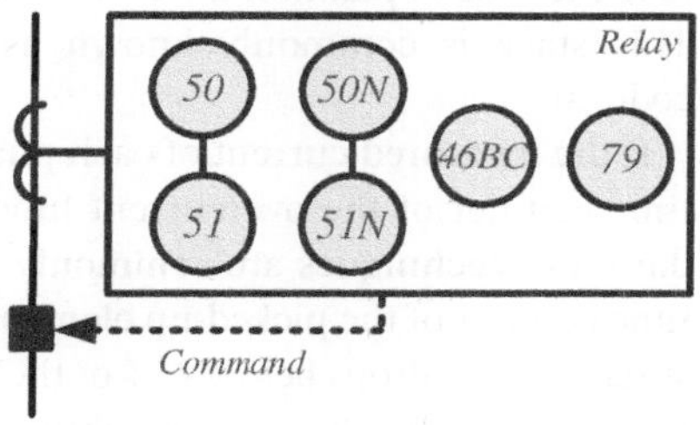

(ANSI code 50/51) measures the current signal of each phase through current transformers (CTs). Then, depending on the exploited signal processing algorithm, the main harmonic component or true-rms value of the measured signal is utilized to detect short-circuit faults on each phase of the collector system feeders. When the measured signal value becomes greater than a predefined threshold, known as the "pickup current," the relay will be triggered and its associated time counter will be started. Afterward, if the amount of timer becomes greater than an intentional time delay, the relay would issue the trip signal. Meanwhile, this intentional time delay (t_{op}) is calculated based on an inverse time–current curve for the ANSI code 51, whose equation is given by [4, 5]

$$t_{op} = \text{TMS} \times \frac{k}{\left(\dfrac{I}{I_{pickup}}\right)^{\alpha} - 1} \tag{4.1}$$

where variable I indicates the measured current signal. Moreover, TMS and I_{pickup} are constant parameters, which represent the time multiplier setting (TMS) and pickup current, respectively. Furthermore, parameters k and α have been used to specify the curve type. Based on the IEC standard, the parameters of widely used operation curves are given in Table 4.1. Various operation stages using different characteristic and parameter settings can be defined in the new digital

Table 4.1 Inverse operation curves of overcurrent protection [4, 5].

Curve \ Value	Parameter α	Parameter k
Standard inverse (SI)	0.02	0.14
Very inverse (VI)	1	13.5
Extremely inverse (EI)	2	80
Long-time inverse (LTI)	1	120

relays. Furthermore, the nondirectional overcurrent protection with constant or instantaneous operation time can also be included in digital OCRs, i.e., $t_{op} = cte.$ This stage is commonly known as the "definite/instantaneous stage (ANSI code 50)."

If the measured current of each phase drops below a predefined threshold, the time counter of the overcurrent function will be reset. Based on the IEC standard, two techniques are commonly exploited in digital OCRs to reset the relay time counter of the picked-up elements [6]. In the first technique, when the measured current drops below 95% of the pickup value, the time counter will be reset instantaneously. The second technique is performed by means of disk emulation. The disk emulation evokes a dropout process (i.e., decrementing of timer counter), which begins after the de-energization. This process can properly simulate the dropout inertia of the disk of conventional electromechanical relays, which is useful for coordination between digital and conventional relays. In this technique, the reset process begins when the measured current becomes less than 90% of the pickup current. Furthermore, the reset time t_{reset} is calculated by [4–6]

$$t_{reset} = \frac{t_r}{\left(\frac{I}{I_{pickup}}\right)^2 - 1} \tag{4.2}$$

where t_r is a constant parameter, whose value should be selected to achieve the most suitable coordination with the downstream conventional relays.

Nondirectional overcurrent functions, i.e., ANSI code 50/51, are applicable for radial and open-loop networks, which are supplied from a single source. For mesh and loop grids, directional overcurrent protection with ANSI code 67 should be utilized. The operation principles, including operation and dropout characteristics, are basically similar to those of the nondirectional overcurrent functions. Nevertheless, the directional overcurrent function provides bidirectional protection by utilizing the measured voltage signal from one phase or various phases. Therefore, it can be utilized for mesh grids in which the power flow direction may change. It should be noted that the setting values for each direction should be adjusted, considering the coordination between the relay and downstream protection devices. Since the directional overcurrent functions require voltage measurements, their implementation may be difficult or not feasible in some cases due to technical and financial challenges. In such a condition, the nondirectional overcurrent functions can be exploited with technical precautions. For SSSG protection purposes, the modified versions of these overcurrent functions are commonly exploited, which are thoroughly explained in Section 4.2.2.2.

Earth Overcurrent Functions (ANSI Code 51N/50N/67N) To detect ground faults at the MG collector system, e.g., single-phase to ground and phase-to-phase to ground faults, nondirectional or directional earth overcurrent functions are widely employed. The earth overcurrent element will be triggered if the amount of zero-sequence current (I_0) becomes greater than a predefined pickup threshold. Similar to the phase overcurrent function, to enhance the security of the earth overcurrent function, an intentional time delay is considered for its operation. For the ANSI code 51N, which represents the inverse nondirectional earth fault protection, the time delay t_{op} is determined by [4–6]

$$t_{op} = \text{TMS} \times \frac{k}{\left(\dfrac{I_0}{I_{pickup}}\right)^{\alpha} - 1} \tag{4.3}$$

where parameters k and α are defined in Table 4.1 for various curve types. Furthermore, constant time delay or instantaneous operation is used for definite nondirectional earth fault protection, which is known by the ANSI code 50N. To detect the direction of earth fault event, displacement voltage (V_0) and I_0 are exploited. Directional earth fault is represented by the ANSI code 67N.

To determine I_0, the following approaches can be exploited:

- **Direct measurement of I_0:** Using core-balance CT or neutral CT (NCT)
- **Indirect measurement of I_0:** Using hardwire summation of three-phase currents, as shown in Fig. 4.4
- **Analytical calculation of I_0:** Using the linear summation of three-phase currents.

Analytical calculation or indirect measurement of I_0 is commonly exploited when core-balance CT or NCT are not applicable at the MG feeders due to technical or financial reasons. The measurement error of each phase CT and also analog inputs of the relay may cause inaccuracy in determining the zero-sequence current through indirect measurement or analytical calculation of I_0. It is worth mentioning that the maximum amount of error in determining zero-sequence current through the indirect measurement and analytical calculation of I_0 is

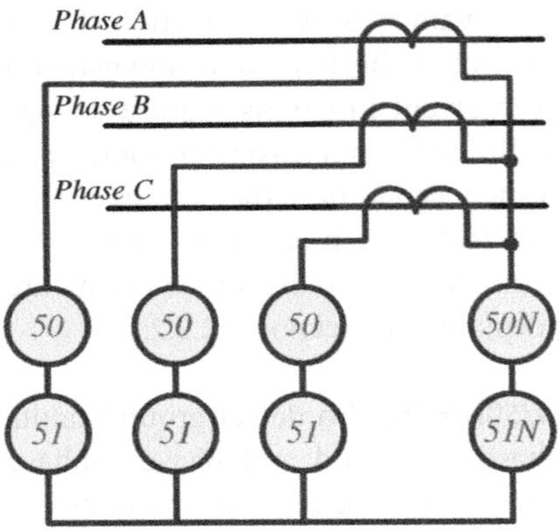

Fig. 4.4 Indirect measurement of I_0 component using hardwire summation.

about 10% and 15% of the line current, respectively [3]. Therefore, for more security, the pickup current of earth fault functions is commonly adjusted greater than 20% when the zero-sequence current is not determined directly. For the applications where more sensitivity is required for detecting high-impedance earth faults events, sensitive earth fault (SEF) protection is commonly exploited. For implementing SEF, direct measurement of I_0 should be utilized through core-balance CT or NCT. In this case, a separate sensitive analog input is exploited for the relay, which has lower error compared to other inputs. The pickup current of SEF function can be considered even less than 5% of the nominal current of NCT or core-balance CT [3–6].

Broken Conductor (ANSI Code 46BC) Open circuit faults may occur in distribution systems and SGBMGs due to various reasons, such as the operation of fuse elements. The open-circuit condition, also known as "broken conductor" condition, may cause various problems, including unbalanced operation of the system, ferroresonance occurrence in downstream transformers, considerable voltage drop, health hazards, etc. Therefore, the broken conductor condition should be detected quickly to prevent the abovementioned issues. The open-circuit faults will not lead to increment of current magnitude; hence, they cannot be detected by the OCRs. The ratio of negative to positive sequence current (I_2/I_1) is exploited to detect open-circuit faults [3].

4.2.1.1.2 Low-Voltage Feeders The primary purpose of LV feeder protection is to detect faults, i.e., short circuits and open circuits, within the LV portion of the system section including cables, connections, sealing ends, and other associated components. In addition, LV-side protection functions can operate as a backup to the DGs, including inverters and SSSGs. Depending on the load current amount, various types of CBs may be exploited to interrupt the fault current at the LV

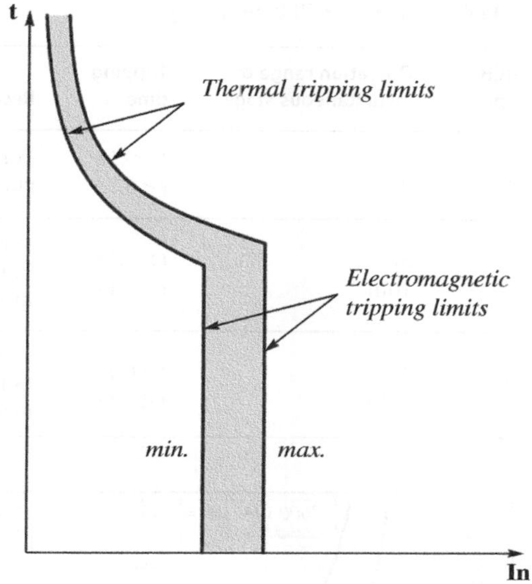

Fig. 4.5 Operation curve of MCB [7–9].

systems including (i) miniature circuit breaker (MCB), (ii) molded case circuit breaker (MCCB), and (iii) air circuit breaker (ACB) [7–9].

MCBs are commonly utilized for feeders whose current magnitudes are less than 125 A. The MCBs are equipped with two operation mechanisms, as shown in Fig. 4.5. The first one is a thermal mechanism, which provides delayed tripping with an inverse time–current characteristic. In addition, the second one provides instantaneous tripping based on an electromagnetic mechanism, which operates for high-magnitude severe short-circuit currents. It is worth mentioning that these stages cannot be adjusted by the users. Therefore, for each MCB, the manufacturer provides the operation band, i.e., the minimum and maximum boundary of the operation curve, as shown in Fig. 4.5.

The operation curves of MCBs are divided into three types as represented in Table 4.2. Based on the type of connected loads, the suitable operation curve of the MCB and its nominal current (I_n) should be selected. For a better visual comparison, different types of MCB curves are depicted in Fig. 4.6. As shown, type B curves operate instantaneously when the current magnitude becomes greater than $5 \times I_n$. Therefore, these curves may not be appropriate for protecting the feeders including high penetration of induction motors (IMs) due to their large starting current. Therefore, for such feeders, type D curves are commonly recommended [7–9].

MCCBs are commonly exploited for the feeders with a larger nominal current. Similar to MCBs, MCCBs have two operation mechanisms, including thermal and

Table 4.2 Various MCB types [7–9].

MCB type	Operation range of instantaneous stage	Tripping time	Recommended application
B	3In 5In	$t > 0.1$ s $t < 0.1$ s	For feeders and loads without inrush currents
C	5In 10In	$t > 0.1$ s $t < 0.1$ s	For feeders with small penetration of induction motors (small-magnitude inrush current)
D	10In 20In	$t > 0.1$ s $t < 0.1$ s	For feeders with high penetration of induction motors (large-magnitude inrush current)

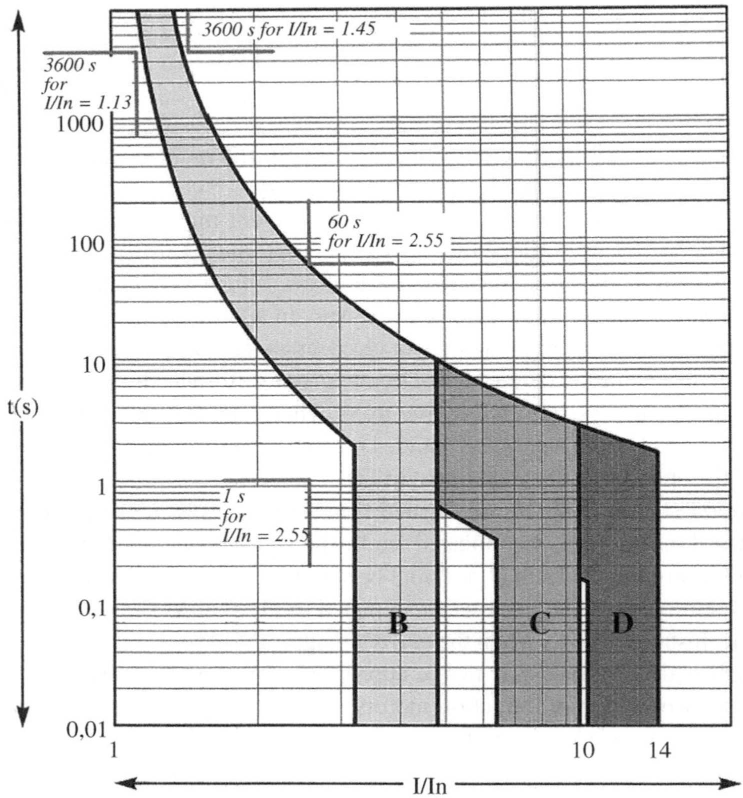

Fig. 4.6 Operation curves of various MCB types [7–9].

electromagnetic, which provide delayed and instantaneous tripping, respectively. The operation characteristic of the delayed tripping stage, which is implemented by a thermal bimetal, can be presented as

$$I^2 \times t_{op} = K \tag{4.4}$$

where K is a relatively constant parameter. In addition, variables I and t_{op} represent the MCCB current and operation time, respectively. Unlike MCBs, the operation characteristics of MCCBs can be adjusted by the user in some cases. In some models, only the pickup current of thermal and electromagnetic stages can be adjusted, while in others, the operation time and parameter K are also adjustable by the user. For instance, a sample type of MCCB in which the pickup current of both stages can be adjusted is depicted in Fig. 4.7. Furthermore, for more flexibility, some types of MCCBs are equipped with an electronic trip unit, which is commonly known as "ETU." The ETU can provide operation curves of IEC-60255-3 standard (instead of $I^2 \times t = K$), including standard inverse, very inverse, extremely inverse, and long-time inverse [7–10]. It should be noted that ETUs are commonly provided for MCCBs whose nominal currents are larger than 800 A.

ACBs are commonly utilized for industrial feeders with large nominal currents up to 6300 A. Various types of ACBs have been developed by the manufacturers. The common type includes an electronic tripping device that provides four protection elements, which is known as "LSIG" type [7–10]. The pickup current and operation time delay of each stage can be adjusted by the user. The sample LSIG operation characteristic is depicted in Fig. 4.8. Stage-L has a large intentional delay time with inverse time–current characteristic, which is suitable for overload protection. Stage-S is provided for short-circuit protection, which its operation curve can be set at both definite and inverse types. Furthermore, stage-I operates instantaneously for high-magnitude short-circuit faults, and finally, stage-G covers the ground faults whose operation curve can be definite or inverse. For some applications where more sensitivity is required, ACBs are commonly exploited along with

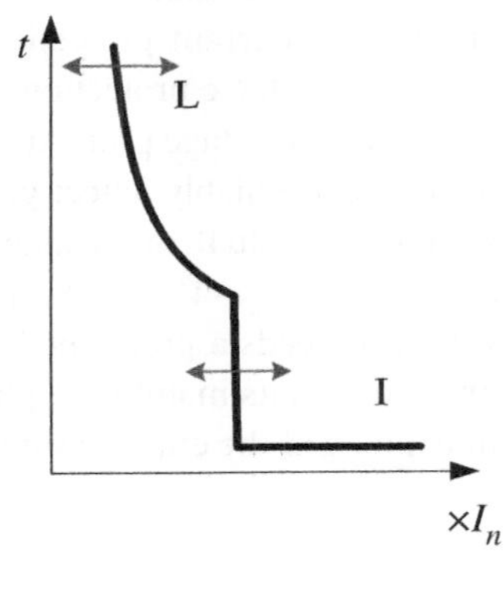

Fig. 4.7 Operation curve of MCCB, including adjustable instantaneous stage (I) and adjustable delayed stage (L).

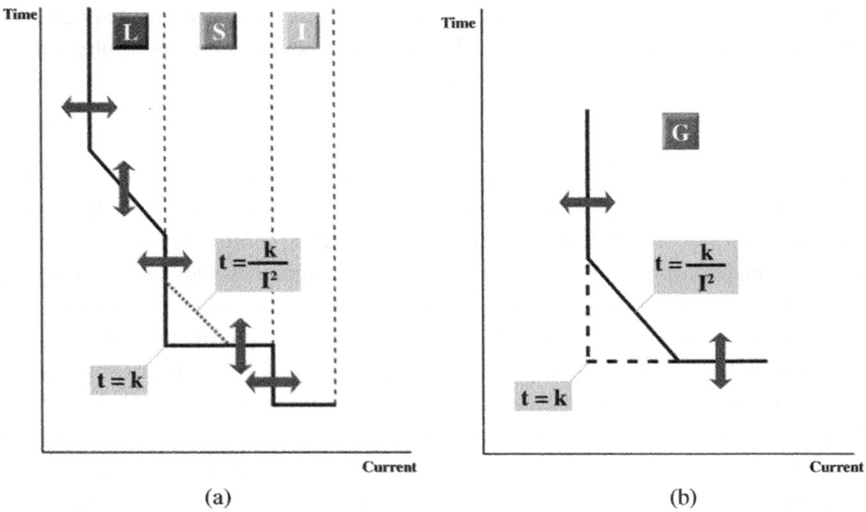

Fig. 4.8 LSIG characteristics of a sample ACB: (a) overload and phase fault protection and (b) ground fault protection [7–10].

digital relays. In these cases, LSIG units are not utilized anymore, and the applied digital relay issues the trip command to the ACB through the trip circuits.

For systems in which the LV system is not isolated and is effectively grounded, overcurrent protection is generally sufficient for fault detection. In such cases, MCCBs or ACBs equipped with appropriate overcurrent units could be used. However, when the LV system is not effectively grounded or is isolated, voltage-based protection functions, such as OV/UV and also residual-voltage protection, should also be employed.

In systems with a floating or noneffectively grounded neutral, the healthy phase voltages increase when a ground fault occurs on one phase. Under these conditions, the fault current may not rise substantially, making phase and ground overcurrent protection insufficient for detecting ground faults. In such cases, overvoltage protection can be employed to identify the fault. Additionally, residual overvoltage protection, also known as neutral voltage displacement protection, can reliably detect ground faults in both grounded and isolated-neutral systems. Residual overvoltage protection monitors the zero-sequence voltage (i.e., $3V_0 = V_a + V_b + V_c$) and issues a trip command when the zero-sequence voltage exceeds a predefined threshold. The primary drawback of voltage-based protection is its inability to pinpoint the specific faulty feeder, typically resulting in tripping of the entire system.

4.2.1.2 Transformer Protection

Transformers are used in MGs for three main purposes: (i) DGs connection to the grid, (ii) supplying the loads, and (iii) connection to the upstream grid. The main protection functions of the transformer MV side are illustrated in Fig. 4.9 [3, 11]. Regardless of the transformer vector group and winding configurations, commonly the overcurrent functions including 51/50/51N/50N/SEF/46BC are exploited to detect the transformer internal short-circuit faults, as shown in Fig 4.9. Since the principles of these functions have been discussed earlier, they are not discussed in this section anymore. Differential protection function is not widely used for MG transformers due to its high capital cost.

On the other hand, overload function (ANSI code 49) can be exploited to prevent probable damage to the transformer caused by thermal overloading. Generally, this overload function determines the transformer over temperatures based on the transformer thermal model using the following differential equation [11]:

$$\frac{d\Theta}{dt} + \frac{1}{\tau}\Theta = \frac{1}{\tau}\left(\frac{I}{I_{max}}\right)^2 + \frac{1}{\tau}\Theta_k \tag{4.5}$$

where Θ represents the transformer temperature, I is the rms value of the transformer current, I_{max} is the maximum permissible phase current, τ is the transformer thermal time constant, and Θ_k is the ambient temperature. The amounts of I_{max} and τ are commonly specified by the transformer manufacturers. However, if the values of these parameters are not available, the I_{max} amount can be properly considered 1.1 times the nominal current and parameter τ can be estimated as

$$\tau = \frac{1}{60} \times \left|\frac{I_{1\,s}}{\mathrm{I}_{max}}\right|^2 \tag{4.6}$$

where $I_{1\,s}$ is the maximum permissible transformer current for 1 s. The overload function can exploit either star- or delta-connected side signals of the transformer.

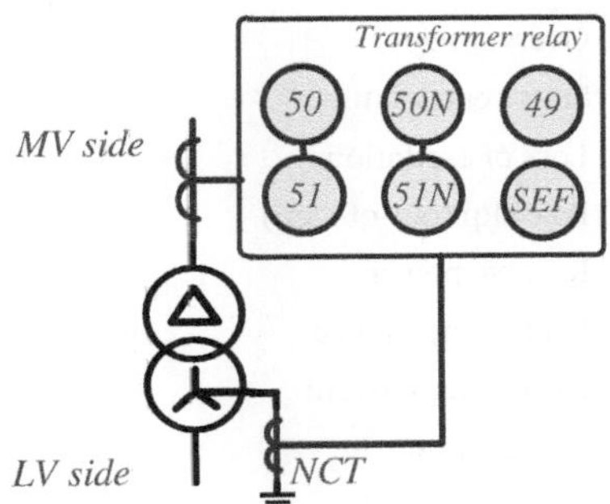

Fig. 4.9 Protection diagram of a distribution transformer. Adapted from [11].

For careful assessment of the transformer thermal condition, exploiting the line current of the star-connected side is more common than the delta-connected side. The main reason is that the zero-sequence current passes through the phase winding of the delta-connected side, while the line current of the delta-connected side does not contain any zero-sequence current. This zero-sequence current can cause thermal overload of the transformer, especially when an unbalanced load is connected to the transformer.

The combination of a digital relay and an ACB is commonly exploited for transformer LV-side protection. It should be noted that the LV- and MV-side protection functions can be integrated into one digital relay. Furthermore, to reduce the expenditures, MCCBs and ACBs may also be utilized to protect the LV side without utilizing digital relays.

4.2.2 SSSG Protection

SSSG protection scheme mainly depends on the corresponding synchronous generator (SG) scale [12–14]. The required functions for each range are presented in Table 4.3. Furthermore, the common protection diagram of SSSGs is illustrated

Table 4.3 Required protection functions of SSSGs [12–14].

Generator capacity / Protection	0–4 MVA	4–15 MVA	15–25 MVA
Rotor earth fault	×	×	×
Interturn fault		×	×
Differential (generator)		×	×
Overall differential (generator and transformer)		×	×
Underfrequency		×	×
Overvoltage	×	×	×
Stator earth fault	×	×	×
Loss of excitation		×	×
Pole slip (out-of-step)			
Reverse power	×	×	×
Under-impedance		×	×
Unbalance current		×	×
Overcurrent	×	×	
Stator overload		×	×
Overcurrent/undervoltage	×	×	
Dead machine			×

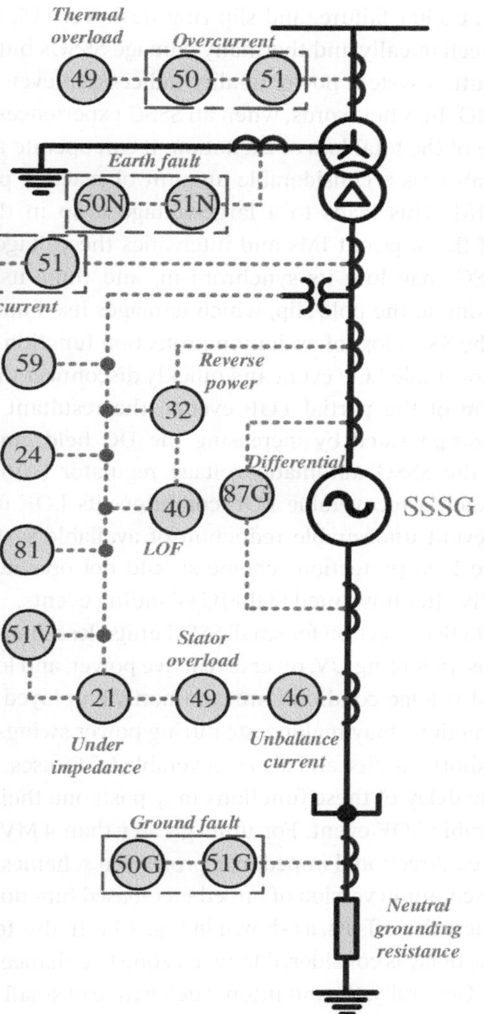

Fig. 4.10 Protection diagram of an SSSG [13].

in Fig. 4.10. Although SSSG relays contain various protection functions, only the main ones that may operate incorrectly during MG perturbations are investigated in the following subsections and the other functions are not covered.

4.2.2.1 LOF Function

One of the major circumstances for which SSSGs require careful attention is SG partial or total loss-of-field (LOF) condition. LOF events may take place due to the generator excitation system short-circuit or open-circuit condition, DC field breaker interruption, exciter failure, and slip ring flashover [15, 16]. This event may not only may mechanically and thermally damage SSSGs but also adversely influence the distribution system power quality indices and even can jeopardize the stability of SGBMG. In other words, when an SSSG experiences an LOF event, especially in the case of the total loss of excitation, it can operate as an induction generator and thus absorbs a considerable amount of reactive power from the upstream grid [15–18]. This leads to a large-voltage drop in the system that can cause stalling of the adjacent IMs and intensifies the voltage drop. In such a condition, the SSSG may lose its synchronism, and thus, its output power becomes oscillatory due to the pole slip, which damages itself and its associated prime mover [19]. The SSSG loss of excitation protection function should operate under such an unrecoverable LOF event and quickly disconnect the SSSG. On the other hand, for some of the partial LOF events, the resultant magnetic field reduction may be compensated by increasing the DC field current using the boost command of the SSSG automatic voltage regulator (AVR). Since SSSG can operate under such a recoverable LOF condition, its LOF function should remain stable to prevent undesirable reduction of available system generation [20]. Meanwhile, the LOF protection scheme should not operate during power oscillations, especially after unwanted SGBMG islanding events.

To provide proper LOF protection for small SSSG units (less than 4 MVA), simple protection algorithms, including UV, reverse reactive power, and low-field current along with abnormal voltage condition are commonly employed [12, 14]. These simple protection functions may maloperate during power swings after non-LOF events, such as grid short circuits, and also recoverable LOF cases. In addition, the long intentional time delay of these functions may postpone their fast operation during an unrecoverable LOF event. For units greater than 4 MVA, the methods used for large SGs, i.e., directional impedance protection schemes, are commonly utilized [12, 14]. The common version of impedance-based functions contains two circular zones with negative offsets, as shown in Fig. 4.11. In this technique, a separate intentional time delay is considered for each zone to enhance security during the non-LOF events. Generally, the intentional delay time of small and large zones is set at 100 and 500 ms, respectively.

4.2.2.2 Backup Protection

An SSSG supplies the out-of-zone faults, which occur at the SGBMG or upstream grids, until they are cleared by their associated main protection systems. These out-of-zone faults are also known as “system faults.” If the associated main protection system fails to clear these faults, the safety of SSSG may be threatened. Therefore, if the system faults cannot be cleared by the main protection system, the backup protection at the SSSG point must operate and disconnect the SSSG to

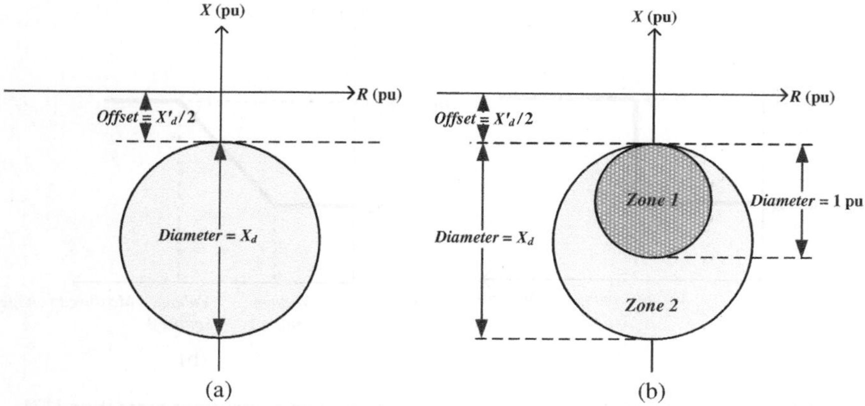

Fig. 4.11 Impedance-based operational characteristics proposed by (a) Mason and (b) Berdy. Adapted from [16].

prevent its mechanical and thermal damages. The widely used backup protection function for system faults is explained as follows.

The SSSG current magnitude varies during system faults due to the AVR response. At the first moment of fault occurrence, AVR sends a boost command to cover the SSSG fault ride-through capability, which maintains the fault current at a relatively high level. After a few seconds, i.e., 2–25 s depending on the SSSG size and AVR technology, the AVR output signal would reach its maximum permissible value, and thus, the fault current would be severely restricted, which results in slow operation of the SSSG backup protection for system faults. As the worst case, the fault current magnitude may drop below the SSSG full-load current, and thus, simple overcurrent protection functions, whose pickup settings are greater than the SSSG full-load current, cannot operate in this case.

System backup protection shall operate quickly during a system fault and must remain stable for nonfault conditions. To achieve these objectives, two techniques are commonly exploited including voltage-dependent overcurrent protection and under-impedance protection. These functions are briefly explained as follows [13–21]:

4.2.2.2.1 Voltage-dependent Overcurrent (ANSI Code 51V) This function provides a phase overcurrent function whose pickup current changes depending on the measured voltage. It may be fed from CTs, which are installed at the terminal or neutral end of the SSSG. Since the SSSG terminal voltage drops during short-circuit faults, a voltage-dependent element is exploited to reduce the pickup current by a factor of K after detecting UV at the terminals. This ensures that the decreased magnitude of SSSG current during system faults,

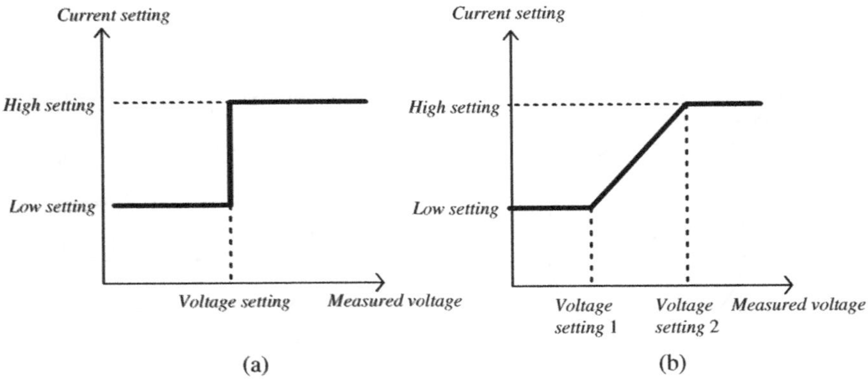

Fig. 4.12 (a) Voltage-controlled and (b) voltage-restrained overcurrent protection [13].

which is caused by the AVR response, is capable of triggering the overcurrent function. To implement the 51V function, a nondirectional overcurrent element is exploited whose operation time delay can be set at inverse time–current or definite time characteristics. Commonly, two types of 51V functions are exploited, which are known as "voltage-controlled" and "voltage-restrained" modes. The operation characteristics of voltage-controlled and voltage-restrained overcurrent functions are depicted in Fig. 4.12(a) and (b), respectively. If the SSSG is directly connected to a busbar, voltage-controlled overcurrent protection may be preferred [13].

4.2.2.2.2 Under-impedance (ANSI Code 21) Under-impedance protection is an alternative to voltage-dependent overcurrent protection, which exploits the impedance seen from the SSSG terminal to detect system faults. If the measured impedance becomes less than a predefined threshold, this function will operate after an intentional time delay. The under-impedance element may contain various stages with different settings and operation time delays. For SSSGs that are connected to the system through a transformer, two stages of impedance protection are commonly exploited. In this case, as recommended by [1], "one stage could be used to reach into the step-up transformer and one stage to reach further into the power system to provide 2 zones of protection." The first stage setting is commonly set at 70% of the maximum load impedance [21]. This provides a sufficient margin for short-time overloads and voltage variations while providing suitable backup protection for SSSG, its transformer, and system faults. In addition, the second stage setting can be set at 50–60% of the SSSG transformer impedance with a smaller time delay. This zone is mainly exploited to provide faster operation for faults closer to the SSSG. The time delay of these stages should be coordinated with the system overcurrent protection devices.

Nevertheless, providing proper coordination between the under-impedance and downstream inverse time overcurrent protections may be difficult in some cases.

4.2.2.3 Reverse Power Protection

In the conditions that the mechanical torque applied by the prime mover drops significantly and thus the delivered mechanical power becomes less than the total losses of the turbo-generator set, the SSSG will absorb some active power from the upstream grid to preserve its synchronism with the system. Therefore, in such a case, the SSSG operates as a synchronous motor. This condition is also known as the reverse power condition. Operation under such a condition does not have considerable adverse impacts on the SSSG if appropriate excitation, cooling, and bearing lubrication are provided. However, the prime mover may be seriously damaged in the reverse power operation of the SSSG. Therefore, the primary concern is the protection of the prime mover, which can be damaged during the reverse power operation [13].

The SSSG relays are commonly equipped with a reverse power function (ANSI code 32) to prevent the motoring operation of SSSG and the related prime mover damages. After detecting the reverse power condition, the responsible relay issues the trip signal to the SSSG main breaker, its field breaker, and prime mover. The amount of active power that the SSSG absorbs during the motoring operation is a function of the load and mechanical losses of the prime mover and generator, and thus, it depends on the type of prime mover, as presented in Table 4.4. Moreover, the damage severity can be different for various types of prime movers. Therefore, the setting values of the reverse power protection should be adjusted based on the prime mover type. For instance, if the prime mover is a gas turbine, the SSSG may consume up to 50% of its nominal rating due

Table 4.4 Motoring power of various prime movers [13].

Prime mover	Motoring power	Possible damage
Diesel engine	5–25%	Explosion and fire risk
Gas turbine	10–15% (split-shaft) >50% (single-shaft)	Mechanical damage due to the reverse torque
Hydraulic turbine	0.2–2% (blades out of water) >2% (blades in water)	Cavitation of blades
Steam turbine	0.5–3% (condensing sets) 3–6% (noncondensing sets)	Thermal stress damage

to the large compressor load of the turbine. In such a condition, the sensitivity of the reverse power function is not critical. On the other hand, the amount of reverse power for the SSSGs equipped with a steam turbine is considerably low and may be even about 0.5% of the nominal power. Therefore, the reverse power function setting should be selected accurately in this case. It should be noted that accurate measurement of low active power levels at low-power factors may not be feasible for some reverse power relays. In addition, for SSSGs coupled with some types of steam turbines, the generator may not be converted to the motoring mode even when the steam turbine valves are fully closed. Therefore, in some applications, a low value forward power function may be exploited instead of the reverse power function [13].

4.2.3 IBDG Protection

Each inverter is equipped with its own protection system on both the AC and DC sides to prevent operation under abnormal conditions and to avoid equipment damage. On the DC side, various protective functions are implemented, such as OV/UV, phase/ground overcurrent, over-temperature, insulation monitoring, and leakage current detection. Meanwhile, on the AC side, basic protection functions, typically OV/UV, phase/ground overcurrent, and OF/UF, are employed. These protection functions are primarily intended to prevent damage to the inverter; however, they should also ensure appropriate ride-through capability by allowing the inverter to remain connected to the grid for the required duration.

4.2.4 PCC Protection

Due to the reliability concerns, the MGs are commonly operated in the grid-connected mode under the normal operation condition. Furthermore, to guarantee the suitable power supply for the local loads during short-circuit faults at the upstream system, the MG will be automatically disconnected from the grid by the PCC protection relay. PCC protection relays should appropriately distinguish the upstream grid faults from the MG faults using proper setting values or directional elements. The PCC relay should be properly adjusted and coordinated with both the MG protection system and the upstream protection system. Considering these explanations, appropriate performance of the PCC relay for the grid and MG faults entails employment of an efficient fault detection method and protection strategy. The widely used protection functions at the PCC include [22]:

1) **Phase and ground overcurrent protection:** Phase and ground overcurrent functions have been explained in the previous sections.

2) **Voltage protection:** Voltage protection includes OV (ANSI code 59) and UV (ANSI code 27) functions. If the measured voltage becomes larger than a predefined threshold, i.e., pickup setting, the OV function will be triggered. In addition, if the measured voltage drops below a pickup setting, the UV function will be activated. To prevent relay maloperation during system transients and improve its security, predefine intentional delays are considered for voltage functions. In addition, residual-voltage protection (59N) is recommended to be used to detect ground faults. As explained earlier, this function is activated when the zero-sequence voltage exceeds a defined threshold during ground faults. This protection is particularly useful for detecting ground faults in situations where the fault current is low and zero-sequence overcurrent protection fails to identify the fault.
3) **Islanding detection protection:** Various islanding detection protection schemes will be thoroughly discussed in the next chapter of this book.

4.2.5 Load-Shedding Protection

Load-shedding protection is commonly exploited to maintain the MG stability and preserve the balance between the DG output power and local load demand by disconnecting a portion of local loads, especially after unwanted islanding events. In this regard, underfrequency load-shedding (UFLS) and undervoltage load-shedding (UVLS) functions are traditionally exploited in MGs due to their simple implementation and cost benefits. These functions are also known as conventional load-shedding functions whose fundamentals have been explained in the following paragraphs [23].

In an islanded MG, if the consumed active power of the local load becomes greater than the active power generation, the MG frequency and voltage drop. The rate of frequency drop and its final value depend on various factors including (i) MG's inertia, (ii) active power mismatch amount, and (iii) dependency of the load active power on the MG frequency. To detect such an abnormal condition and prevent frequency instability of the MG, UFLS relays have been traditionally exploited in MGs. In this method, when the MG frequency becomes lower than a predefined threshold, the UFLS relay will be triggered and its associated time counter is started. Afterward, if the amount of timer becomes greater than an intentional time delay, the relay would issue the trip signal and disconnect some prespecified loads. This intentional time delay is mainly considered to avoid UF relay maloperation during power swings [23, 24].

The MG voltage magnitude may also drop when the amount of reactive or active power generation becomes less than the load demand. Therefore, UVLS relays may also be exploited to prevent severe voltage drop and maintain the MG voltage stability. If the voltage drops below a predefined threshold, the UVLS relay will be triggered, and after an intentional time delay, the tripping command is sent to some prespecified loads. For both the UVLS and UFLS functions, which are also known as conventional load-shedding functions, the locations of the candidate loads are specified by the MG operator, considering the various factors such as the importance of loads and penalty cost. For instance, crucial loads should not be disconnected by load-shedding functions.

The algorithm flowchart of conventional approaches is depicted in Fig. 4.13. If the first shed is insufficient and thus voltage and frequency are not recovered

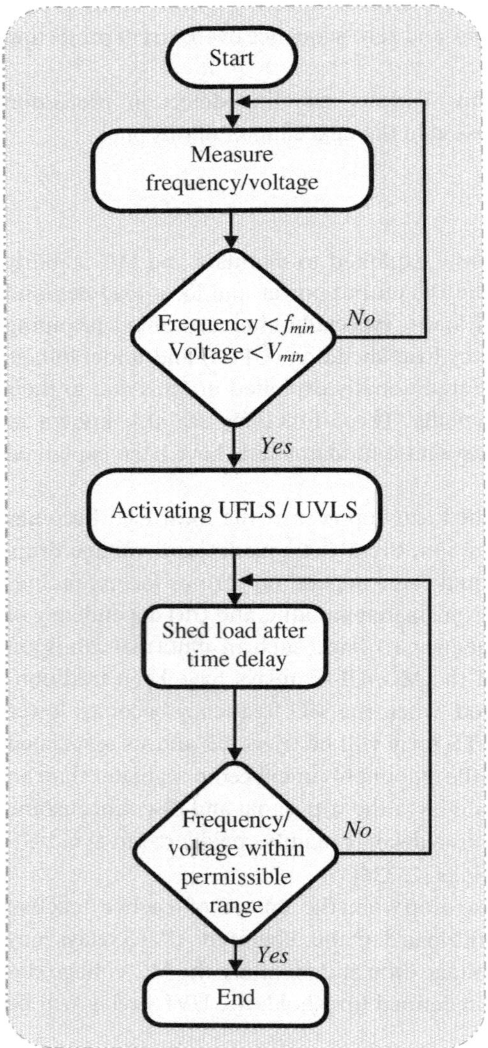

Fig. 4.13 Conventional UVLS and UFLS flowchart.

to their permissible ranges, the load-shedding procedure should be repeated. Furthermore, the conventional load-shedding functions may contain various stages with different thresholds and intentional time delays. For instance, the UVLS function commonly includes two stages, where the first and second ones, respectively, operate after 1 and 1.5 s for $V < 0.85$ pu. It should be noted that each stage can trip different predefined loads. For instance, four stages are commonly considered for UFLS function, in which 35%, 30%, 20%, and 15% of the islanded MG power imbalance will be disconnected when the frequency drops below 49, 48.8, 48.6, and 48.4 Hz, respectively [23, 24]. To minimize the amount of load-shedding and enhance the MG reliability, the load-shedding schemes should be properly coordinated with the MG controllers and regulators. The conventional load-shedding functions can be properly integrated with PCC or feeder protection relays due to their simplicity.

Although the conventional load-shedding functions may help MGs to preserve the voltage and frequency in a permissible range in the presence of static loads through a repetitive load-shedding procedure, they may fail for MGs including dynamic loads such as IMs. In addition, the actual power imbalance value during the islanding event cannot be specified by the traditional techniques, which may finally lead to over-shedding or under-shedding. The shortcomings of these conventional methods are discussed in Section 4.3.4. To diminish the drawbacks of these conventional methods, various novel methods have been proposed in the literature. The main purpose of these new algorithms is to find an optimal solution for load-shedding problem, considering the following questions [23, 24]:

- When should the load-shedding function operate?
- Which loads should be disconnected?
- How much load should be disconnected?

Based on these explanations, an appropriate load-shedding function should have the following features:

- Fast operation to preserve MG stability and balance between the consumption and generation.
- Prevention of DG instability, especially during severe voltage drop conditions.
- High security against stable events, such as power swings.
- Optimal determination of load-shedding amount and load candidates.

For instance, an adaptive procedure has been proposed by several research works to resolve the issues of conventional methods and meet the aforementioned features. For conventional power systems and distribution networks, the UFLS and UVLS are designed to operate separately and make the load-shedding decision individually. However, since in MGs the voltage and the frequency are considerably coupled with each other, separate operation of UVLS and UFLS may lead to over-shedding and also instability problems. Therefore, some research studies recommend exploiting combinational UVLS and UFLS for modern power systems, which are dependent on both voltage and frequency signals simultaneously [25, 26]. This combinational scheme will increase the security and effectiveness of the load-shedding function. The proper load-shedding algorithms and the mentioned combinational schemes have been discussed in Chapter 6.

4.3 Microgrid Protection Challenges

4.3.1 Feeder Protection

After the connection of MG to the upstream system, various protection challenges may occur for feeder protection of both the upstream system and MG. In fact, the connection of MG to the upstream grid increases the complexity of feeder protection schemes due to different reasons such as various operation modes of MG, bidirectional power flow, dynamic behavior of DGs and their associated controller during short-circuit faults, and variation of the short-circuit level.

The major feeder protection challenges include sympathetic tripping, loss of coordination between protection devices, unintentional islanding, reduction of sensitivity of the protection system, and variation in the short-circuit level. Furthermore, dynamic behavior of MG local loads, especially after islanding events, may lead to a semifault condition that may result in feeder protection maloperation. These problems are investigated in the following subsections.

4.3.1.1 Sympathetic Tripping

"Sympathetic tripping" is the false operation of healthy feeder relays after the fault occurrence on the adjacent feeders. This condition may occur due to various reasons, which are described as follows.

4.3.1.1.1 MG Back-Feed As explained earlier, the protection system of upstream distribution feeders and even MG feeders are commonly designed based on the unidirectional power flow (i.e., radial configuration) due to the higher cost of directional relays and their required infrastructures. On the other hand, the setting value of OCR pickup current is commonly determined based on the maximum load current, thermal margins of equipment and distribution lines, and short-circuit current magnitude. As a rule of thumb, the amount of this pickup current is normally adjusted between 1.2 and 1.7 pu. Moreover, considering the radial configuration, its operation time is coordinated only with the downstream protection devices, e.g., overcurrent protection relays, ACBs, MCCBs, and, fuses.

During the grid-connected operation of MG, the power flow direction of distribution feeders may be reversed, especially during short-circuit faults at the adjacent distribution feeders [27]. Figure 4.14 depicts a typical configuration of a

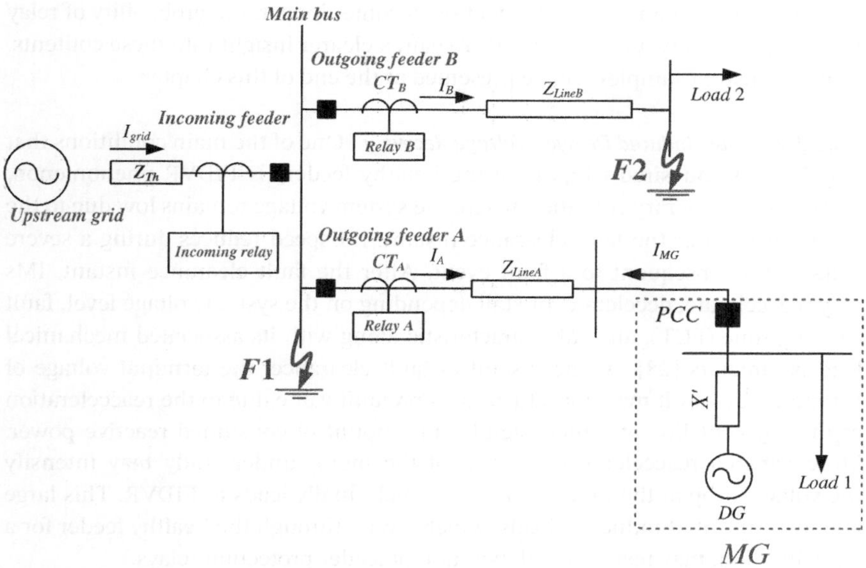

Fig. 4.14 Sympathetic tripping due to the SGBMG back-feed. [27] / with permission of IEEE.

distribution system to which an MG is connected. Considering this figure, after the short-circuit fault occurrence at F2, the MG injects reactive fault current into the fault point and thus a large current may pass through relay A. Since this relay is coordinated only with the downstream protection devices, relay A and relay B have not been selected such that to be coordinated with each other, and thus, relay A may operate incorrectly even before the operation of faulty feeder protection system, i.e., relay B. Therefore, in such a condition, the healthy feeder relay may incorrectly operate for the adjacent feeder faults due to the back-feeding of MG.

The likelihood of healthy feeder maloperation due to the adverse impact of MG back-feeding strongly depends on the type of DG. For example, SSSGs produce significantly higher fault currents compared to IBDGs and, therefore, have a much greater impact on distribution system protection. In addition, the transformer's neutral configuration, such as effectively grounded or isolated neutral, substantially influences the MG's contribution to ground faults. This impact becomes even more pronounced as the capacity of grid-connected DGs increases. Based on the common setting rules of distribution system relays, the false operation of healthy feeders due to back-feeding impact is not expected if the total installed capacity of DGs in a distribution feeder would be less than 1 MW. It is worth mentioning that the physical location of MG has a considerable impact on the probability of relay A false operation. In other words, if the electrical length between the fault location and the point of MG connection becomes longer, the probability of relay A false operation would be reduced. To gain a clearer insight into these contents, comprehensive examples will be presented at the end of this chapter.

4.3.1.1.2 Fault-Induced Delayed Voltage Recovery One of the main conditions that may lead to undesired tripping of the healthy feeders is FIDVR phenomenon. FIDVR is a temporary condition where the system voltage remains low due to the IM behavior after the fault clearance instant. IM speed reduces during a severe voltage dip subsequent to a fault event. After the fault clearance instant, IMs may reaccelerate, decelerate, or stall depending on the system voltage level, fault clearing time (FCT), and IM characteristic along with its associated mechanical load parameters [28]. At the instant of fault clearance, the terminal voltage of IM increases, but it may not return to its prefault value due to the reacceleration of the adjacent IMs and their significant amount of consumed reactive power. Moreover, the reacceleration current of the motor under study may intensify the voltage drop at the motor terminal, which finally leads to FIDVR. This large reactive current of inductive loads, which passes through the healthy feeder for a long duration, may result in maloperation of feeder protection relays.

FIDVR phenomenon commonly occurs in MGs with high penetration of IMs, especially after a fault-initiated islanding event. Fault-initiated islanding is a condition in which the PCC relay disconnects the MG from the upstream system due to a system short-circuit fault. Since the islanded MG has limited resources of reactive power, IMs cannot properly reaccelerate after the fault clearance, which leads to a considerable voltage drop at the MG buses. In such a condition, large reactive currents will pass through the MG OCRs for a long period, which may finally result in relay maloperation of healthy feeders. The distribution system shown in Fig. 4.15 is utilized in this section for investigation of the OCR performance against FIDVR. The distribution network parameters are given in [29].

Various scenarios including short-circuit faults and fault-initiated islanding events have been analyzed in this system. The information of these scenarios is given in Table 4.5. For short-circuit cases, all fault types at two locations, F1 and F2, have been tested. Furthermore, security of OCRs during islanding after a short-circuit event is also discussed. The performance of the common commercialized OCR is given in Table 4.5. The simulation results are discussed in the following subsections [29].

Short-circuit Events (Cases 1–5) When a short-circuit fault takes place on a feeder, the related protection device should operate and clear the fault. Moreover, other protection relays should remain stable during the fault and after its clearance. To evaluate the performance of OCR during FIDVR, various types of faults occurred at $T_f = 8$ s have been investigated in scenarios 1–5. Tripping instants of OCRs are

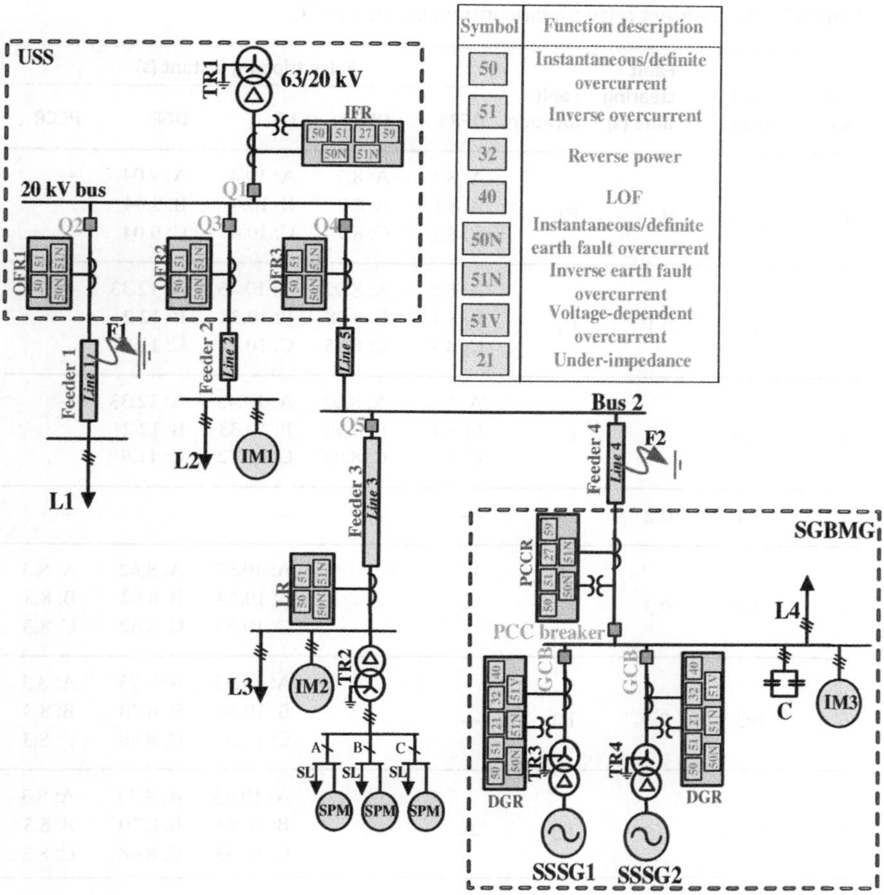

Fig. 4.15 Sample test system.

given in Table 4.5, which indicates their maloperation due to FIDVR. To clarify these results, the waveforms related to few scenarios are explained in the following paragraphs.

In scenario 1 (case no. 1 in Table 4.5), a three-phase short-circuit occurs on feeder 1. After 100 ms, OFR1 operates as the main protection and clears the fault. Figure 4.16 depicts the rotor speed of IMs. As shown, IM speeds decrease during the fault. After the fault clearance, the speed of IM1 and IM3 recovers to their nominal values. However, stall condition occurs for IM2 and SPMs due to their large electrical distances from the USS main bus. Current magnitude of feeder 1, feeder 2, and feeder 3 are illustrated in Fig. 4.17. As shown, the current

Table 4.5 Overcurrent relay tripping instant during FIDVR.

Case no.	Fault type	Fault clearing time (s)	Fault location	False tripping instant (s)				
				OFR1	OFR2	LR	DGR	PCCR
1	ABC	0.1	F1	A: 8.1 B: 8.1 C: 8.1	A: 8.7 B: 8.7 C: 8.7	A: 10.4 B: 10.4 C: 10.4	A: 9.04 B: 9.04 C: 9.04	—
2	BC	0.1	F1	A: 8.1 B: 8.1 C: 8.1	A: 8.92 B: 8.98 C: 8.95	A: 10.45 B: 10.33 C: 10.12	A: 12.33 B: 12.2 C: 11.9	—
3	BCG	0.1	F1	A: 8.1 B: 8.1 C: 8.1	A: 8.92 B: 8.98 C: 8.95	A: 10.45 B: 10.33 C: 10.12	A: 12.33 B: 12.2 C: 11.9	—
4	BG	0.2	F1	—	—	—	—	—
5	ABC	0.3	F2	—	—	A: 10.53 B: 10.53 C: 10.53	A: 8.62 B: 8.62 C: 8.62	A: 8.3 B: 8.3 C: 8.3
6	BC	0.3	F2	—	—	A: 10.63 B: 10.56 C: 10.53	A: 8.75 B: 8.70 C: 8.68	A: 8.3 B: 8.3 C: 8.3
7	BCG	0.3	F2	—	—	A: 10.63 B: 10.56 C: 10.53	A: 8.75 B: 8.70 C: 8.68	A: 8.3 B: 8.3 C: 8.3

magnitude of faulty feeder (feeder 1) increases after the fault occurrence. Albeit, current magnitude of adjacent feeders decreases for a short duration after the fault occurrence due to severe voltage drop. After decreasing the speed of IMs and increasing their slip, the current magnitude of healthy feeders increases. When the faulty feeder is tripped by OFR1, the voltage of USS bus increases but is not recovered to its nominal value due to FIDVR condition, as shown in Fig. 4.18. Moreover, the current magnitude of healthy feeders including DG feeder increases at the fault clearing instant due to large slip of IMs. In such a condition, OFR2, LR, and DGR relays operate incorrectly due to the FIDVR phenomenon, as presented in Table 4.5.

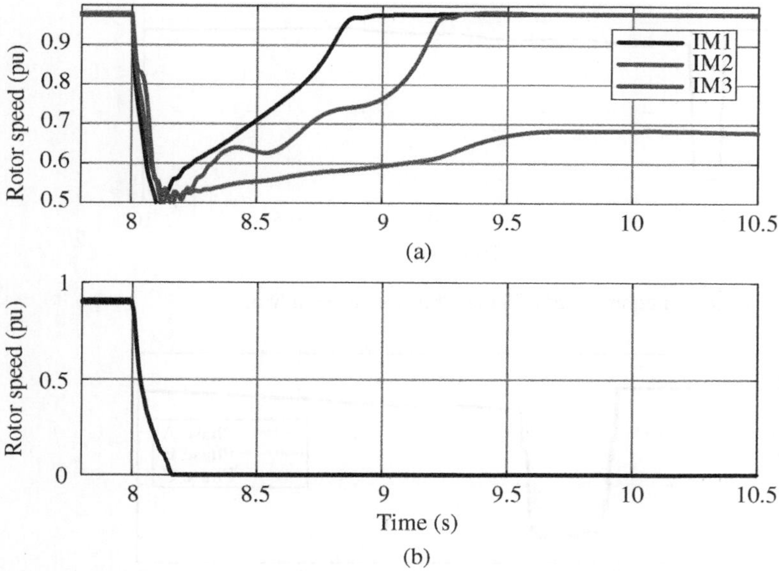

Fig. 4.16 Rotor speed of induction motors for case study no. 1: (a) three-phase IMs and (b) single-phase IMs.

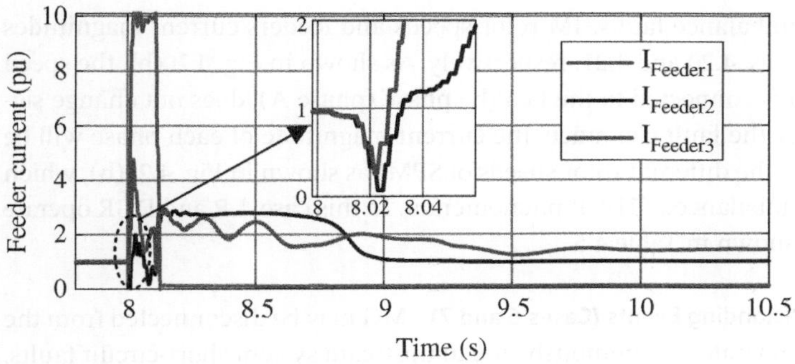

Fig. 4.17 Current magnitude of distribution feeders for case study no. 1.

In case 2, a two-phase fault (B-C) on feeder 1 is investigated. The voltage signal of USS bus is depicted in Fig. 4.19. The voltage dips at phases B and C are almost similar at the high-voltage side of distribution transformer. However, voltage dips at the LV side of the transformer with a Dyn connection are different for each phase. Thus, the amount of speed decrement of SPMs will also be different at each

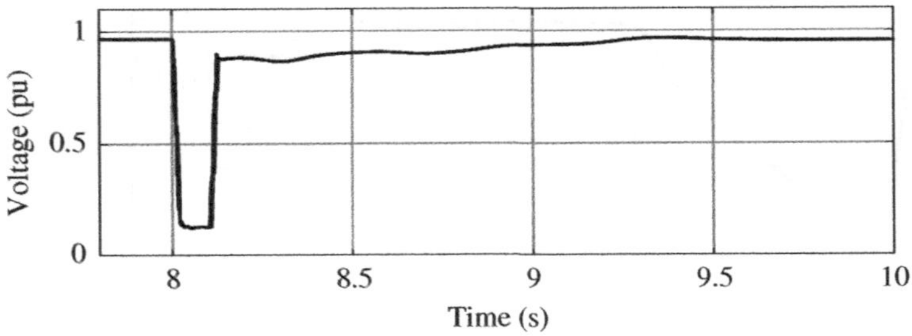

Fig. 4.18 Voltage magnitude of USS main bus for case study no. 1.

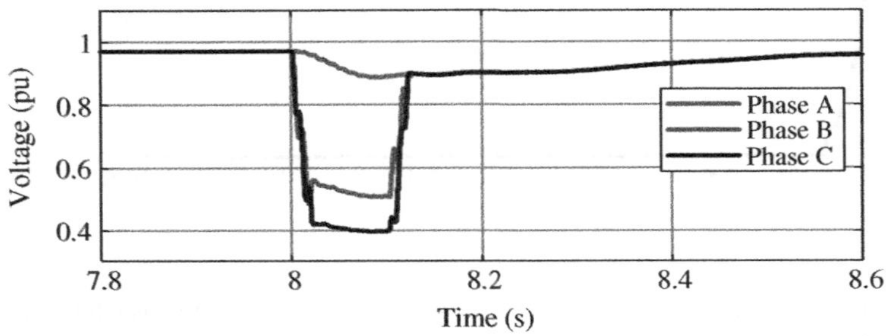

Fig. 4.19 Voltage magnitude of USS main bus for case study no. 2.

phase during unbalance faults. IM rotor speeds and feeders current magnitudes are plotted in Figs. 4.20 and 4.21, respectively. As shown in Fig. 4.20(b), the speed of SPM, which is connected to the healthy phase (phase A), does not change significantly. After the fault clearance, the current magnitude of each phase will be different due to the different rotor speeds of SPMs as shown in Fig. 4.21(b), which represents an unbalanced FIDVR phenomenon. In this case, LR and DGR operate incorrectly as shown in Table 4.5.

Fault-initiated Islanding Events (Cases 6 and 7) MG may be disconnected from the main grid and operate autonomously due to upstream system short-circuit faults. Its voltage and frequency deviate during short-circuit faults and after the islanding. The IMs located in MG decelerate during the faults. Large slip of IMs at the islanding instant may result in a severe FIDVR phenomenon in the MG after its disconnection. In such a condition, proper control and protection actions should be performed automatically to retain the MG stability. In addition, DGRs should remain secure during FIDVR to stabilize the MG and recover its voltage and frequency using appropriate control strategies [30]. To investigate this, two islanding events subsequent to a short-circuit fault are simulated at cases 6 and 7 in Table 4.5.

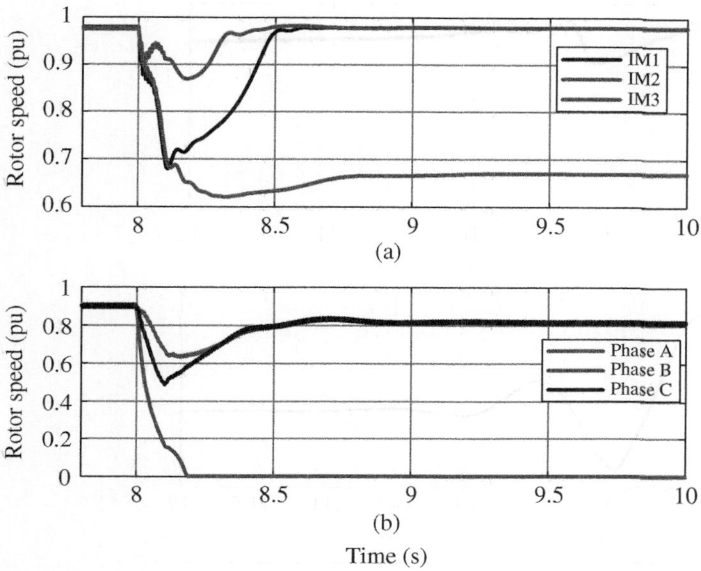

Fig. 4.20 Rotor speed of induction motors for case study no. 2: (a) three-phase IMs and (b) single-phase IMs.

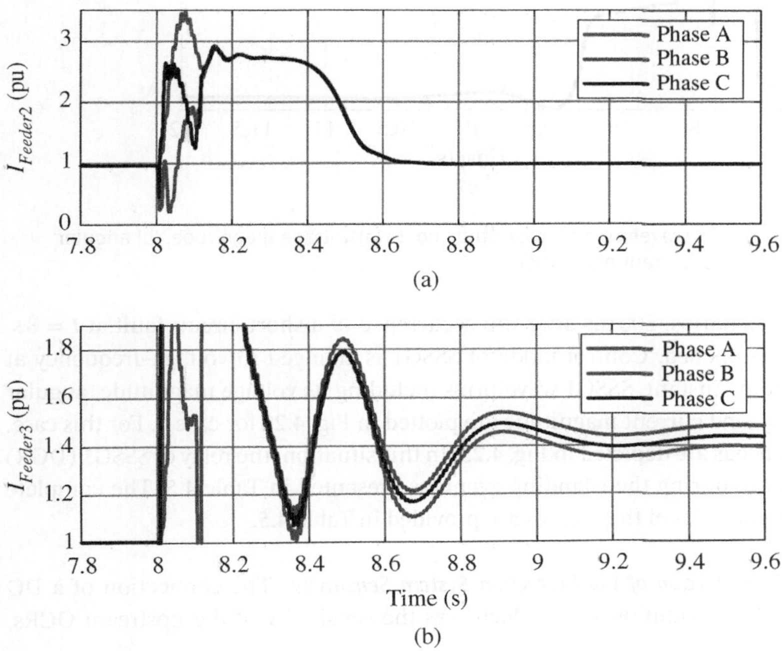

Fig. 4.21 Current magnitudes of distribution feeders for case study no. 2: (a) feeder 2 and (b) feeder 3.

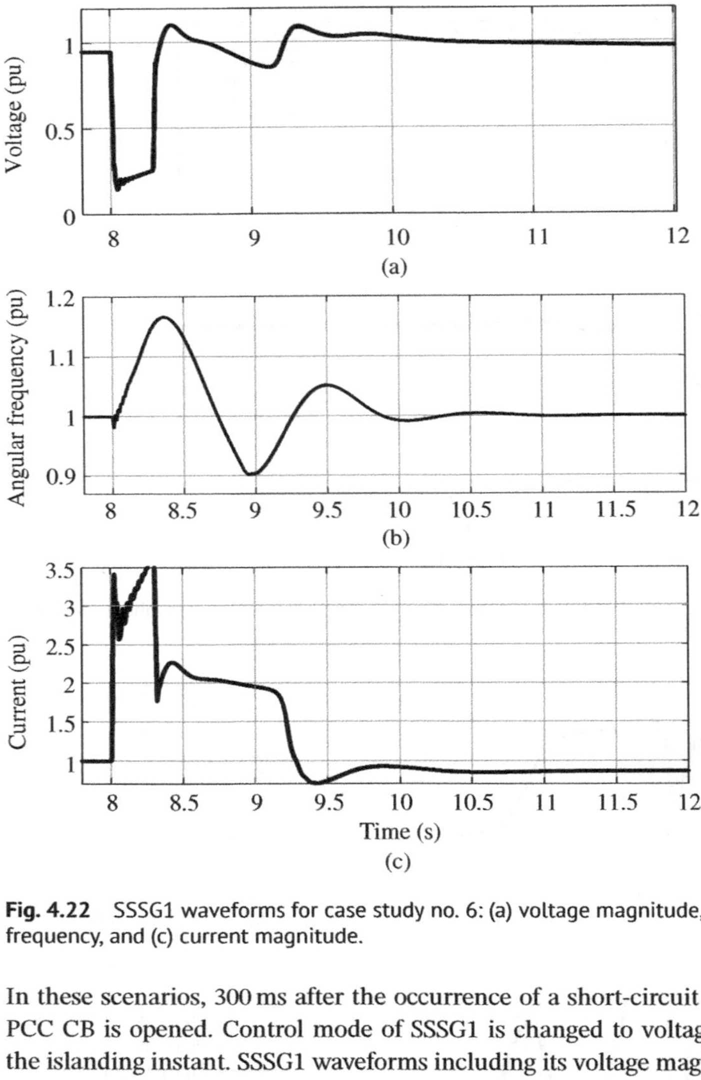

Fig. 4.22 SSSG1 waveforms for case study no. 6: (a) voltage magnitude, (b) angular frequency, and (c) current magnitude.

In these scenarios, 300 ms after the occurrence of a short-circuit fault at $t = 8$ s, PCC CB is opened. Control mode of SSSG1 is changed to voltage–frequency at the islanding instant. SSSG1 waveforms including its voltage magnitude, angular frequency, and current magnitude are plotted in Fig. 4.22 for case 6. For this case, MG IM speeds are depicted in Fig. 4.23. In this situation, the relay of SSSG1 (DGR) maloperates during the islanding event, as presented in Table 4.5. The complete simulation results of these cases are provided in Table 4.5.

4.3.1.1.3 Reduction of the Protection System Sensitivity The connection of a DG or MG to a distribution feeder decreases the sensitivity of the upstream OCRs,

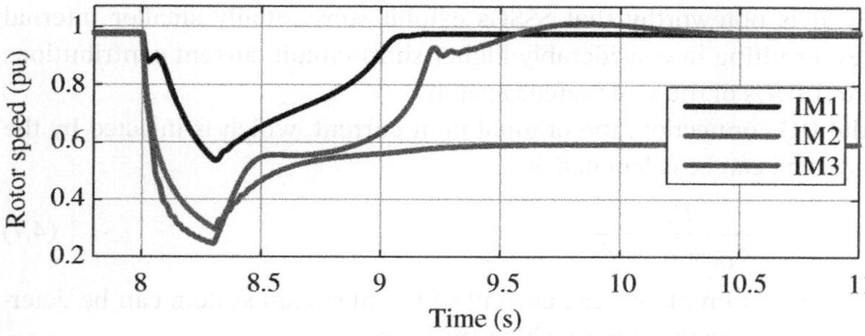

Fig. 4.23 Rotor speed of three-phase induction motors for case study no. 6.

due to the reduction of upstream system contribution to the downstream faults. To justify this, the sample distribution system of Fig. 4.24 is considered. In this configuration, relay B protects the whole line, i.e., $Z_1 + Z_2$, and relay A protects the *Load* feeder. The DG illustrated in Fig. 4.24, represented as a voltage source in series with an impedance, may correspond to either an SSSG or an IBDG. SGs are conventionally modeled as internal voltage sources in series with their subtransient or transient impedances. In addition, as discussed in Chapter 3, grid-following (GFL)-IBDGs are typically modeled as controlled current sources in parallel with an impedance; through the Norton–Thevenin equivalent transformation, this representation can be converted to an equivalent voltage-source model in series with an impedance, as shown in Fig. 4.24. Grid-forming IBDGs can also be represented using a voltage-source model with a corresponding series

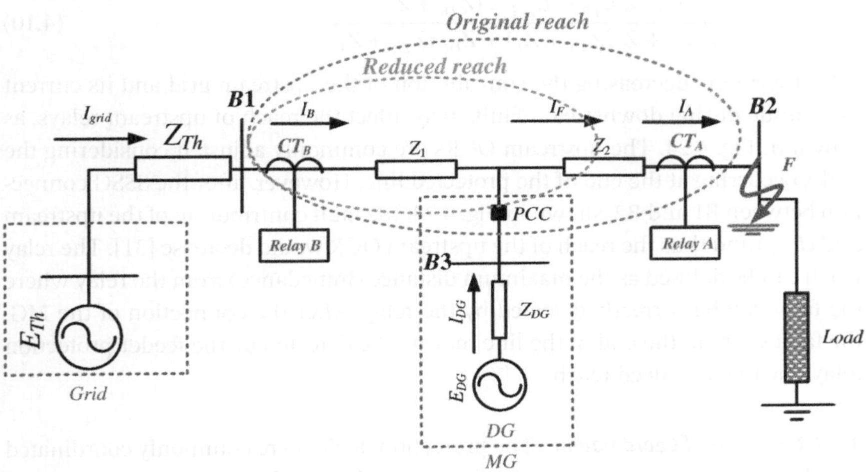

Fig. 4.24 Reduced reach of feeder protection after grid connection of SSSG. [31] / with permission of Elsevier.

impedance. It is noteworthy that SSSGs exhibit substantially smaller internal impedances, resulting in considerably higher short-circuit current contributions compared to IBDGs of the same rated capacity.

Before the DG connection, the original fault current, which is injected by the upstream system, can be calculated as

$$I_F = I_{grid} = \frac{E_{Th}}{Z_{Th} + Z_1 + Z_2} \tag{4.7}$$

After the connection of DG, the current of the upstream system can be determined as follows using the superposition theorem:

$$I_{grid} = \frac{E_{Th}}{Z_{DG} \| Z_2 + Z_{Th} + Z_1} - \frac{E_{DG}}{(Z_{Th} + Z_1) \| Z_2 + Z_{DG}} \times \frac{Z_2}{Z_1 + Z_{Th} + Z_2} \tag{4.8}$$

Comparing (4.7) and (4.8) reveals that the fault contribution of the upstream grid decreases after the connection of DG. Since the fault current decreases, the FCT for OCRs equipped with the inverse time functions increases. Therefore, the sensitivity of relay B decreases.

On the other hand, the total fault current I_F increases after the connection of DG, which can be determined by

$$I_F = \frac{E_{Th}}{Z_{DG} \| Z_2 + Z_{Th} + Z_1} \times \frac{Z_{DG}}{Z_2 + Z_{DG}} + \frac{E_{DG}}{(Z_{Th} + Z_1) \| Z_2 + Z_{DG}} \times \frac{Z_{Th} + Z_1}{Z_2 + Z_{Th} + Z_1} \tag{4.9}$$

For more simplification, this equation can be rewritten as

$$I_F = \frac{Z_{DG} \times E_{Th} + E_{DG} \times (Z_{Th} + Z_1)}{Z_{DG} Z_2 + Z_2 (Z_{Th} + Z_1) + Z_{DG} (Z_{Th} + Z_1)} \tag{4.10}$$

Furthermore, decreasing the contribution of the upstream grid and its current magnitude during downstream faults may affect the reach of upstream relays, as shown in Fig. 4.24. The upstream OCRs are commonly adjusted considering the faults occurring at the end of the protected line. However, after the SSSG connection between B1 and B2, shown in Fig. 4.24, the fault contribution of the upstream grid (I_{grid}) and thus the reach of the upstream OCR would decrease [31]. The relay reach can be defined as the maximum distance (impedance) from the relay where the fault can be correctly detected by the relay. After the connection of the MG, the fault event at the end of the line may not be detected by the feeder protection relay due to its reduced reach.

4.3.1.1.4 *Loss of Coordination* The protection devices are commonly coordinated based on the maximum short-circuit level of the distribution system. However, after the connection of MGs to the system, the short-circuit level can considerably change, and thus, the coordination between protection devices may be lost in such a condition. The main reasons for changing the short-circuit level include:

- Changing the number of DGs connected to the grid
- Various operation modes of DGs and control loops
- Changing the configuration of distribution system
- Decreasing the short-circuit level of MG after the islanding event

The loss of coordination may occur between various protection devices, which are studied as follows.

1) **Relay-relay:** OCRs are coordinated with each other based on the maximum fault current experienced at the downstream relay location. For instance, in the distribution system of Fig. 4.25, relay B operates as the backup protection for relay A, and thus, these relays are coordinated for the maximum fault current (I_{max}) at the relay A position [32]. Suppose that at the absence of DG, a proper coordination time interval (CTI) is selected between the operation times of relay A and relay B at $I_{max(1)}$, as shown in Fig. 4.26. After the DG connection at the main upstream bus, the fault current passing through relays A and B increases to $I_{max(2)}$, and thus, the difference between the operation times of main and backup relays, represented by TD in Fig. 4.26, becomes smaller than CTI. This may lead to incorrect operation of the backup relay B, which may endanger the security of the protection system.
2) **Relay-fuse:** The operation curve of fuse elements includes two parts: (i) MMT and (ii) maximum clearing time (MCT) [33]. The fuse characteristics are commonly selected, considering (i) the thermal damage curve of downstream loads and power equipment and (ii) proper coordination with the other protection devices. On the other hand, the upstream adjustable protection relays should be coordinated with the maximum melting time of the downstream fuses.

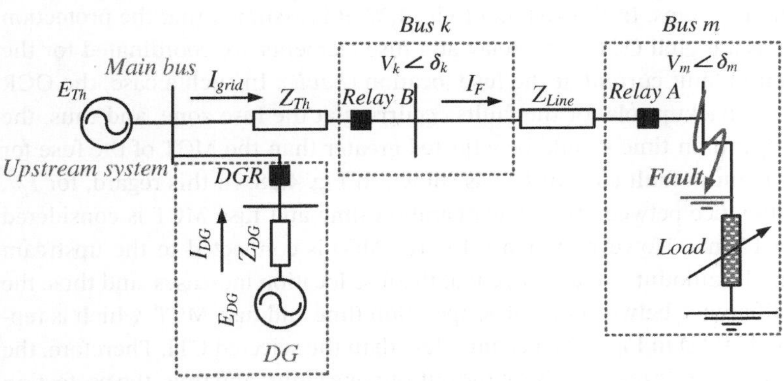

Fig. 4.25 Contribution of DG during short-circuit fault.

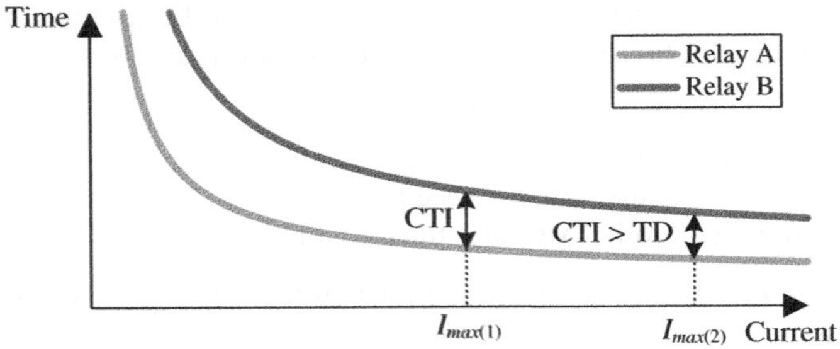

Fig. 4.26 Loss of coordination between two relays after DG connection.

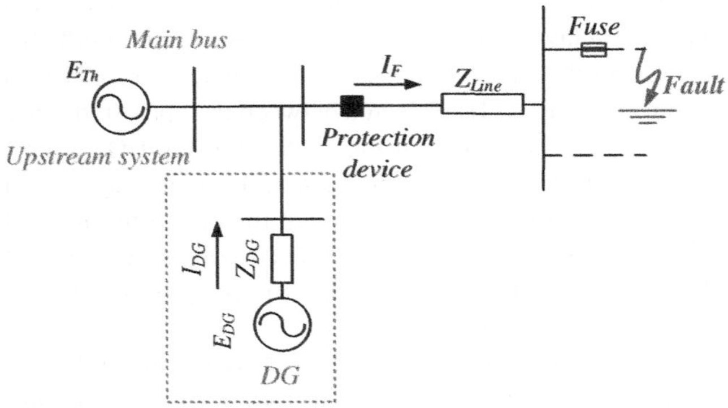

Fig. 4.27 Contribution of DG during short-circuit fault at the fuse protection zone.

A typical configuration of a distribution system including fuses is depicted in Fig. 4.27. As shown, fuse elements are normally utilized in the laterals of distribution systems. In the system of Fig. 4.27, it is assumed that the protection device is a digital OCR. This relay and fuse elements are coordinated for the maximum fault current at the fuse location (*Fault*). In such a case, the OCR plays the backup role for the faults occurring at the fuse zone, and thus, the relay operation time should be adjusted greater than the MCT of the fuse for the maximum fault current I_{F1}, as shown in Fig. 4.28. In this regard, for I_{F1}, the difference between the relay operation time and fuse MCT is considered CTI = 120 ms. However, when a DG (or MG) is connected to the upstream region, the amount of fault current at the fuse location increases, and thus, the time difference between the OCR operation time and fuse MCT, which is represented by TD in Fig. 4.28, becomes less than the selected CTI. Therefore, the relay may operate incorrectly for the out-of-zone faults, and thus, the protection system loses its selectivity in this case.

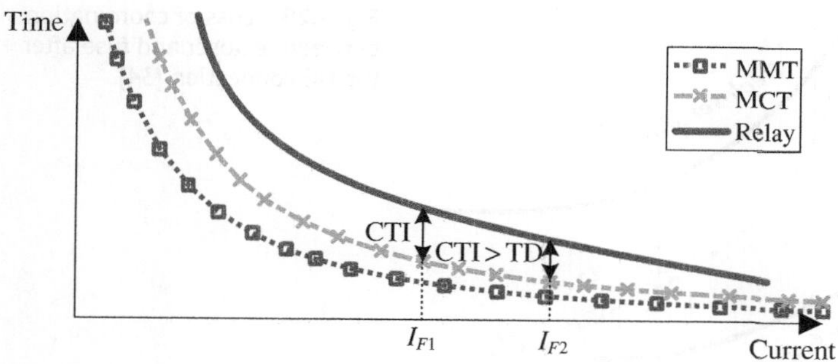

Fig. 4.28 Loss of coordination between relay and fuse after the SSSG connection.

3) **Recloser-fuse:** To decrease the additional cost of replacing fuses and the outage time of system laterals, fuse-saving schemes are widely utilized in radial conventional distribution networks. For investigating fuse-saving schemes and coordination between recloser and fuse, a simple conventional distribution system depicted in Fig. 4.27 is considered, and a protection device is assumed to be a recloser. The main purpose of fuse-saving methods is the prevention of fuse melting during temporary short-circuit faults. To do so, one recloser should be installed at the main distribution feeder, as shown in Fig. 4.27. If a short-circuit fault occurs at the protection zone of the fuse (for instance, "Fault" shown in Fig. 4.27), at first, the recloser should operate before the fuse melting, which is known as the first stage operation of the recloser [31]. In such a condition, the whole feeder will be de-energized for a while. After a predefined dead-time, the feeder will be energized by the recloser command. After reconnection of the feeder, if the fault was temporary, the feeder and downstream laterals can continue their steady-state operation, while if the fault is permanent, it should be cleared by its associated protection device, i.e., its associated fuse, and thus, the fuse should melt before the second stage operation of the recloser [31]. Based on these explanations, to implement a fuse-saving strategy, the recloser should be equipped with two operation curves, including fast and slow curves. The recloser and fuse characteristics are shown in Fig. 4.29. As depicted, for a range of fault current $I_{min} < I_F < I_{max}$, the operation time of the recloser first stage t_{1R} (i.e., recloser fast curve) should be less than the MMT of the fuse, which can be calculated by

$$t_{1R}\big|_{@I_{min}} = t_{MMT}\big|_{@I_{min}} - \text{CCT} \tag{4.11}$$

where t_{MMT} represents the MMT of the fuse at I_{min} and CCT is a constant margin.

For a range of fault current $I_{min} < I_F < I_{max}$, the operation time of the second stage of recloser t_{2R} (i.e., recloser slow curve) should be greater than the MCT

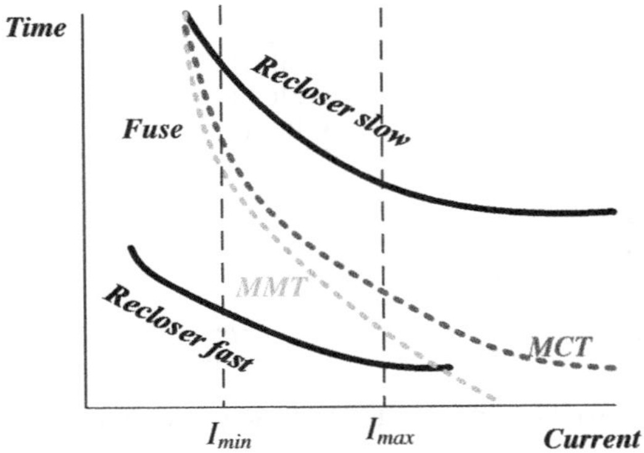

Fig. 4.29 Loss of coordination between recloser and fuse after the DG connection [34].

of the fuse, which can be calculated by

$$t_{2R}\big|_{@I_{max}} = t_{MCT}\big|_{@I_{max}} + \text{CCT} \tag{4.12}$$

where t_{MCT} represents the MCT of the fuse at I_{max}. It should be noted that the range of fault parameters I_{min} and I_{max} is commonly determined based on the off-line simulation studies of the distribution network.

After connecting a DG to the network, the structure of the distribution system will considerably change where the power flow becomes bidirectional. Therefore, during short-circuit fault conditions, the DG (or MG) contributes to feed the short-circuit fault, and thus, a significant current may pass through the fuse for upstream faults. On the other hand, the number of grid-connected DGs and their operation control modes (e.g., voltage, droop, and constant power modes) have a considerable impact on the fault current magnitude. In such a condition, the amount of I_{min} and I_{max} may frequently change depending on the number of grid-connected DGs and the operation modes, and thus, the coordination between recloser and fuse may be lost.

4.3.2 False Operation of Fault Indicators

Fault indicators (FIs) are widely exploited in distribution networks to help field crews in detecting the fault region and decrease the outage duration. The conventional distribution system FIs are not directional. Since the fault point is supplied by the grid-connected DGs, the FIs between the DGs and the fault location operate incorrectly. In such a condition, the section of the fault location cannot be determined appropriately.

As an example, a part of a real distribution system is depicted in Fig. 4.30. The FIs of this system are represented by FI_1 to FI_9. If a short-circuit fault takes place at F_1 and the DG is not connected to the grid, FI_1, FI_2, and FI_3 operate correctly and the

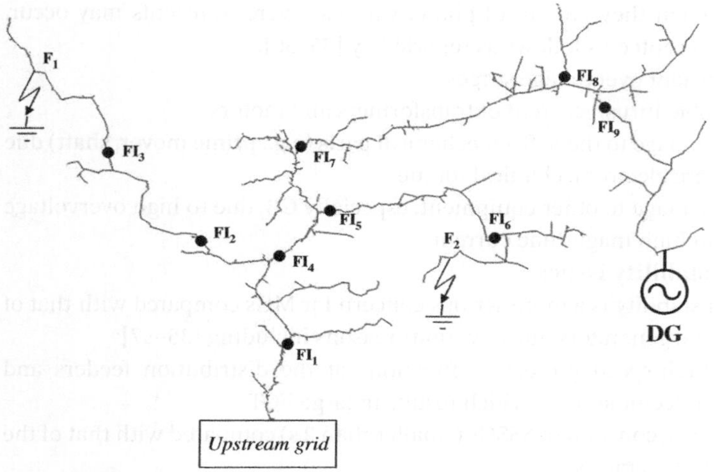

Fig. 4.30 A sample distribution system.

fault section can be properly determined by the field crews. After the connection of DG, FI_4, FI_7, and FI_9 may also operate for the fault at the F1 point. In this condition, the actual fault section cannot be specified correctly by the field crews.

4.3.3 Recloser Issues

After connection of an MG/DG to the distribution system, the operation of system reclosers may face various problems, which are discussed as follows.

1) **Preventing arc extinction**
   The primary concern for reclosing distribution feeders in the presence of DGs or MGs is endangering the system reliability. If a short-circuit fault has a temporary nature and the DGs remain connected to the system after the first stage operation of the recloser, the arc extinction time considerably increases and thus the arc current may still exist after elapsing the reclosing dead-time. In other words, the fault arc does not extinguish prior to the reclosing and thus the reclosing attempt will not be successful. This can endanger the automatic restoration of the whole feeder, including healthy laterals, which finally leads to a power outage for an enormous number of customers.
2) **Out-of-phase reclose**
   After the disconnection command of the recloser, the islanded MG including its DGs, especially SSSGs, is likely to drift out of synchronism with the upstream grid. If the MG and upstream grid are reconnected to each other by the recloser

command when they are out of phase, various severe transients may occur, which are presented as follows as reported by [35, 36]:

- Sever transient overvoltage surges
- Considerable inrush current of transformers and motors
- Possible damage to the SSSG mechanical parts (e.g., prime mover, shaft) due to the severe electromechanical torques
- Possible damage to other equipment, especially CB, due to high overvoltage surges and high magnitude current.

3) **Transient stability issues**

   Transient instability is a more serious concern for MGs compared with that of the large-scale generators due to various reasons including [35–37]:

   - Lack of high-speed protection functions at the distribution feeders and SGBMG collector system, which results in large FCT
   - Lower inertia constant of SSSGs (smaller than 2 s) compared with that of the large-scale generators
   - Instability of the phase-locked loop (PLL) in GFL-IBDGs, particularly in weak distribution systems such as MGs, as discussed in Chapter 3
   - Higher reactance value of SSSGs compared with that of the large-scale generators
   - Higher Thevenin impedance of upstream distribution system compared with that of the transmission system
   - Higher capability of large-scale generators to provide a better fault ride-through feature than that of the SSSGs.

The occurrence of transient instability in either SSSGs or IBDGs can lead to severe oscillations in voltage, current, torque, and frequency within the MG. Furthermore, these voltage oscillations may have adverse impacts on the adjacent dynamic loads and equipment such as IMs. On the other hand, unlike large-scale generators, DGs are not equipped with the out-of-step function, and thus, the transient instability of the DGs cannot be detected quickly. Such prolonged transient instability in an MG can reduce equipment lifetime and may even cause severe mechanical damage to rotating machines, such as SSSGs, and their associated prime movers. Therefore, appropriate protection and control schemes should be exploited to prevent transient instability against short-circuit faults.

Most of the short-circuit faults that occur in distribution systems are temporary. Therefore, to restore service with minimum interruption time, an auto-reclosing algorithm with a fixed dead-time is conventionally exploited in distribution feeders to reenergize the feeder. In the conventional distribution systems, to ensure de-ionization of arc and successful reclosing, the fixed dead-time is adjusted sufficiently long. However, in modern MGs and distribution systems that contain DGs, using this predefined fixed dead-time may lead to transient instability. If a

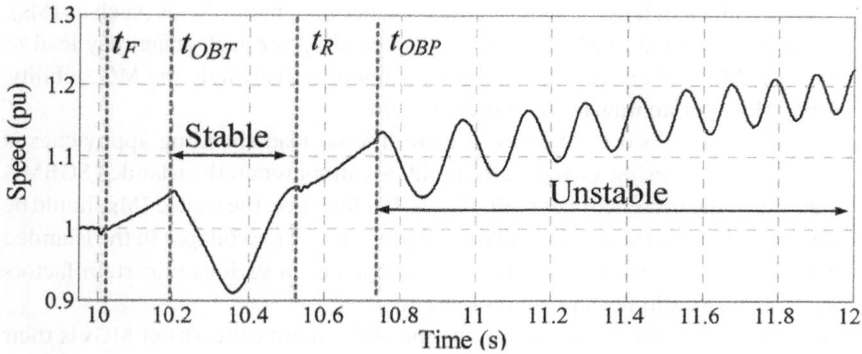

Fig. 4.31 Transient stability of a sample SSSG after an unsuccessful reclosing. [37] / with permission of John Wiley & Sons.

permanent short-circuit fault occurs at an adjacent distribution feeder of the DG, the reclosing attempt at this feeder will be unsuccessful, and it intensifies the DG energy level, which may result in transient instability of the MG. For instance, Fig. 4.31 illustrates the speed signal for an SSSG, which is connected to a sample distribution system where the parameters of the SSSG and distribution system are presented in [37]. In this case, a permanent short-circuit fault occurs at $t_F = 10$ s, and the faulty feeder breaker is opened at $t_{OBT} = 10.2$ s. After the fault clearing, the SSSG remains stable. However, considering the 320 ms recloser dead-time, and after an unsuccessful reclosing at $t_R = 10.52$ s, the SSSG loses its stability. It is worth mentioning that if the value of dead-time changes in this case, the SSSG may remain stable even after an unsuccessful reclosing attempt. In fact, instability of SSSG would take place if the reclosing attempt is made during the SSSG accelerating swing cycle. Since the accelerating swing cycle depends on various factors such as loading level, using the traditional fixed dead-time may not be an appropriate practice to stabilize the SSSG after the short-circuit faults.

4.3.4 Load-Shedding Relay Problems

As expressed earlier, UFLS and UVLS methods are traditionally exploited for load-shedding in conventional power systems. However, these approaches may not be suitable for modern distribution systems and MGs, which include SSSGs and IBDGs. In the following paragraphs, the main shortcomings of these load-shedding relays are discussed [24, 38].

- For both the aforementioned traditional schemes, the load-shedding amount in each stage is commonly determined based on the real power mismatch that passes through the PCC breaker. However, voltage instability of MG commonly occurs due to the lack of reactive power. On the other hand, since the amount of power mismatch may change in the presence of dynamic loads (such as IMs), using a predefined fixed amount of load shedding for each stage may lead to over-shedding or under-shedding and cannot effectively help the MG stability, especially after an unwanted islanding event.

- The location of load candidates in conventional load-shedding approaches is fixed. However, using a fixed load candidate cannot retain the islanded SGBMG stability in the presence of dynamic loads. For instance, the stalled IMs should be disconnected after short-circuit faults to prevent voltage collapse in the islanded MG where the location of stalled motor depends on various uncertain factors such as the location of short-circuit event.
- As expressed in the previous section, one of the main concerns of MGs is their transient stability during different perturbations. However, this concern is not considered in the aforementioned traditional load-shedding schemes. Therefore, the fixed improper operation time and location of these methods may result in MG transient instability, especially after a fault-initiated islanding event. For a sample MG whose parameters are given in [24] and contains SSSGs, IMs, and static loads, the maximum permissible voltage recovery time (t_S), which can prevent transient instability of the SSSGs after clearing a three-phase fault, is illustrated in Fig. 4.32. In this simulated scenario, the fault is cleared by opening of the PCC breaker at $t = t_{FCT}$, which leads to unwanted islanding of the MG. For instance, for $t_{FCT} = 0.3$ s and postfault voltage equal to 0.7 pu, if the voltage does not recover to its nominal value within $t_S = 600$ ms, the MG loses its transient stability. Therefore, the intentional time delay of traditional schemes cannot be greater than 600 ms in this case. As shown in this figure, for various

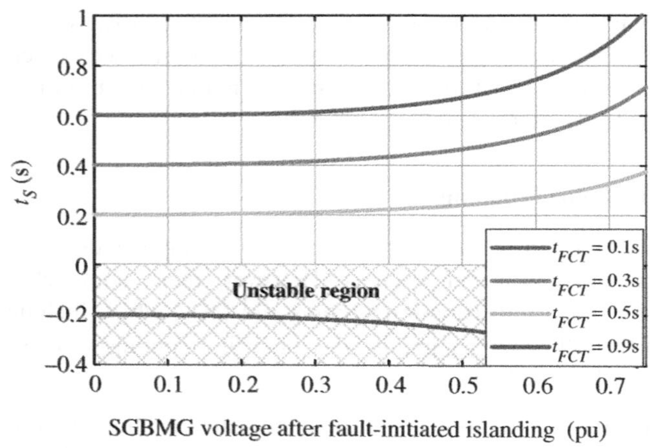

Fig. 4.32 The maximum permissible voltage recovery time (t_S) for various fault clearing time.

cases, the amount of t_S is different and thus using a fixed intentional operation time for load-shedding schemes is not suitable for MGs. Meanwhile, the conventional load-shedding approaches have a large fixed operation delay, and thus they may fail to recover system frequency and voltage.

- One of the main drawbacks of conventional load-shedding strategy is its false operation against stable events. For instance, the UFLS method may maloperate during stable power swings. Besides, the UVLS technique may disconnect the loads incorrectly during the short-circuit fault and stable FIDVR phenomena. To enhance the security of conventional strategies, their sensitivity and operation speed may be reduced. However, this can cause their prolonged operation during unstable transients.

4.3.5 SSSG Protection Challenges

4.3.5.1 LOF Protection

In this section, some LOF and system disturbance cases are studied, which lead to nonproper operation of the widely used impedance-based LOF method. To do so, the distribution system shown in Fig. 4.15 is utilized for investigating the LOF protection challenges. Some LOF and system disturbance cases, which may result in malfunction of LOF relays, are scrutinized and simulated as follows.

4.3.5.1.1 Delay in Detection of Unrecoverable LOF Event For the studied system of Fig. 4.15, SSSG1 is considered to operate in the constant active/reactive power control mode, and SSSG2 is out of service. SSSG1 loses its excitation system due to a solid short circuit across its field winding terminals at $t = 8.3$ s, and the system voltage (V_{bus2}) is illustrated in Fig. 4.33(a). In such a condition, SSSG1 operates as an induction generator, whose reactive power is supplied by the upstream grid. In addition, the speed of the adjacent IM load, i.e., IM3, decreases due to its associated bus voltage drop, as shown in Fig. 4.33(b). If SSSG1 is not disconnected from the system, IM3 loses its suitable operation at $t = 11.42$ s, which leads to drawing a large amount of locked-rotor current. Finally, voltage instability occurs at $t = 11.5$ s due to the incapability of the network to restore equilibrium between the load demand and supply. In this condition, as shown in Fig. 4.33(c), SSSG1 frequency continuously increases, which will result in the operation of the corresponding overspeed protection function.

Analysis of conventional impedance-based LOF function for this case shows that LOF occurrence is realized at 11.74 s, while the system is under an adverse operating condition after 11.42 s. In other words, the studied unrecoverable LOF case should be quickly detected to prevent further stress on the SSSG and other system equipment.

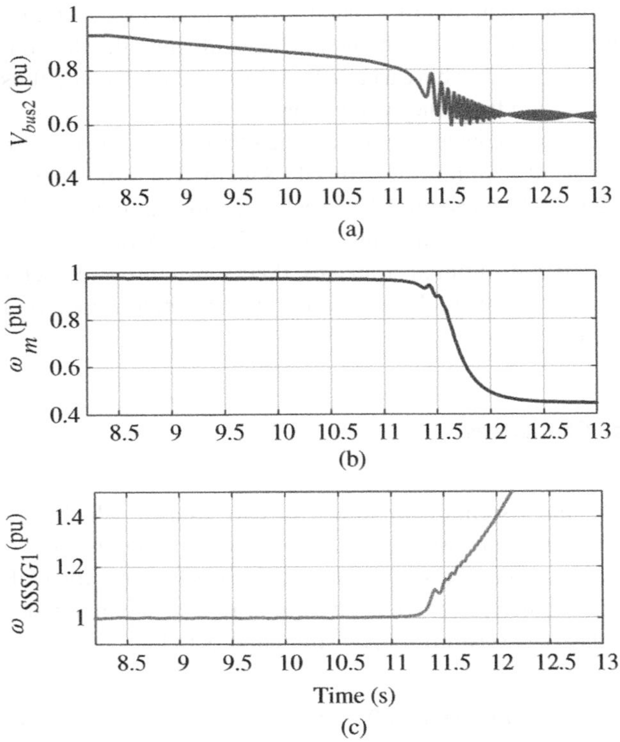

Fig. 4.33 System behavior after LOF event for SSSG1: (a) system voltage, (b) IM3 speed, and (c) SSSG1 angular frequency.

For further investigations, it is assumed that SSSG2 is in service and operates under the active/reactive power control mode during the total LOF event occurrence of SSSG1. Similar to the previous case, after such an LOF event, SSSG1 operates as an induction machine and draws considerable reactive power from the upstream grid, which leads to SGBMG voltage drop. Since SSSG2 operates under the active/reactive power control mode, its reactive power is not boosted during such a severe voltage drop, and thus, this generator cannot participate in compensating the reactive power and restoring the equilibrium between SGBMG reactive power consumption and generation. Finally, SSSG1 loses its synchronism with the upstream grid, and the conventional impedance-based LOF detects the LOF event of SSSG1 after its loss of synchronism. It should be noted that when SSSG2 operates under the voltage control mode, the AVR excitation system will increase its injected reactive power to restore the SGBMG voltage drop during the total LOF event occurrence of SSSG1.

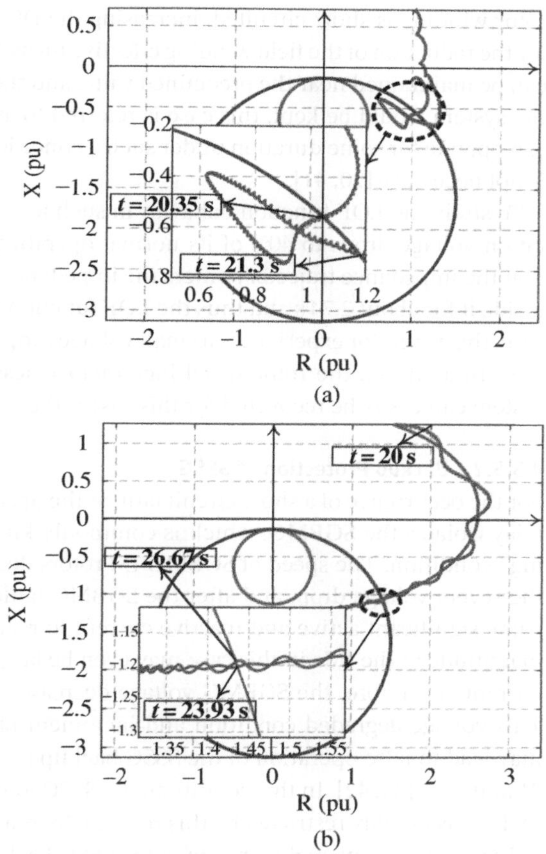

Fig. 4.34 Impedance trajectory for (a) islanding event and (b) recoverable LOF. [39] / with permission of IEEE.

4.3.5.1.2 LOF Maloperation During Islanding Event In this case, SSSGs operate in the grid-connected mode in the system depicted in Fig. 4.15, and both SSSGs work under the constant active/reactive power control strategy. Moreover, SSSGs are equipped with the widely used traditional impedance-based LOF protection function. After the islanding occurrence at $t = 20$ s by opening of the PCC breaker, the operation mode of SSSG1 is instantly changed to the voltage/frequency control strategy and thus some reactive power is absorbed by SSSG1 to maintain the SGBMG voltage [39]. The impedance trajectory seen from the SSSG1 terminal after the islanding instant is depicted in Fig. 4.34(a). As shown, this trajectory remains inside the LOF protection zone for about 1 s, resulting in issuing the false trip command.

4.3.5.1.3 LOF Maloperation During Recoverable LOF Partial LOF event of SGs may be recoverable by fast operation of AVR [40]. In other words, if some portion of the rotor winding is short-circuited, increasing the DC field current can compensate for the reduction of the field winding effective turns. Since the SSSG output voltage can be maintained near the predefined value and the generator synchronism with the system would be kept, this case is referred to as a recoverable LOF. As SSSG can operate for some duration under such a condition, its LOF function operation is not necessary [20, 41].

To study the LOF function behavior in such a case, it is assumed that the excitation voltage drops to 40% of its normal operating value. Figure 4.34(b) shows that the impedance trajectory enters the impedance-based LOF zone and remains inside it for about 2.74 s and thus the LOF function maloperates. Whereas for this case, the generator experiences a small voltage drop of 0.1 pu only for a short duration. In addition, the rotor speed increment is less than 0.2%. Accordingly, the system can easily be recovered for this case if the LOF function remains stable.

4.3.5.2 Backup Protection of SSSG

For the occurrence of a short-circuit fault at the upstream grid, the PCC protection relay isolates the SGBMG, which is commonly known as "fault-initiated islanding" condition. The speed of SGBMG IM rotors decreases during the fault due to the severe voltage drop, and after the SGBMG fault-initiated islanding, IMs consume significant active and reactive current to reaccelerate [29, 42]. The current magnitude of the IMs in this case may even be larger than 350% of their nominal current. Therefore, the SGBMG voltage decreases during the IM reacceleration. This voltage-degraded condition and increment of the IMs current magnitudes may lead to false operation of the SSSG backup protection system, e.g., functions 21 and 51 V [29, 42]. In this condition, the SSSG low-voltage ride-through capability is considerably restricted by this relay maloperation.

To investigate the maloperation of the SSSG backup protection, the distribution system shown in Fig. 4.15 is utilized. Meanwhile, it is assumed that SSSG1 operates in the constant active/reactive power control mode. Moreover, SSSG2 and the capacitor bank are out of service. Furthermore, the amount of SGBMG local load consumption is equal to the SSSG1 generation. The SGBMG voltage and current are depicted in Fig. 4.35 when a fault occurs at F2, as shown in Fig. 4.15. In addition, parameters t_1, t_2, and t_3 represent the fault occurrence instant, fault clearing instant, and recovery instant, respectively. The impedance seen from the SSSG1 terminals is illustrated in Fig. 4.36. As shown in this figure, the measured impedance during the voltage-degraded condition becomes less than its threshold value, and thus, the under-impedance function 21 operates incorrectly in this case. Furthermore, the 51 V function may also operate falsely due to the voltage reduction and current increment, as shown in Fig. 4.37.

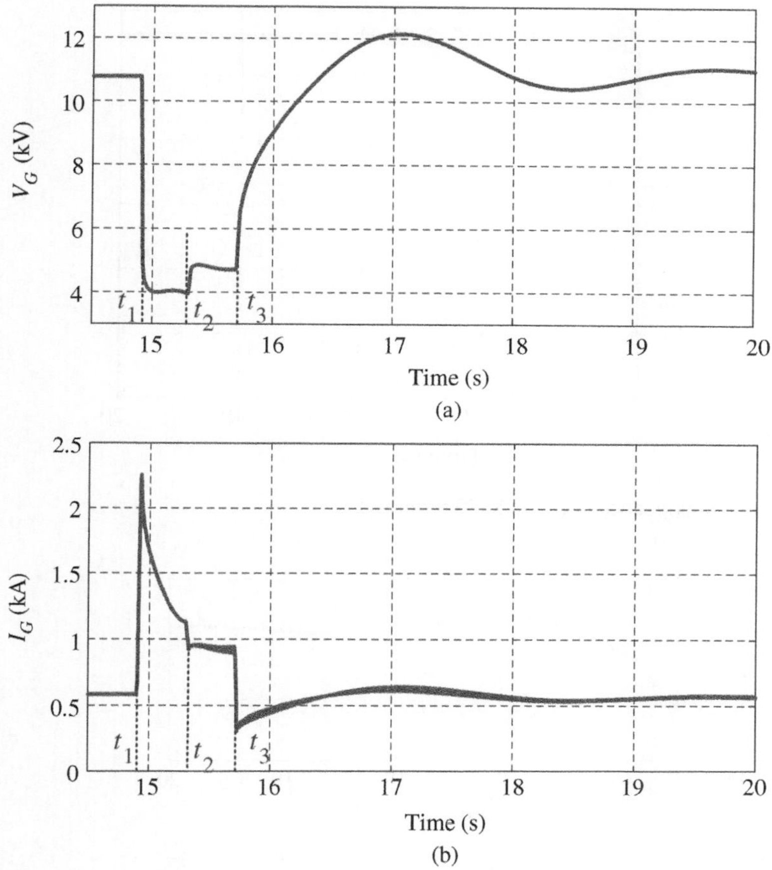

Fig. 4.35 Voltage and current of SSSG.

Some examples are provided to better comprehend the subjects discussed in the previous sections. These examples try to summarize the key points of this chapter and provide a clearer insight into the challenges of the SGBMG protection system.

Example 4.1

A sample distribution system, which includes two outgoing feeders, is depicted in Fig. 4.38. The nominal voltage of this system is 20 kV, and the maximum three-phase short-circuit current at its main bus is $I_F^{max} = 10$ kA. Both of the outgoing feeders are similar, and their protection relays, i.e., relays A and B, are adjusted as given in Table 4.6.

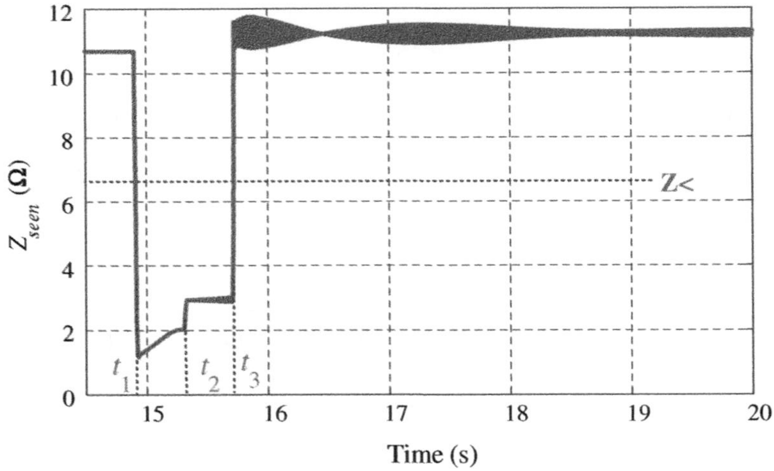

Fig. 4.36 Seen impedance from the SSSG terminals.

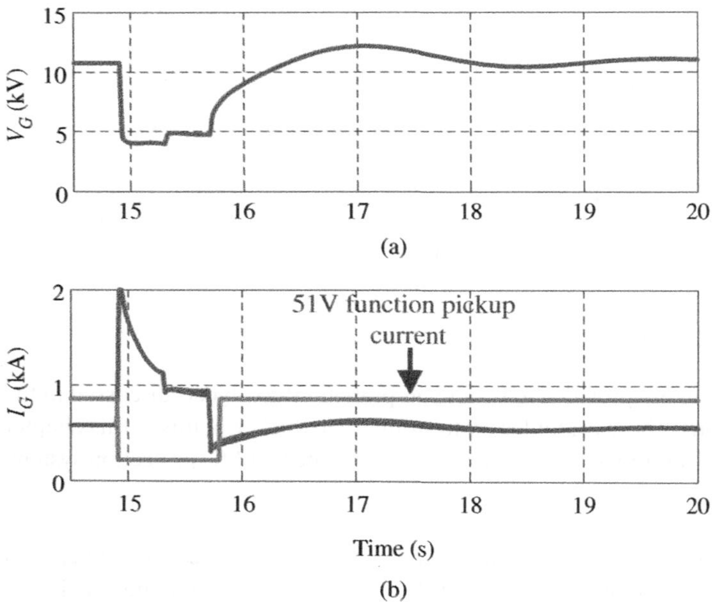

Fig. 4.37 51V function maloperation.

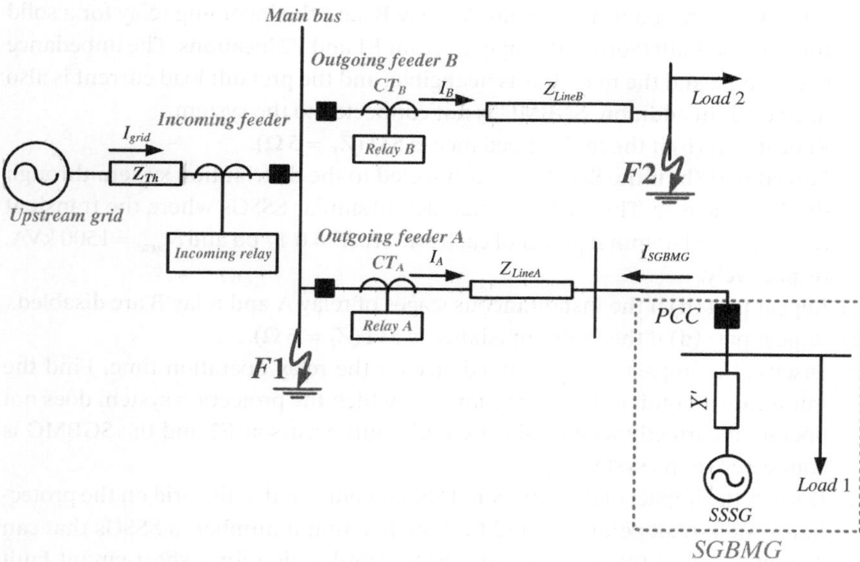

Fig. 4.38 Sample distribution system (Example 4.1). [27] / with permission of IEEE.

Table 4.6 Setting values of relay A and relay B.

ANSI code	Parameter	Value
51	Curve	NI
	Pickup $\left(I_{51}^{Outgoing}\right)$	1.2 A
	TMS	0.05 s
50	Pickup $\left(I_{50}^{Outgoing}\right)$	6 A
	Delay	0 s

In addition, the CT ratio of both outgoing feeders (CT_A and CT_B) is selected as 400/1 A, and the CT ratio of incoming feeder is considered 1000/1 A. Furthermore, for simplicity, the load current during the short-circuit fault and also the line impedances (Z_{LineA} and Z_{LineB}) are neglected.

a) Find proper settings for the incoming OCR while the SGBMG is not connected to the system. The maximum load current of the incoming feeder ($I_{loadmax}$) is considered 1000 A, and its CT ratio is 1000/1 A.

b) Find the operation times of relay A, relay B, and the incoming relay for a solid three-phase fault (zero fault impedance) at F1 and F2 locations. The impedance between F2 and the main bus is negligible, and the prefault load current is also neglected. In addition, SGBMG is not connected to the system.
c) Repeat part (b) if the fault impedance is 5 Ω ($Z_f = 5\,\Omega$).
d) Repeat part (b) if the SGBMG is connected to the distribution system through the PCC breaker. This SGBMG includes 10 similar SSSGs where the transient reactance and nominal power of each one are $X' = 0.17$ pu and $S_{base} = 1500$ kVA, respectively.
e) Repeat part (d) if the instantaneous stages of relay A and relay B are disabled.
f) Repeat part (d) if the fault impedance is 5 Ω ($Z_f = 5\,\Omega$).
g) Discuss the impact of fault impedance on the relay operation time. Find the minimum amount of fault impedance at which the protection system does not operate incorrectly when a short-circuit fault occurs at F2 and the SGBMG is connected to the system.
h) Discuss the impact of the number of SSSGs connected to the grid on the protection system maloperations, and find the maximum number of SSSGs that can be connected to the grid without losing coordination for a short-circuit fault at F2.

Solution

a) The nondirectional OCRs are traditionally coordinated with each other, considering the radial configuration for the distribution system. The CTI between the incoming relay and the outgoing relays is considered 300 ms, i.e., CTI = 300 ms. The setting values of functions 50 and 51 of the incoming relay are determined in the following paragraphs using conventional methodology.

   Function 50: Function 50 commonly refers to the instantaneous operation of the overcurrent protection. However, in some of the new digital relays, this function can be adjusted at either definite or instantaneous operation time. For adjusting function 50 with instantaneous operation time and without exploiting an intentional time delay, at least one of the following conditions should be fulfilled:

   1) No protection devices would be present in the downstream region. In such a condition, the operation time of function 50 can be adjusted instantaneously without considering coordination requirements.
   2) The maximum fault current magnitude at the upstream relay point should be at least 120% greater than that at the downstream relay point. In other words, a considerable amount of impedance (distance) should be present between these two points. In such a condition, the pickup current of the instantaneous overcurrent function is adjusted at least 120% greater than the maximum fault current at the downstream relay location.

In this example, due to the short distance of feeder A, the maximum fault current at the incoming relay location is almost equal to that at the outgoing relay location, and therefore, the current grading coordination method cannot be applied. Consequently, the operation time of function 50 cannot be adjusted instantaneously and the time grading coordination technique is utilized in which this operation time is adjusted at CTI = 300 ms. Furthermore, the pickup current of incoming relay function 50 at the CT primary side should be considered equal to that of the outgoing relays. The value of outgoing relay function 50 pickup at the primary side of CT, i.e., $I_{primary}$, can be calculated as

$$I_{Primary} = I_{50}^{Outgoing} \times \text{CT ratio} = 6\ \text{pu} \times 400 = 2.4\ \text{kA}$$

Consequently, the pickup current of the incoming relay function 50 is calculated as

$$I_{50}^{Incoming} = \frac{I_{Primary}}{\text{CT ratio}} = \frac{2.4\ \text{kA}}{1000} = 2.4\ \text{A}$$

Function 51: The pickup current of function 51 $\left(I_{51}^{incoming}\right)$ is considered 120% of the maximum load current:

$$I_{51}^{Incoming} = \frac{1.2 \times I_{loadmax}}{\text{CT ratio}} = \frac{1.2 \times 1000}{1000} = 1.2\ \text{A}$$

To determine the TMS value of function 51, the CTI between the incoming relay and the outgoing relays is considered 300 ms at the maximum fault current for which function 51 operates and function 50 is not triggered. This fault current magnitude, corresponding to point A in Fig. 4.38, is equal to 2.4 kA. At this fault current magnitude, the operation time of outgoing relays $\left(t_{op}^{Outgoing}\right)$ can be calculated as

$$t_{op}^{Outgoing}\Big|_{@I_F=2.4\ \text{kA}} = \text{TMS} \times \frac{0.14}{\left(\frac{I_F/\text{CT Ratio}}{I_{51}^{Outgoing}}\right)^{0.02} - 1} = 0.05 \times \frac{0.14}{\left(\frac{2400/400}{1.2}\right)^{0.02} - 1} = 214\ \text{ms}$$

Therefore, the operation time of incoming relay $\left(t_{op}^{Incoming}\right)$ at this fault current magnitude, i.e., 2.4 kA, can be determined as

$$t_{op}^{Incoming}\Big|_{@I_F=2.4\ \text{kA}} = t_{op}^{Outgoing}\Big|_{@I_F=2.4\ \text{kA}} + \text{CTI} = 214\ \text{ms} + 300\ \text{ms} = 514\ \text{ms}$$

Consequently, the TMS value of the incoming relay can be calculated as

$$t_{op}^{Incoming}\Big|_{@I_F=2.4\ \text{kA}} = 514\ \text{ms} = \text{TMS} \times \frac{0.14}{\left(\frac{I_F/\text{CT Ratio}}{I_{51}^{Incoming}}\right)^{0.02} - 1}$$

$$= \text{TMS} \times \frac{0.14}{\left(\frac{2400/1000}{1.2}\right)^{0.02} - 1}$$

$$\Rightarrow \text{TMS} = 0.068\ \text{s}$$

The operation curves of both incoming and outgoing relays are depicted in Fig. 4.39. As shown, for all various amounts of fault current magnitudes, the operation time difference between the incoming and outgoing relays is greater than the selected CTI = 300 ms.

b) Before the connection of SGBMG to the network, the distribution system remains radial, and thus, for fault F1 that occurs at the main bus, the current magnitude that passes through both outgoing feeders is zero. Therefore, relay A and relay B remain stable in this case. In addition, the fault current at F1 is 10 kA, which triggers the definite stage of the incoming relay. In this case, the incoming relay operates at $t = 300$ ms.
Fault F2 occurs at the main protection zone of relay B, resulting in the short-circuit fault current passing through CT_B. The current measured by CT_B (I_B) and injected into relay B can be determined as

$$I_B = \frac{10\ \text{kA}}{\text{CT ratio}} = \frac{10\,000}{400} = 25\ \text{A}$$

Since the obtained I_B current is greater than the pickup value of the instantaneous stage $\left(I_{50}^{Outgoing}\right)$, i.e., 6 A, relay B issues the trip command instantaneously. For better demonstration, the operation curves of relays are

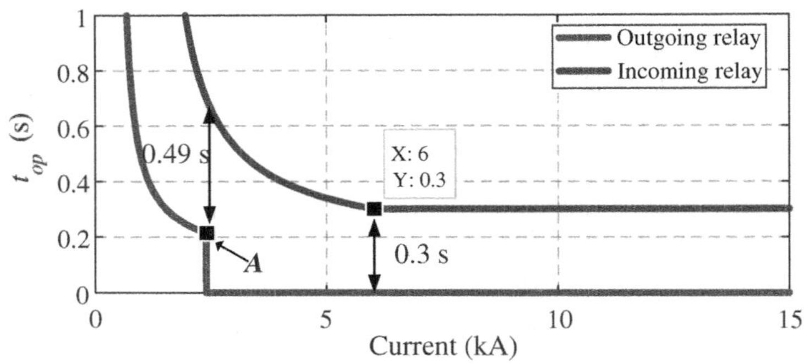

Fig. 4.39 Operation curves of incoming and outgoing relays.

illustrated in Fig. 4.39. Due to the radial configuration of the distribution system before the connection of SGBMG, no fault current passes through relay A, and thus, this relay remains stable. In addition, although the incoming relay has been triggered, the fault is cleared properly by relay B before the operation of the incoming relay.

c) The short-circuit fault current (I_F) for $Z_f = 5\,\Omega$ can be calculated as

$$I_F = \frac{V_s/\sqrt{3}}{Z_{Th} + Z_f}$$

where V_s is the nominal voltage of the system, i.e., 20 kV, and Z_{Th} is the upstream system impedance. Parameter Z_{Th} can be determined as

$$Z_{Th} = \frac{V_s/\sqrt{3}}{I_F^{max}} = \frac{20\text{ kV}/\sqrt{3}}{10\text{ kA}} = j1.15\,\Omega$$

Consequently, the I_F magnitude for $Z_f = 5\,\Omega$ can be calculated as

$$I_F = \left|\frac{V_s/\sqrt{3}}{Z_{Th} + Z_f}\right| = \left|\frac{20\text{ kV}/\sqrt{3}}{1.15j + 5}\right| = 2.25\text{ kA}$$

If the fault occurs at F1, similar to the previous parts, neither relay A nor relay B senses the fault current, and both remain stable properly. In such a condition, the fault current passes through the incoming feeder, and the measured current at its CT secondary is 1.875 A. This current is injected into the incoming relay and only triggers its function 51. In this case, the operation time of the incoming relay can be calculated as

$$t_{op}^{Incoming}\Big|_{@I_F=2.25\text{ kA}} = \text{TMS} \times \frac{0.14}{\left(\dfrac{I_F/\text{CT Ratio}}{I_{51}^{Incoming}}\right)^{0.02} - 1}$$

$$= 0.068 \times \frac{0.14}{\left(\dfrac{2250/1000}{1.2}\right)^{0.02} - 1} = 752\text{ ms}$$

For the fault F2, the current measured by CT_B (I_B) and injected into relay B can be determined as

$$I_B = \frac{2.25\text{ kA}}{\text{CT ratio}} = \frac{2250}{400} = 3.75\text{ A}$$

In this case, $I_B = 3.75$ A is smaller than $I_{50}^{Outgoing}$ and thus, the instantaneous stage of relay B is not triggered. However, its inverse stage (51) is activated, whose operation time can be determined as

$$t_{op}^{Relay\ B} = \text{TMS} \times \frac{0.14}{\left(\dfrac{I_B}{I_{51}}\right)^{0.02} - 1} = 0.05 \times \frac{0.14}{\left(\dfrac{3.75}{1.2}\right)^{0.02} - 1} = 303\text{ ms}$$

Similar to part (a), no fault current passes through relay A due to the radial configuration of distribution system before the connection of the SGBMG. Furthermore, before operation of the incoming relay, the fault has been properly cleared by relay B. If, due to a failure, relay B does not operate properly, the incoming relay will operate as the backup protection and clear the fault properly. The operation time of the incoming relay in this case can be calculated as

$$t_{op}^{Incoming}\Big|_{@I_F=2.25\text{ kA}} = \text{TMS} \times \frac{0.14}{\left(\frac{I_F/\text{CT Ratio}}{I_{51}^{Incoming}}\right)^{0.02} - 1}$$

$$= 0.068 \times \frac{0.14}{\left(\frac{2250/1000}{1.2}\right)^{0.02} - 1} = 752\text{ ms}$$

Based on these calculations, if the main protection relay, i.e., relay B, fails to clear the fault, the incoming relay properly operates at $t = 752$ ms as the backup and de-energizes the main bus. The time difference (TD) between the operation of incoming relay and relay B for this short-circuit fault is equal to TD = 752–303 = 449 ms, which is greater than the required CTI = 300 ms. This reveals that these relays have a proper time grading coordination for the radial configuration condition of the distribution system.

d) After the connection of SGBMG, the fault points will be supplied by two sources, i.e., the upstream grid and SGBMG.
First, it is assumed that the fault occurs at F1 and its impedance is zero, i.e., $Z_f = 0\,\Omega$. In such a condition, the 10 kA fault current is injected by the upstream grid. Moreover, the value of the SGBMG back-feed fault current (I_{SGBMG}) can be determined by

$$I_{SGBMG} = N \times \frac{E_{SSSG}}{X'}\text{(pu)}$$

where N is the number of grid-connected SSSGs, and E_{SSSG} is the internal voltage of each SSSG, which is considered equal to 1 pu for simplicity. Therefore, the rms value of I_{SGBMG} can be determined as

$$I_{SGBMG} = N \times \frac{E_{SSSG}}{X'} \times \frac{S_{base}}{\sqrt{3} \times V_s} = 10 \times \frac{1}{0.17} \times \frac{1500\text{ kVA}}{\sqrt{3} \times 20\text{ kV}} = 2.54\text{ kA}$$

Therefore, the current I_{SGBMG} passes through feeder A, and its associated relay senses the following current (I_A):

$$I_A = \frac{I_{SGBMG}}{\text{CT ratio}} = \frac{2.54\text{ kA}}{400} = 6.35\text{ A}$$

Since $I_A = 6.35$ A is greater than $I_{50}^{Outgoing}$, the instantaneous stage of relay A operates incorrectly. Since no fault current passes through feeder B, its associated relay, i.e., relay B, does not maloperate. Moreover, the current magnitude that passes through the incoming relay remains 10 kA, and thus, as expressed in part (b), the incoming relay operates after 300 ms.

For the fault occurring at F2 with $Z_f = 0\,\Omega$, the summation of SGBMG and upstream grid currents passes through relay B, which can be calculated as

$$I_B = I_{SGBMG} + I_{grid} = 10\text{ kA} + 2.54\text{ kA} = 12.54\text{ kA}$$

$$I_B = \frac{12.54\text{ kA}}{\text{CT ratio}} = \frac{12\,540}{400} = 31.35\text{ A}$$

Since the amount of I_B is greater than the pickup value of instantaneous stage $\left(I_{50}^{Outgoing}\right)$, i.e., 6 A, relay B issues the trip command instantaneously. In such a condition, $I_{SGBMG} = 2.54$ kA passes through feeder A, and its associated relay measures the following current (I_A):

$$I_A = \frac{I_{SGBMG}}{\text{CT ratio}} = \frac{2.54\text{ kA}}{400} = 6.35\text{ A}$$

As $I_A = 6.35$ A is greater than $I_{50}^{Outgoing}$, the instantaneous stage of relay A maloperates incorrectly. In addition, $I_{grid} = 10$ kA passes through the incoming feeder, which triggers its associated relay. Although the incoming relay has been triggered, relay B operates instantaneously before the incorrect operation of this relay at $t = 752$ ms.

e) In some applications, the instantaneous stage of outgoing relays, such as relay A and relay B, may be disabled due to their probable maloperations during system transients and nonfault events. In this part, the protection system performance without considering the instantaneous stage is investigated.

For fault F1, $I_{SGBMG} = 2.54$ kA passes through feeder A, and its operation time can be determined as

$$t_{op}^{Relay\,A} = \text{TMS} \times \frac{0.14}{\left(\frac{I_A}{I_{51}}\right)^{0.02} - 1} = 0.05 \times \frac{0.14}{\left(\frac{2.54\text{ kA}/400}{1.2}\right)^{0.02} - 1} = 207\text{ ms}$$

For fault F1, after 207 ms, relay A issues the trip signal incorrectly. Since no fault current passes through feeder B, its associated relay, i.e., relay B, remains stable in this case.

For fault F2, the operation times of relay A and relay B can be calculated as

$$t_{op}^{Relay\,A} = \text{TMS} \times \frac{0.14}{\left(\frac{I_A}{I_{51}}\right)^{0.02} - 1} = 0.05 \times \frac{0.14}{\left(\frac{2.54\text{ kA}/400}{1.2}\right)^{0.02} - 1} = 207\text{ ms}$$

$$t_{op}^{Relay\,B} = \text{TMS} \times \frac{0.14}{\left(\frac{I_B}{I_{51}}\right)^{0.02} - 1} = 0.05 \times \frac{0.14}{\left(\frac{12.54\text{ kA}/400}{1.2}\right)^{0.02} - 1} = 103\text{ ms}$$

Therefore, for the fault at F2, relay B issues the trip signal 104 ms before relay A operation. However, the interval between the operation times of these relays is smaller than the required CTI = 300 ms. Therefore, in such a condition, relay A may still operate incorrectly.

f) The SGBMG fault current magnitude (I_{SGBMG}) when $Z_f = 5\,\Omega$ can be calculated as

$$I_{SGBMG} = \left| N \times \frac{E_{SSSG}}{X' + Z_f} \times \frac{S_{base}}{\sqrt{3} \times V_s} \right|$$

$$= \left| 10 \times \frac{1}{j0.17 + 5} \times \frac{1500\,\text{kVA}}{\sqrt{3} \times 20\,\text{kV}} \right| = 86\,\text{A}$$

Therefore, the SGBMG back-feed current considerably decreases in this case, and thus, relay A is not triggered. Since the SGBMG current is negligible in this case, the operation times of other relays are almost similar to those obtained in part (c).

g) The maximum amount of Z_f that can trigger relay A can be calculated as

$$I_{SGBMG} = \left| N \times \frac{E_{SSSG}}{X' + Z_f} \times \frac{S_{base}}{\sqrt{3} \times V_s} \right| = I_{51} \times \text{CT ratio}$$

Therefore,

$$\left| 10 \times \frac{1}{j0.17 + Z_f} \times \frac{1500\,kVA}{\sqrt{3} \times 20\,kV} \right| = 1.2 \times 400 \rightarrow Z_f \approx 0.89\,\Omega$$

Therefore, for $Z_f > 0.89\,\Omega$, relay A will not be triggered. For $Z_f < 0.89$, relay A is triggered, and the operation times of relay A and relay B are depicted in Fig. 4.40. As shown, for $Z_f < 0.06\,\Omega$, the instantaneous stage of relay A operates incorrectly. Furthermore, the difference between the operation times of relay A and relay B is less than CTI = 300 ms for $0.06\,\Omega < Z_f < 0.229\,\Omega$. However, for $Z_f > 0.229\,\Omega$, this difference is greater than CTI = 300 ms, and thus, the fault will be cleared by relay B before the incorrect operation of relay A. Based on these explanations, the SGBMG back-feed current can cause relay A maloperation for $Z_f < 0.229\,\Omega$.

These studies properly reveal that the higher fault impedances lead to a decrement in the magnitude of the SGBMG back-feed current. Therefore, the adverse impact of SGBMG back-feed on relay operation considerably decreases with increasing the fault impedance. In other words, for high values of fault impedance, the relay maloperation due to the SGBMG back-feed is not expected. For instance, based on the calculations of this part, the SGBMG back-feed leads to relay A maloperation only for low values of impedance faults, i.e., $Z_f < 0.229\,\Omega$.

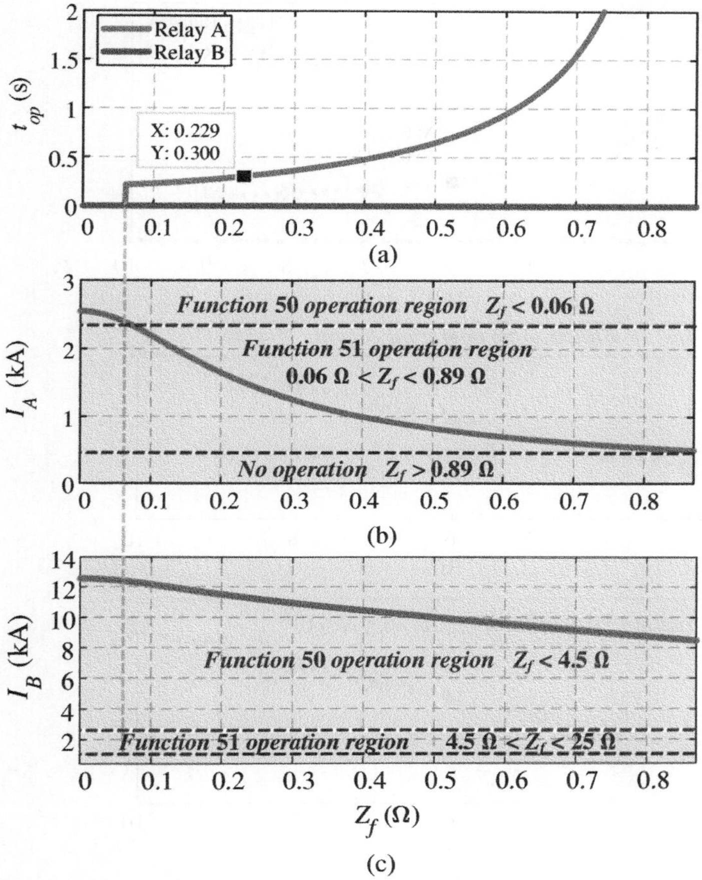

Fig. 4.40 (a) Operation times of relay A and relay B for various amount of Z_f. (b) Relay A current for various amounts of Z_f. (c) Relay B current for various amounts of Z_f.

h) Assuming $Z_f = 0$, the minimum number of SSSGs that can trigger relay A can be determined as

$$I_{SGBMG} = \left| N \times \frac{E_{SSSG}}{X' + Z_f} \times \frac{S_{base}}{\sqrt{3} \times V_s} \right| = I_{51} \times \text{CT ratio}$$

Therefore,

$$N \times \frac{1}{j0.17} \times \frac{1500\ \text{kVA}}{\sqrt{3} \times 20\ \text{kV}} \geq 1.2 \times 400 \rightarrow N = 2$$

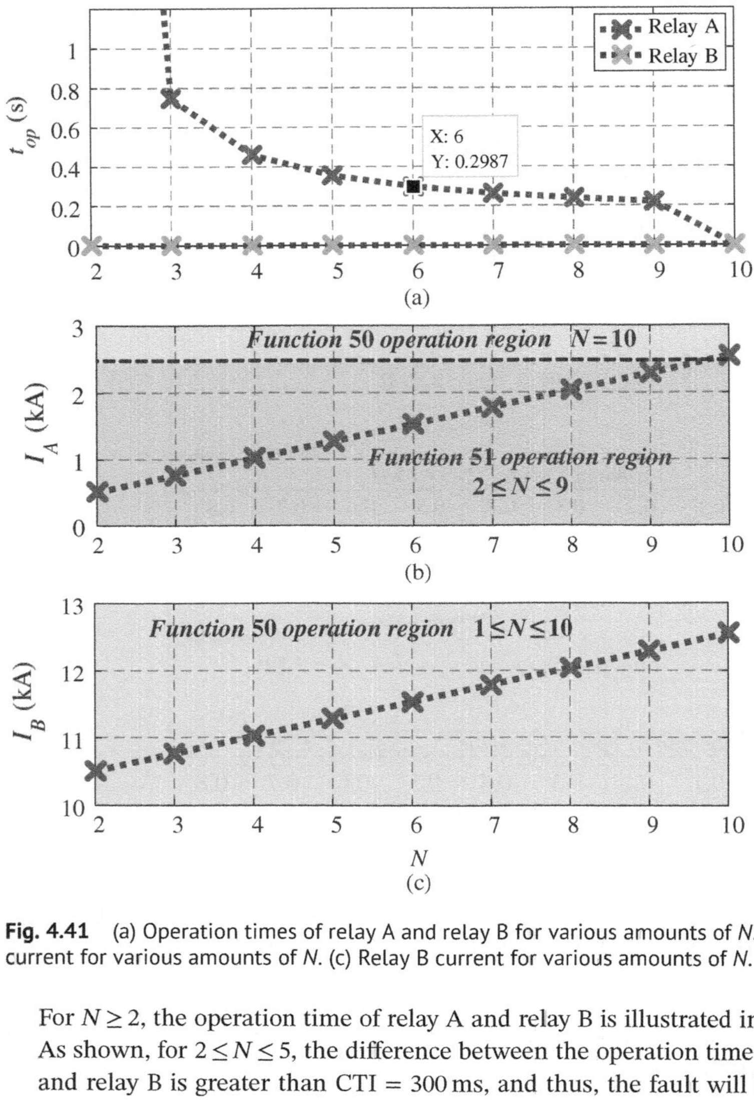

Fig. 4.41 (a) Operation times of relay A and relay B for various amounts of *N*. (b) Relay A current for various amounts of *N*. (c) Relay B current for various amounts of *N*.

For $N \geq 2$, the operation time of relay A and relay B is illustrated in Fig. 4.41. As shown, for $2 \leq N \leq 5$, the difference between the operation time of relay A and relay B is greater than CTI = 300 ms, and thus, the fault will be cleared by relay B before the incorrect operation of relay A. Therefore, the maximum number of SSSGs that can be connected to the grid without losing coordination between relay A and relay B is 5. For $N > 5$, the difference between operation time of relay A and relay B becomes less than CTI = 300 ms, and thus, relay A may maloperate before the fault occurred at the outgoing feeder B would be cleared by relay B. These calculations reveal that the probability of relay miscoordination due to SGBMG back-feed considerably increases when the number of connected SSSGs and their nominal power increases. For instance, in this example, when the total power of grid-connected SSSGs is 1500 kVA ($N = 1$), the SGBMG back-feed does not trigger the healthy feeder relay, and thus, the miscoordination is not expected. Generally, for various configurations of MV distribution feeders, the relay miscoordination is not expected when the total capacity of grid-connected SSSG is less than 1 MW.

Based on these studies, to adjust the relay setting and minimize the probability of relay maloperation due to the SGBMG back-feed injected current, the following guidelines can be considered:

- If the maximum amount of SGBMG back-feed current during the fault is greater than the instantaneous stage of the associated healthy feeder relay, it might be better that this protection stage would be disabled or its pickup value should be adjusted greater than the maximum SGBMG back-feed current.
- The magnitude of SGBMG back-feed current considerably decreases when the fault impedance is not zero. For instance, for a fault with $Z_f = 5\,\Omega$, the SGBMG back-feed current is about 86 A, which does not trigger the healthy feeder relay. Since in the distribution systems the probability of having high value of fault impedance is higher than that of low impedance ones, in most cases, the probability of maloperation is reduced considerably. In this direction, and to consider the random nature of fault impedance, a probabilistic approach can be utilized to determine the optimal setting values.
- For radial configuration of distribution systems and before connection of any SGBMG, the feeder protection relays should be coordinated only with their downstream protection devices. After the SGBMG connection, due to the SSSG back-feed current, function 51 of the feeders to which an SGBMG is connected should also be coordinated with the relays of the other feeders.

Example 4.2

A sample distribution system is depicted in Fig. 4.42. The nominal voltage of the distribution system is 20 kV, and the maximum amount of three-phase short-circuit current injected by the upstream grid is $I_F^{max} = 10$ kA. An SGBMG that contains 10 similar SSSGs is connected to this system through the PCC breaker. The nominal capacity of each SSSG is 1500 kVA. Furthermore, the transient reactance of each SSSG is considered 0.17 pu. The line length between B1 and B2 is considered 1 km. Furthermore, the amount of line impedances, i.e.,

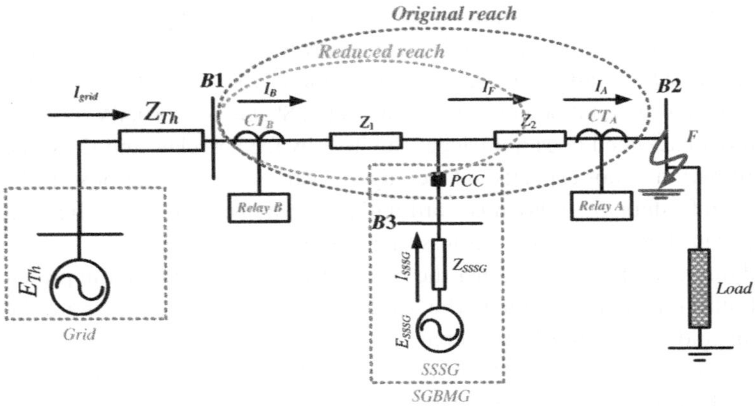

Fig. 4.42 Sample distribution system (Example 4.2). [31] / with permission of Elsevier.

Table 4.7 Setting values of relay A.

ANSI code	Parameter	Value
51	Curve	NI
	Pickup $\left(I_{51}^{Outgoing}\right)$	1.2 A
	TMS	0.05 s

Z_1 and Z_2 are similar to each other and equal to 0.46 Ω. The load current during the short-circuit fault is neglected. Furthermore, the setting values of relay A are given in Table 4.7.

a) Find the proper settings for relay B without considering the SGBMG connection. The maximum load current of the incoming feeder ($I_{loadmax}$) is considered 400 A.
b) Find the reach of relay B before and after the connection of SGBMG.
c) Find the difference between relay A and relay B operation times after the connection of SGBMG.
d) Discuss the impact of the Thevenin impedance of the upstream grid and Z_1 (SGBMG connection point impedance) on the reach of relay B.

Solution

a) To coordinate relay A and relay B, the maximum magnitude of three-phase short-circuit fault at the location F should be determined. To do so, the

Thevenin impedance of the upstream grid should be calculated as

$$Z_{Th} = \frac{V_s/\sqrt{3}}{I_F^{max}} = \frac{20\ \text{kV}/\sqrt{3}}{10\ \text{kA}} = j1.15\ \Omega$$

Then, the short-circuit fault current (I_F) is obtained as

$$I_F = \frac{V_s/\sqrt{3}}{Z_{Th} + Z_1 + Z_2} = \left|\frac{20\ \text{kV}/\sqrt{3}}{j1.15 + j0.92}\right| = 5.57\ \text{kA}$$

The operation time of relay A for $I_F = 5.57$ kA can be calculated as

$$t_{op}^{Relay\ A}\Big|_{@I_F=5.57\ \text{kA}} = \text{TMS} \times \frac{0.14}{\left(\frac{I_F/\text{CT Ratio}}{I_{51}^{Relay\ A}}\right)^{0.02} - 1}$$

$$= 0.05 \times \frac{0.14}{\left(\frac{5.57\ \text{kA}/400}{1.2}\right)^{0.02} - 1} = 139\ \text{ms}$$

Therefore, considering CTI = 300 ms, the operation time of relay B $\left(t_{op}^{Relay\ B}\right)$ at this fault current magnitude, i.e., 5.57 kA, can be determined as follows:

$$t_{op}^{Relay\ B}\Big|_{@I_F=5.57\ \text{kA}} = t_{op}^{Relay\ A}\Big|_{@I_F=5.57\ \text{kA}} + \text{CTI}$$

$$= 139\ \text{ms} + 300\ \text{ms} = 439\ \text{ms}$$

Consequently, considering the relay B pickup current equal to 1.2 A, i.e., $I_{51}^{Relay\ B} = 1.2$ A, the TMS value of relay B can be calculated as follows:

$$t_{op}^{Relay\ B}\Big|_{@I_F=5.57\ \text{kA}} = 439\ \text{ms} = \text{TMS} \times \frac{0.14}{\left(\frac{I_F/\text{CT Ratio}}{I_{51}^{Relay\ B}}\right)^{0.02} - 1}$$

$$= \text{TMS} \times \frac{0.14}{\left(\frac{5.57\ \text{kA}/400}{1.2}\right)^{0.02} - 1}$$

$$\Rightarrow \text{TMS} = 0.157\ \text{s}$$

b) Before connection of the SGBMG to the system, the maximum reach of relay B where $I_F \geq I_{51}^{Relay\ B}$ can be calculated as

$$I_F = \frac{V_s/\sqrt{3}}{Z_{Th} + Z_{reach}} = \left|\frac{20\ \text{kV}/\sqrt{3}}{j1.15 + Z_{reach}}\right| \geq 480\ \text{A} \Rightarrow |Z_{reach}| \leq 22.9\ \Omega$$

After the connection of SGBMG to the system, the maximum reach of relay B where $I_B \geq I_{51}^{Relay\ B}$ can be calculated as

$$I_B = \frac{E_{Th}}{Z_{SGBMG} \| (Z_{reach} - Z_1) + Z_{Th} + Z_1}$$

$$-\frac{E_{SGBMG}}{(Z_{Th}+Z_1)\|(Z_{reach}-Z_1)+Z_{SGBMG}}\times\frac{Z_{reach}-Z_1}{Z_{reach}+Z_{Th}}\geq I_{51}^{Relay\ B}$$

where

$$Z_{SGBMG}={}^{X'}\!/_{N}\times Z_{base}={}^{0.17}\!/_{10}\times{}^{(20\ \text{kV})^2}\!/_{1500\ \text{kVA}}=j4.53\ \Omega$$

$$E_{Th}=E_{SGBMG}=20\ \text{kV}/\sqrt{3}$$

Therefore,

$$I_B=\frac{{}^{20\ \text{kV}}\!/_{\sqrt{3}}}{j4.53\|(Z_{reach}-j0.46)+j1.15+j0.46}$$

$$-\frac{{}^{20\ \text{kV}}\!/_{\sqrt{3}}}{(j1.15+j0.46)\|(Z_{reach}-j0.46)+j4.53}\times\frac{Z_{reach}-j0.46}{Z_{reach}+j1.15}\geq 480\ \text{A}$$

$$\Rightarrow |Z_{reach}|\leq 17\ \Omega$$

Consequently, the reach of relay B is reduced from 22.9 to 17 Ω after the SGBMG connection. In other words, after the SGBMG connection in the middle of the line, the reach of relay B and its sensitivity are decreased by 26%. The main negative consequences of this reduction are:

- The relay cannot detect the short-circuit faults at medium and long length distribution lines, i.e., when Z_2 is greater than 17 Ω. Depending on the type of line conductor, the protected length of the relay may be different. For conductors with $|Z| = 1\ \Omega/\text{km}$, the relay covers only the faults up to 17 km.
- Since the relay reach is reduced to 17 Ω, the faults with relatively high impedances may not be detected by the relay.
- The relay cannot detect the faults that occur at the downstream lines. In such a condition, the relay may not properly operate as the backup protection for out-of-zone faults when the downstream relay fails to clear them.
- Reduction of the relay reach and its sensitivity is mainly caused by the reduction of the upstream grid share in the fault current. In other words, the current that passes through relay B is decreased after the SGBMG connection. Therefore, considering the inverse time–current operation curve of relay B, the operation time of this relay and also FCT will increase due to the decrement of the current passing through the relay after the SGBMG connection.

The percentage of reach reduction significantly depends on the location of the SGBMG connection and the upstream grid Thevenin impedance (short-circuit capacity). Therefore, in part (d) of this example, the impact of these parameters on relay reach is discussed.

c) After the connection of SGBMG, and using the superposition theorem, the fault current at location F can be calculated as follows:

$$I_F=\frac{E_{Th}}{Z_{SGBMG}\|Z_2+Z_{Th}+Z_1}\times\frac{Z_{SGBMG}}{Z_{SGBMG}+Z_2}$$

$$+\frac{E_{SGBMG}}{(Z_{Th}+Z_1)||Z_2+Z_{SGBMG}}\times\frac{Z_1+Z_{Th}}{Z_1+Z_{Th}+Z_2}$$

Therefore, I_F is obtained as 7 kA after the SGBMG connection, based on the following calculations:

$$I_F=\frac{20\text{ kV}/\sqrt{3}}{j4.53||j0.46+j1.15+j0.46}\times\frac{j4.53}{j4.53+j0.46}$$
$$+\frac{20\text{ kV}/\sqrt{3}}{(j1.15+j0.46)||j0.46+j4.53}\times\frac{j0.46+j1.15}{j0.46+j1.15+j0.46}=7\text{ kA}$$

In this case, the current that passes through relay B (I_B) can be calculated by

$$I_B=\frac{E_{Th}}{Z_{SGBMG}||Z_2+Z_{Th}+Z_1}$$
$$-\frac{E_{SGBMG}}{(Z_{Th}+Z_1)||Z_2+Z_{SGBMG}}\times\frac{Z_2}{Z_1+Z_2+Z_{Th}}$$

Consequently, I_B is calculated as 5.17 kA based on this equation. Therefore, after connection of the SGBMG, the relay A current (downstream relay of the SGBMG connection point) increases while the current passing through relay B (upstream relay of the SGBMG connection point) decreases. As such, the operation time between relay A and relay B increases, which can be calculated as

$$t_{op}^{Relay\ A}\Big|_{@I_F=7\text{ kA}}=\text{TMS}\times\frac{0.14}{\left(\frac{I_F/\text{CT Ratio}}{I_{51}^{Relay\ A}}\right)^{0.02}-1}$$
$$=0.05\times\frac{0.14}{\left(\frac{7\text{ kA}/400}{1.2}\right)^{0.02}-1}$$
$$=127\text{ ms}$$

$$t_{op}^{Relay\ B}\Big|_{@I_F=5.17\text{ kA}}=\text{TMS}\times\frac{0.14}{\left(\frac{I_F/\text{CT Ratio}}{I_{51}^{Relay\ A}}\right)^{0.02}-1}$$
$$=0.157\times\frac{0.14}{\left(\frac{5.17\text{ kA}/400}{1.2}\right)^{0.02}-1}=451\text{ ms}$$

Therefore, the time difference between the operation of relay A and relay B increases to 324 ms after the SGBMG connection.

d) Figure 4.43(a) shows the variations of relay B reach (Z_{reach}) before and after the SGBMG connection for various amounts of Z_{Th}. As shown, after and before SGBMG connection, the relay reach decreases with an increment of Z_{Th}. In fact, if the short-circuit capacity of upstream system reduces, the reach of relay B decreases. On the other hand, after the SGBMG connection, the relay reach is reduced. However, the amount of reach reduction is different for various values of Z_{Th}. As shown in Fig. 4.43(b), the percentage of reach reduction considerably increases with an increment of Z_{Th}. For instance, the amount of reach reduction for $Z_{Th} = 0\,\Omega$ is about 10%, while for $Z_{Th} = 6\,\Omega$, it is about 60%. Therefore, the percentage of reach reduction is smaller for higher short-circuit capacities of the upstream grid.

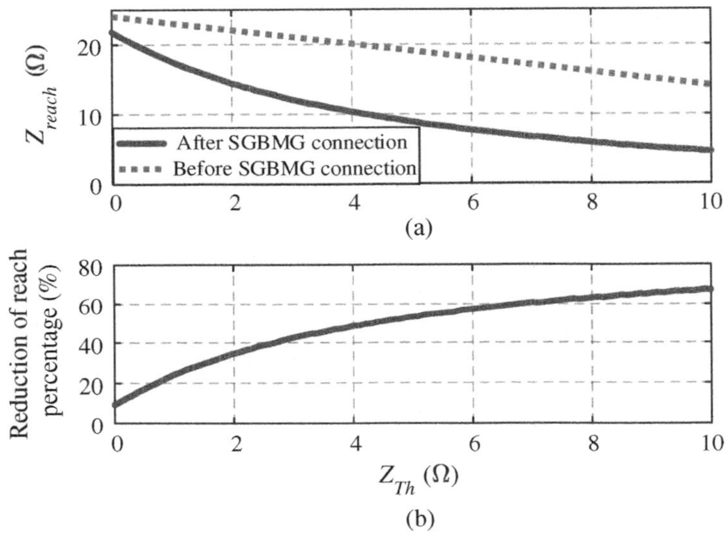

Fig. 4.43 Reach of relay B for various amounts of upstream grid impedance: (a) Z_{reach} and (b) reduction of reach percentage.

Figure 4.44(a) shows the variations of relay B reach (Z_{reach}) before and after the SGBMG connection for various amounts of Z_1. As shown, before the SGBMG connection, the relay reach is not affected by Z_1. However, after the SGBMG connection, the Z_1 value has a considerable impact on the amount of relay reach. In other words, the location of SGBMG connection has a significant impact on the upstream relay reach. As shown in Fig 4.44(b), the percentage of reach reduction considerably increases when Z_1 increases from 0 to 5 Ω. For instance, the amount of reach reduction for $Z_1 = 0\,\Omega$ is about 20%, while for $Z_{Th} = 5\,\Omega$, it is about 45%.

4.3.6 IBDG Protection Challenges

The protection relays used for IBDGs mostly include frequency, voltage, and current relays. Although these protection schemes are simple and straightforward,

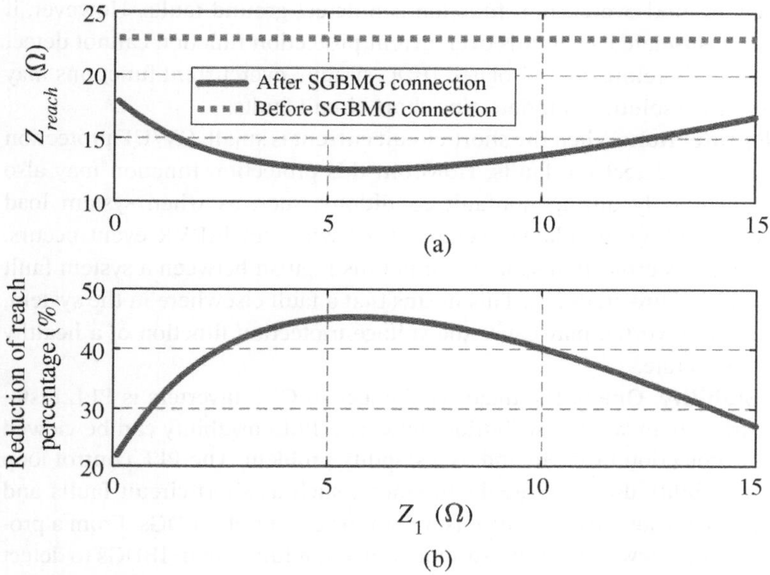

Fig. 4.44 Reach of relay B for various amounts of Z_1, (a) Z_{reach}, and (b) reduction of reach percentage.

they may cause the inverter to trip wrongly or fail to detect a fault correctly. In this section, the protection challenges of IBDGs are discussed briefly.

- **OF/UF protection:** This function issues the trip signal when the system frequency lies outside its permissible range to prevent the IBDG from operating at an improper frequency. However, in some cases, this protection relay may cause a false trip. In an HMG including an SSSG and GFL inverters operating in the islanded mode, the inverters frequency functions are highly sensitive. Because the MG has a low inertia, small load changes, nonlinear varying power loads, and changes in generation can result in frequency variations. These variations may trigger the frequency protection and disconnect the inverter by mistake. Therefore, proper adjustment of the frequency protection settings is crucial. In addition, this protection should be set so that the inverter can meet grid requirements, support system frequency, and cover ride-through faults. If the settings are too strict, the inverter may trip quickly during normal frequency swings and, therefore, would not be able to support the system frequency.
- **Overcurrent protection:** In IBMGs, the short-circuit current is usually limited to about 1.5 times the inverter's rated current. Because of this low fault current level, phase overcurrent functions may not work properly and their sensitivity befalls very low. On the other hand, if the system neutral is effectively

grounded, ground overcurrent function can detect ground faults. However, if the neutral is isolated, even this overcurrent protection function cannot detect these faults. Therefore, in an isolated IBDG system, overcurrent functions may not be effective solutions for detecting short-circuit faults.

- **OV/UF protection:** When the short-circuit current is small, OV/UF protection can be used to detect the faults. However, this protection function may also operate incorrectly during nonfault conditions, such as when system load increases, resulting in voltage decrement, or when an FIDVR event occurs. In addition, this protection scheme cannot distinguish between a system fault and an internal inverter fault. This means that a fault elsewhere in the system, or in another inverter, may cause the voltage protection function of a healthy inverter to operate.
- **PLL instability:** One of the major challenges in GFL inverters is PLL instability, especially in weak distribution networks. PLL instability can be viewed both as a protection problem and as a stability problem. The PLL control loop may lose stability during large disturbances, such as short-circuit faults and reclosing events, and lose synchronism with the grid or other DGs. From a protection point of view, there is no specific protection function in IBDGs to detect this problem promptly and disconnect the unstable unit quickly. If the unstable inverter is not isolated fast enough, the PLL instability can be spread to other DGs, including SSSGs, and endanger the stability of the entire MG.
- **Fault type classification:** The control systems of IBDGs should be designed and implemented to meet the grid code requirements, which results in dynamic behavior that differs significantly from that of SSSGs. Therefore, the values of electrical parameters such as voltage and current magnitude and phase angle are governed by the IBDG control systems, and these values can deviate considerably from those in conventional SG-based systems. For instance, IBDG control system typically limits fault current to only 120–150% of the rated value, and the phase angles of the positive- and negative-sequence currents can differ substantially from those observed in conventional SSSGs. Therefore, this can negatively affect the performance of fault type classification schemes designed based on the characteristics of conventional SGs. As reported in [43], traditional methods, including current-angle-based and current-magnitude-based techniques, may yield incorrect decisions in the presence of IBDGs in MGs.

4.4 Summary

In this chapter, various problems and challenges of the MG protection systems have been raised and discussed through both the analytical and simulation studies. Based on these studies, some of the main challenges of these protection systems include (i) miscoordination between the relays and other protection devices, (ii) maloperation during MG transients such as the unwanted transition between the grid-connected and islanded modes, and (iii) delayed operation during severe fault conditions. These challenges may endanger the security and dependability of the following protection systems:

- Collector system protection, including OCRs, reclosers, and fuse elements
- SSSG protection system, including LOF and backup protection functions
- IBDG protection system
- Load-shedding relays, including UVLS and UFLS relays
- PCC protection relay.

To resolve the shortcomings of the MG protection system and enhance its security, especially during the transition between the grid-connected and islanded modes, some practical solutions are provided in the following chapters.

Problems

4.1 A sample distribution system, which includes two outgoing feeders, is depicted in Fig. 4.45. The nominal voltage of this system is 20 kV, and the maximum three-phase short-circuit current at its main bus is $I_F^{max} = 7$ kA.

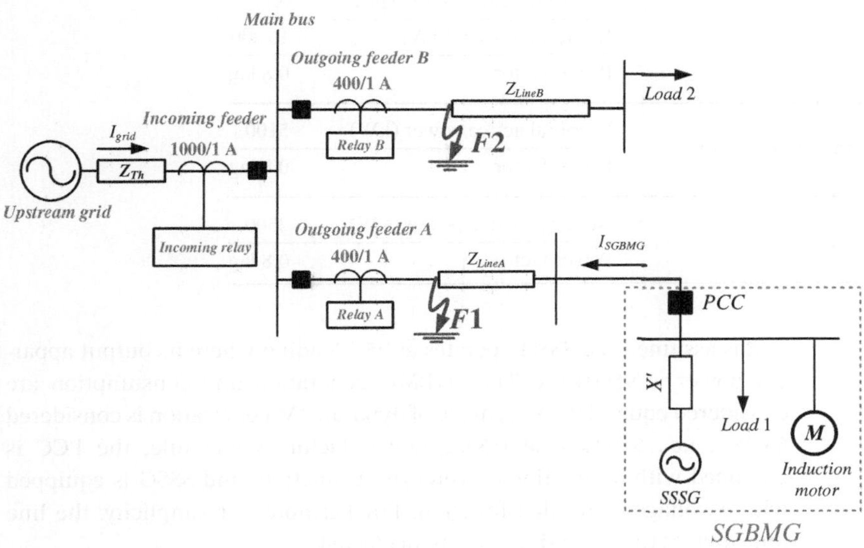

Fig. 4.45 Sample distribution system (Problem 4.1).

Before connection of the SGBMG, relay A and relay B are adjusted as shown in Table 4.8:

Table 4.8 Setting values of relay A and relay B.

ANSI code	Parameter	Value
51	Curve	NI
	Pickup $\left(I_{51}^{Outgoing}\right)$	1.2 A
	TMS	0.1 s
50	Pickup $\left(I_{50}^{Outgoing}\right)$	6 A
	Delay	0 s

The SGBMG is connected to the distribution system through the PCC breaker, as shown in Fig. 4.45. This SGBMG includes one SSSG, and static and dynamic loads whose specifications are presented in Table 4.9.

Table 4.9 SGBMG parameters.

Description	Parameter	Value
SSSG	Transient reactance X' (pu)	0.17
	Nominal power (kVA)	10 000
	Power factor	0.8 lag
Induction motor	Nominal active power (kW)	5100
	Power factor	0.8 lag
Static load	Nominal active power (kW)	2500
	Power factor	0.8 lag

It is assumed the SSSG operates at 95% loading where its output apparent power is 9500 kVA. The SGBMG generation and consumption are considered equal. The percentage of dynamic IM penetration is considered 63.75%, i.e., 5100 kW at 0.8 lag power factor. Meanwhile, the PCC is equipped with a directional overcurrent function, and SSSG is equipped with a voltage-controlled function. Furthermore, for simplicity, the line impedances (Z_{LineA} and Z_{LineB}) are neglected.

a) Find proper setting values for PCC directional overcurrent function and voltage-controlled function of the SSSG. Meanwhile, revise the setting values of relay A.
b) A short-circuit fault occurs at F1. The PCC relay operates and isolates the SGBMG. After fault-initiated islanding of the SGBMG, the IM current magnitude becomes 3.5 pu at 0.55 lag power factor due to its reacceleration. Calculate the voltage magnitude at the IM terminal in this case. If the motor reacceleration time is 6 s, specify which relay maloperates in this case. Find the operation time of these relays.
c) Find the minimum percentage of the IM penetration for which the SSSG relay may maloperate. Discuss the impact of the percentage of the IM penetration on the SSSG relay maloperation.
d) A short-circuit fault occurs at F2. Relay B operates and isolates the fault. After the fault isolation, the IM current magnitude becomes 4 pu at 0.6 lag power factor due to its reacceleration. If the motor reacceleration time is 5 s, specify which relay maloperate in this case.
e) Modify the relay settings to prevent relay maloperation during motor reacceleration.

4.2 Figure 4.46 illustrates a sample distribution system to which an SGBMG is connected. The parameters of this system are given in Table 4.10. The outgoing feeder is equipped with a recloser whose dead-time is considered 100 ms. A three-phase temporary short-circuit fault occurs at the middle of the line (point F). The recloser de-energizes the feeder after

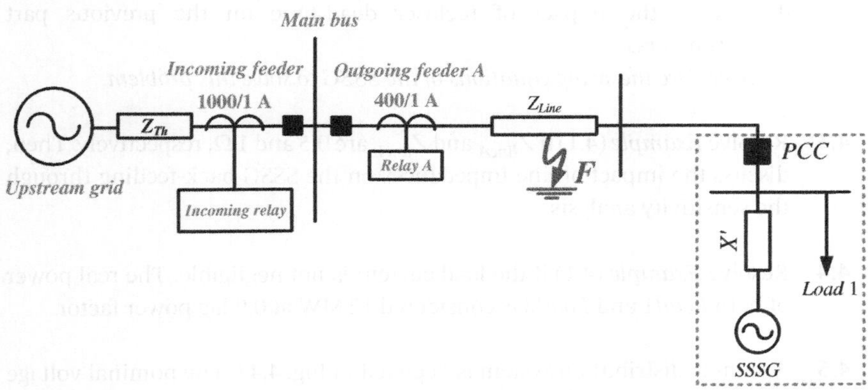

Fig. 4.46 Sample distribution system (Problem 4.2).

Table 4.10 SGBMG and distribution system parameters.

Description	Parameter	Value
Upstream grid	Maximum short-circuit fault (kA)	7
	Nominal voltage (kV)	20
SSSG	Transient reactance X' (pu)	0.17
	Nominal power (kVA)	10 000
	Loading	90%
	Power factor	0.8 lag
	Inertia (s)	1
Line	Length (km)	2
	Impedance (Ω/km)	0.4
Static load	Nominal active power (kW)	5000
	Power factor	0.8 lag

100 ms. Meanwhile, the FCT is considered 200 ms. After reenergizing the feeder, calculate the following parameters.

a) Find the angular frequency and power angle of the SSSG.
b) Find the voltage difference between SSSG and upstream grid.
c) Discuss the impact of SSSG inertia and its loading on the previous items.
d) Discuss the impact of recloser dead-time on the previous part parameters.

Note: Use the swing equations of the SSSG to solve this problem.

4.3 Resolve *Example* (4.1) if Z_{lineA} and Z_{lineB} are 0.5 and 1 Ω, respectively. Then, discuss the impact of line impedances on the SSSG back-feeding through the sensitivity analysis.

4.4 Resolve *Example* (4.1) if the load current is not negligible. The real power of both *Load*1 and *Load*2 is considered 12 MW at 0.9 lag power factor.

4.5 A sample distribution system is depicted in Fig. 4.47. The nominal voltage of this system is 20 kV, and the maximum three-phase short-circuit current at its main bus is $I_F^{max} = 5$ kA. The transient reactance, nominal power, and inertia of the SSSG are $X' = 0.17$ pu, $S_{base} = 5000$ kVA, and $H = 0.75$ s,

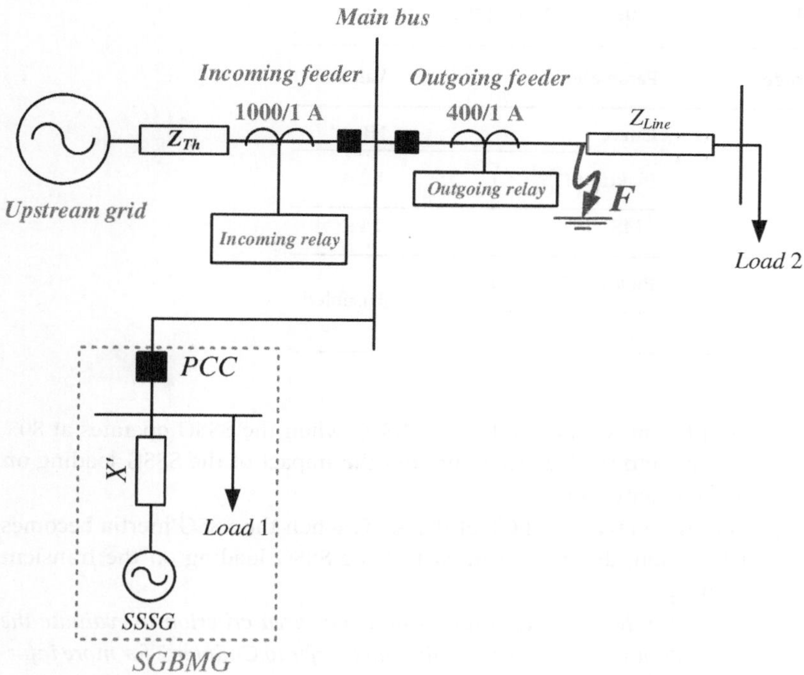

Fig. 4.47 Sample distribution system (Problem 4.5).

respectively. It is assumed the SSSG operates at 95% loading and 0.9 lag power factor. Furthermore, the setting values of the outgoing feeder relay are given in Table 4.11. The impedance of the line is neglected. Furthermore, the load parameters are:

Load 1: 3.2 MW at 0.9 lag power factor
Load 2: 5 MVA at 0.95 lag power factor

a) A three-phase short-circuit fault occurs at F. If function 50 of outgoing relay is disabled, specify whether the SSSG becomes unstable or not.
b) Find the maximum FCT for which the SSSG remains stable.
c) If function 50 of the outgoing relay is disabled, revise the setting values of function 51 of this relay to guarantee the stability of SSSG after fault occurrence at F.
d) If function 50 of outgoing relay is enabled, find the proper pickup current of this function to guarantee the stability of SSSG after fault occurrence at F.

Table 4.11 Setting values of outgoing feeder relay.

ANSI code	Parameter	Value
51	Curve	NI
	Pickup $\left(I_{51}^{Outgoing}\right)$	1.2 A
	TMS	0.1 s
50	Pickup $\left(I_{50}^{Outgoing}\right)$	Disabled
	Delay	

e) Find the maximum FCT of the SSSG when the SSSG operates at 80% loading and 0.9 lag. Then, discuss the impact of the SSSG loading on the transient stability.

f) Find the maximum FCT of the SSSG when the SSSG inertia becomes 1.1 s. Then, discuss the impact of the SSSG loading on the transient stability.

Note: Use the swing equations and equal area criterion to evaluate the SSSG stability after short-circuit faults. Refer to Chapter 2 for more information about transient stability of SSSGs.

4.6 A sample islanded SGBMG whose parameters are given in Table 4.12 is depicted in Fig. 4.48.

a) Specify proper protective functions for protection of SSSG, feeders, and PCC.

b) Determine proper setting values for overcurrent functions of feeders and PCC assuming the islanded operation mode of the SGBMG.

c) Determine proper setting values for voltage-dependent OCR of SSSG.

d) Determine the setting values for the functions mentioned in parts (b) and (c) for the SGBMG grid-connected condition. Discuss the impact of grid connection on the setting values.

4.7 For the sample system depicted in Fig. 4.42, discuss the impact of SSSG reactance value on the upstream relay reach and sensitivity.

4.8 For the sample system depicted in Fig. 4.38, discuss the impact of fault type on the SGBMG back-feeding.

4.9 Resolve *Example* (4.2) considering both functions 51 and 50.

Table 4.12 SGBMG and distribution system parameters.

Description	Parameter	Value
Upstream grid	Maximum short-circuit fault (kA)	4
	Nominal voltage (kV)	20
SSSG	Transient reactance X' (pu)	0.17
	Nominal power (kVA)	5000
	Loading	90%
	Power factor	0.8 lag
	Inertia (s)	1
Line	Length (km)	1
	Impedance (Ω/km)	0.5
Static load	Nominal active power (kW)	3600
	Power factor	0.8 lag

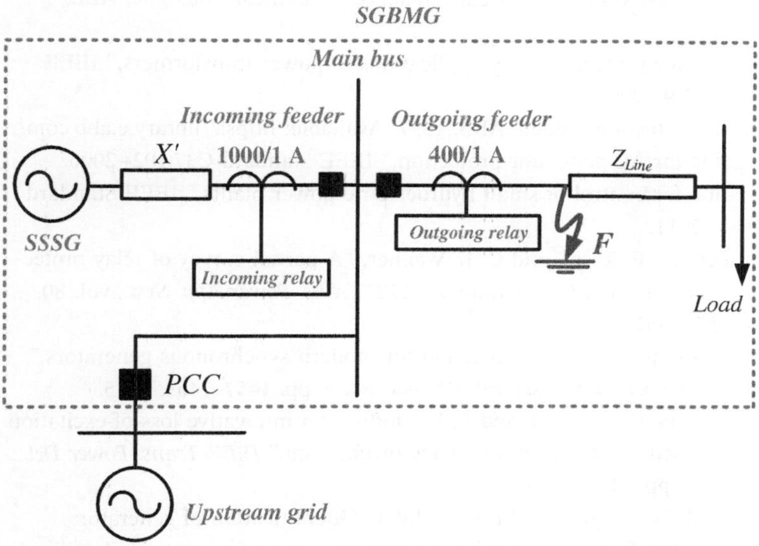

Fig. 4.48 Sample distribution system (Problem 4.6).

References

1 R. Razzaghi, M. Davarpanah, and M. Sanaye-Pasand, "A novel protective scheme to protect small-scale synchronous generators against transient instability," *IEEE Trans. Ind. Electron.*, vol. 60, no. 4, pp. 1659–1667, 2013.
2 J. H. Teng, W. H. Huang, and S. W. Luan, "Automatic and fast faulted line-section location method for distribution systems based on fault indicators," *IEEE Trans. Power Syst.*, vol. 29, no. 4, pp. 1653–1662, 2014.
3 "Technical guidelines for protection of the distribution networks," Iran, Tavanir, 2020. Available: https://wamp.tavanir.org.ir/rule/index.php.
4 "IEEE recommended practice for protection and coordination of industrial and commercial power systems (IEEE Buff Book)," IEEE Standard 242-2000.
5 "Measuring relays and protection equipment," IEC-60255 Standard, 2022.
6 SIPROTEC overcurrent protection 7SJ8 manual, SIEMENS, Available: https://www.siemens.com/siprotec.
7 "Low-voltage switchgear and controlgear - Part 2: Circuit-breakers," IEC 60947-2, 2016.
8 "Miniature circuit breaker and residual current protection devices," Technical catalogue, Schneider Electric, 2012.
9 "ABB molded case circuit breakers," Technical catalogue, ABB, 2016.
10 "SACE Emax low voltage air circuit-breakers," Technical catalogue, ABB, 2013.
11 "IEEE guide for protective relay applications to power transformers," IEEE Standard C37.91-2000.
12 "Generator protection guide," ABB, 1997, Available: https://library.e.abb.com/.
13 "IEEE guide for AC generator protection," IEEE Standard C37.102–2006.
14 "IEEE guide for control of small hydroelectric power plants," IEEE Standard 1020-1988, 2011.
15 J. M. Bisbee, M. F. Rosol, and C. L. Wagner, "A partial survey of relay protection of steam-driven A-C generators," *IEEE Trans. Power App. Syst.*, vol. 80, pp. 954–957, 1962.
16 J. Berdy, "Loss of excitation protection for modern synchronous generators," *IEEE Trans. Power App. Syst.*, vol. PAS-94, no. 5, pp. 1457–1463, 1975.
17 A. P. de Morais, G. Cardoso, and L. Mariotto, "An innovative loss-of-excitation protection based on the fuzzy inference mechanism," *IEEE Trans. Power Del.*, vol. 25, no. 4, pp. 2197–2204, 2010.
18 E. Pajuelo, R. Gokaraju, and M. S. Sachdev, "Identification of generator loss-of-excitation from power-swing conditions using a fast pattern classification method," *IET Gener. Transm. Distrib.*, vol. 7, no. 1, pp. 24–36, 2013.

19 A. Hasani and F. Haghjoo, "A secure and setting-free technique to detect loss of field in synchronous generators," *IEEE Trans. Energy Convers.*, vol. 32, no. 4, pp. 1512–1522, 2017.

20 J. K. Park, C. H. Kim, K. H. Kim, Y. S. Lyu, and J. J. Yang, "Development of an adaptive underexcitation limiter in excitation system," *IEEE Trans. Power Del.*, vol. 33, no. 5, pp. 2135–2142, 2017.

21 "MiCOM P343 technical manual," GE Solutions, 2014, Available: https://www.gegridsolutions.com.

22 S. F. Zarei and M. Parniani, "A comprehensive digital protection scheme for low-voltage microgrids with inverter-based and conventional distributed generations," *IEEE Trans. Power Del.*, vol. 32, no. 1, pp. 441–452, 2017.

23 N. N. A. Bakar *et al.*, "Microgrid and load shedding scheme during islanded mode: A review," *Renew. Sust. Energ. Rev.*, vol. 71, pp. 161–169, 2017.

24 R. Bekhradian, M. Sanaye-Pasand, and M. Davarpanah, "Innovative load shedding scheme to restore synchronous generator-based microgrids during FIDVR," *IEEE Trans. Smart Grid*, vol. 14, no. 1, pp. 388–399, 2023.

25 A. Saffarian and M. Sanaye-Pasand, "Enhancement of power system stability using adaptive combinational load shedding methods," *IEEE Trans. Power Syst.*, vol. 26, no. 3, pp. 1010–1020, 2011.

26 A. P. Ghaleh, M. Sanaye-Pasand, and A. Saffarian, "Power system stability enhancement using a new combinational load-shedding algorithm," *IET Gener. Transm. Distrib.*, vol. 5, no. 5, pp. 551–560, 2011.

27 T. D. Firouzabadi, D. A. Zarchi, M. Mazid, H. Safdarkhani, and H. Nafisi, "Overarching preventive sympathetic tripping approach in active distribution networks without telecommunication platforms and additional protective devices," *IEEE Access*, vol. 10, pp. 28411–28421, 2022.

28 J. V. Milanovic, M. T. Aung, and S. C. Vegunta, "The influence of induction motors on voltage sag propagation—Part I: Accounting for the change in sag characteristics," *IEEE Trans. Power Del.*, vol. 23, no. 2, pp. 1063–1071, 2008.

29 R. Bekhradian, M. Davarpanah, and M. Sanaye-Pasand, "Current-based blocking scheme to stabilize distribution network relays against FIDVR," *Int. J. Electr. Power Energy Syst.*, vol. 132, p. 107205, 2021.

30 "IEEE recommended practice for excitation system models for power system stability studies," IEEE Standard 421.5-2005, 2006.

31 N. K. Sharma and S. R. Samantaray, "Issues and challenges in microgrid protection." *Microgrid Cyberphysical Systems*. Elsevier, pp. 233–254, 2022.

32 A. Chandra, G. K. Singh, and V. Pant, "Protection of AC microgrid integrated with renewable energy sources–A research review and future trends," *Electr. Power Syst. Res.*, vol. 193, p. 107036, 2021.

33 H. Bisheh, B. Fani, G. Shahgholian, I. Sadeghkhani, and J. M. Guerrero, "An adaptive fuse-saving protection scheme for active distribution networks," *Int. J. Electr. Power Energy Syst.*, vol. 144, p. 108625, 2023.

34 M. Bollen and F. Hassan, *Integration of Distributed Generation in the Power System*, vol. 80. Hoboken, New Jersey: John Wiley & Sons, Inc., 2011.

35 "IEEE standard for interconnection and interoperability of distributed energy resources with associated electric power systems interfaces," IEEE Std 1547-2018 (Revision of IEEE Std 1547-2003), 2018.

36 S. Teimourzadeh, M. Davarpanah, F. Aminifar, and M. Shahidehpour, "An adaptive auto-reclosing scheme to preserve transient stability of microgrids," *IEEE Trans. Smart Grid*, vol. 9, no. 4, pp. 2638–2646, 2018.

37 M. Abedini, M. Sanaye-Pasand, M. Davarpanah, H. Lesani, and M. Shahidehpour, "Predictive auto-reclosure approach to enhance transient stability of grid-connected DGs," *IET Gener. Transm. Distrib.*, vol. 13, no. 14, pp. 3011–3019, 2019.

38 S. Nourollah and G. B. Gharehpetian, "Coordinated load shedding strategy to restore voltage and frequency of microgrid to secure region," *IEEE Trans. Smart Grid*, vol. 10, no. 4, pp. 4360–4368, 2018.

39 R. Bekhradian, M. Sanaye-Pasand, M. Davarpanah, and M. Abedini, "Innovative loss of field protection algorithm for small-scale synchronous generators," *IEEE Trans. Power Del.*, vol. 38, no. 4, pp. 2661–2672, 2023.

40 A. Hasani, X. Liang, M. Sanaye-Pasand, and M. Abedini, "Challenges and remedy for synchronous generators asynchronous operation following a fast loss of field detection," *IEEE Trans. Ind. Appl.*, 2025.

41 M. Abedini, M. Sanaye-Pasand, and M. Davarpanah, "An analytical-based method to postpone generator tripping during loss-of-excitation events," in *2018 IEEE 12th International Conference on Compatibility, Power Electronics and Power Engineering*, Qatar, 2018.

42 "Technical Reference Paper Fault-Induced Delayed Voltage Recovery," NERC, 2009. Available: https://www.nerc.com.

43 A. Hooshyar, E. F. El-Saadany, and M. Sanaye-Pasand, "Fault type classification in microgrids including photovoltaic DGs," *IEEE Trans. Smart Grid*, vol. 7, no. 5, pp. 2218–2229, 2016.

5

Microgrid Islanding Detection Challenges and Solutions

5.1 Introduction

Among the various operational issues in microgrids (MGs), islanding is one of the most critical events that requires careful consideration [1, 2]. An islanding event, which is also known as "loss of main" [3], refers to a condition in which the MG becomes isolated from the main grid, either intentionally (planned) or unintentionally (unplanned). This isolation may occur symmetrically in three phases or asymmetrically in one or two phases, typically as a result of upstream breakers tripping, fuse melting, and broken conductors.

One of the most crucial requirements for stable, uninterruptible, and safe islanded operation of the synchronous generator-based microgrids (SGBMGs) and inverter-based microgrids (IBMGs) is a quick and secure detection of islanding events [4–8]. Following the disconnection of the MG from the main distribution system, its controlling systems should be activated automatically to preserve its stability during the islanded operation. Therefore, the unintentional islanding event of MGs should be detected quickly and accurately to ensure an uninterruptible power supply for local essential loads and avoid adverse impacts on the main distribution system. Rapid identification of unintentional islanding also enables appropriate control and protection actions to preserve the stable operation of the MG [9].

Various islanding detection schemes (IDSs) have been presented in the past decades, which include remote and local techniques. Remote techniques use information from remote nodes and equipment, and the operating status of remote lines and generators, obtained through appropriate communication links, to detect islanding conditions, whereas local techniques rely solely on the locally measured data and information and, therefore, do not require communication links. Although remote techniques provide better performance, local techniques are commonly preferred due to their simplicity and lower capital expenditure.

Local techniques can be classified into passive and active methods. Passive methods monitor the MG signals (such as voltage, frequency, and impedance) or the indices derived from them in real time to detect islanding conditions without injecting any perturbation into the system. In contrast, active methods require the deliberate injection of a perturbation for islanding detection. To choose the most suitable IDS, various features and technical factors should be considered, including the type of distributed generation (DG) and its technology. The existing IDSs offer different performances and features depending on the MG configuration and DG types.

The dynamic behaviors of SGBMGs and IBMGs differ following an islanding event; therefore, a specific IDS should be used for each type and configuration, and thus, the IDS for SGBMGs may be entirely different from that for IBMGs. Due to the inherent mechanical inertia of small-scale synchronous generators (SSSGs), the SGBMG frequency would vary after islanding events, even for small values of power mismatch. Hence, the SGBMG frequency becomes greater or lower than the permissible range after a while and the SGBMG becomes unstable, even without inserting any intentional disturbance. However, for the IBMGs, the frequency may not change significantly, and thus, intentional perturbations may be required to identify the islanding condition [10–13]. This disturbance is mainly exploited to provide an intentional power mismatch, which results in voltage and frequency deviations from their nominal values. Based on these explanations, active methods are commonly adopted by IBMGs, while these methods are not recommended for SGBMGs [14]. Therefore, the appropriate IDS should be employed depending on the dynamic behavior of the DGs and MGs. In addition, some other factors may also be important for selecting suitable IDS, such as grid codes/standards and control/protection requirements. In this regard, various features of IDS should be evaluated, including simplicity, dependability, detection speed, implementation cost, and security.

In this chapter, the appropriate IDSs that can be practically applied to the MGs are thoroughly scrutinized. To do so, at first, some proper indices are presented, which can be appropriately employed by researchers and also power system engineers to compare various IDSs and select the most suitable one. Afterward, various types of existing IDSs are carefully investigated using the presented indices. Based on these studies, it can be concluded that passive techniques are more suitable for SGBMGs due to their technical and economic advantages. However, for inverter-based DGs (IBDGs), passive techniques may not guarantee proper performance. Furthermore, to enhance the existing IDSs, several novel ideas are suggested in this chapter, which can be applied in future research works.

5.2 Proper Indices for Performance Evaluation of IDSs

To evaluate the performance of different IDSs, suitable indices and criteria should be used to quantify the effectiveness and features of IDSs. In this regard, various criteria have been presented by different research works and standards. For instance, the required islanding detection time (IDT) for different types of MGs provided by different standards are presented in Table 5.1 [1, 15]. As presented, only the IEEE 1547 standard is applicable to SGBMGs, whereas various other standards address the requirements of IBMGs. To evaluate the IDS of MGs, some suitable indices and techniques are presented in this section, considering the requirements of the international standards.

5.2.1 Nondetection Zone

The most widely used graphical technique to evaluate the performance of the IDSs is exploiting the nondetection zone (NDZ) [16–18]. NDZ presents the region of active and reactive power mismatches where the IDS cannot identify the islanding event within the permissible detection time [19, 20]. Therefore, NDZ is exploited to adjust the relay thresholds, especially for conventional techniques [16–21].

To extract the boundary of NDZ, simulation-based studies and analytical techniques can be adopted [18, 21]. It is worth mentioning that the NDZ extraction of SGBMGs may be different from that of IBMGs [21]. The IBMGs have no inherent

Table 5.1 International standards for islanding detection [1, 15].

Standard	MG type	Load quality factor	Frequency range (Hz)	Voltage range (pu)	Required detection time (s)
IEEE 1547	All types	1	59.3–60	0.88–1.1	<2
IEEE 929	PV	2.5	59.3–60.5	0.88–1.1	<2
IEC 62116	PV	1	59.3–60	0.88–1.15	<2
German VDE 0126-1-1	IBMG (small inverters less than 4.6 kVA)	1–2.5	47.5–50.2	0.88–1.15	<0.2
German VDE-AR-N 4105	IBMG (low-voltage grid)	1–2.5	47.5–50.2	0.88–1.15	<0.2
German VDE-AR-N 4110	IBMG (medium-voltage grid)	1–2.5	47.5–50.2	0.88–1.15	<0.2
Canadian C22.2 No. 107-1	IBMG	2.5	59.5–60.5	0.88–1.06	<2

mechanical inertia, and thus, their NDZ boundary is commonly affected by the quality factor of the local RLC loads and their resonant frequency [21]. Hence, the detection time of IBMGs is less important for their NDZ extraction. However, due to the inherent mechanical inertia of SGBMGs, their IDT should be considered for determining the NDZ boundary. After the SGBMGs islanding event, their frequency does not change instantaneously; it takes some time for that to reach the predefined thresholds of the relays. This time period depends significantly on the SGBMG inertia, and its value will be enlarged by increasing the inertia of the SGBMG [22]. Therefore, the operation time of the islanding detection relay is strongly influenced by the SGBMG's inertia and its dynamic behavior following the islanding event. Hence, unlike IBMGs, the IDT should be considered for SGBMG NDZ determination. The NDZs for a sample SGBMG and IBMG are presented below. For this purpose, the postislanding frequency and voltage behavior of both SGBMG and IBMG are examined.

5.2.1.1 SGBMG

Fig. 5.1(a) illustrates the SGBMG IDT versus its active and reactive power imbalances (ΔP, ΔQ). In addition, the (ΔP, ΔQ) points for which the IDT is larger than a predefined threshold are depicted in Fig. 5.1(b) [23]. The amount of this threshold, i.e., the maximum IDT for SGBMGs, is commonly considered equal to 2 s based on the recommendations of the IEEE 1547 standard [9].

The SGBMG NDZ region may be influenced by the SSSG controllers, such as the governor and exciter controllers. On the other hand, the dynamic behavior of the local static loads, and their dependency on SGBMG voltage/frequency, as well as dynamic loads behavior, such as induction motors (IMs), should be considered for NDZ extraction [24]. Commonly, numerous transient electromagnetic simulations considering various case studies should be performed for determining the NDZ region of IDS [23–25]. For example, to determine the NDZ region, about 195

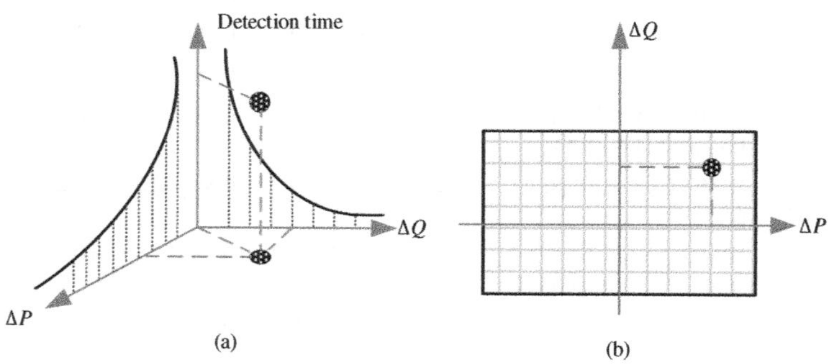

Fig. 5.1 (a) Islanding detection time versus active and reactive power imbalances. (b) NDZ. [10] / with permission of IEEE.

various scenarios are investigated by [25] around the perfect power imbalance through transient simulation studies. Using the simulation-based methods for determining NDZ region may be time-consuming in some cases due to the required iterative computations. In this regard, some analytical techniques have been recommended by [21–24] to estimate the NDZ boundary of widely used conventional IDSs including under/over-voltage (UV/OV), under/over-frequency (UF/OF), and rate of change of frequency (ROCOF) schemes. Although these techniques are quite simple and straightforward, they neglect the response of the SSSG controllers, including governor and automatic voltage regulator (AVR) controllers, as well as the dynamic behavior of the loads. Furthermore, the dynamics of SSSG is only modeled by the second-order swing equation.

In the following, a simple analytical technique for determining the SGBMG NDZ region for the abovementioned conventional IDSs is explained, which is developed by [21]. To conduct the analysis, a typical distribution system to which an SGBMG is connected, shown in Fig. 5.2, has been considered. As illustrated, the SGBMG is equipped with an islanding detection relay. During the steady-state grid-connected operation, the SSSG supplies active power P_{SSSG} to both the local load and the upstream grid. The active power consumed by the load is denoted by P_{Load}, while the active power exchanged with the upstream system is denoted by P_{Sys}. An islanding event is assumed to occur when breaker CB is opened. Following islanding, the resulting active power imbalance, i.e., ΔP, induces transient behavior in the SSSG, governed by the swing equation:

$$\begin{aligned} \frac{2H}{\omega_0}\frac{d\omega}{dt} &= P_{SSSG} - P_{Load} = -\Delta P \\ \frac{d\delta}{dt} &= \omega - \omega_0 \end{aligned} \tag{5.1}$$

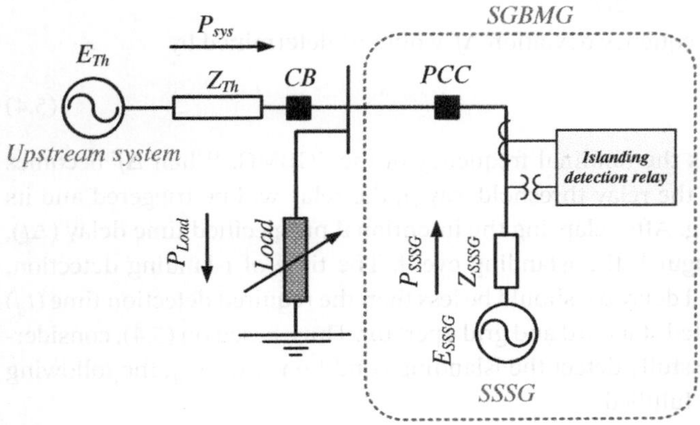

Fig. 5.2 A typical distribution system containing an SGBMG.

where H is the SSSG inertia, ω_0 is the synchronous angular frequency, δ is the SSSG rotor angle, and ω is the SSSG angular frequency. In (5.1), the machine's damping coefficient is neglected because its value is relatively small. Furthermore, the mechanical power of the SSSG (P_M) is assumed to remain constant and equal to its steady-state active power output, i.e., P_{SSSG}. According to (5.1), the rotor speed and frequency vary in response to the active power imbalance (mismatch power), i.e., ΔP, which is equal to P_{sys}, assuming a constant power load model and neglecting its dependence on voltage and frequency. The swing equation of the SSSG can, therefore, be effectively used to determine the NDZ region of conventional IDSs, as described in the following:

5.2.1.1.1 *Under/Over-Frequency Technique* In this technique, the frequency of the SGBMG is used to identify an islanding event. If the frequency deviates above or below a predefined threshold for a specified duration, the relay recognizes the islanding condition and trips the PCC breaker. To prevent the relay maloperation during nonislanding events, an intentional delay (Δt) is applied in this approach. Since this method is widely adopted for islanding detection worldwide, its NDZ is analytically examined in this section. The discussion is particularly useful for engineers who need to properly adjust frequency relays for effective islanding detection.

Based on swing equation, expressed in (5.1), the angular frequency of the SSSG, ω, is obtained as follows [21]:

$$\omega = -\frac{\omega_0 \Delta P}{2H} t + \omega_0 \tag{5.2}$$

Thus, the variation of angular speed could be determined by

$$\omega - \omega_0 = \Delta\omega = -\frac{\omega_0 \Delta P}{2H} t \tag{5.3}$$

Consequently, the frequency deviation Δf would be determined by

$$\Delta f = -\frac{f_0 \Delta P}{2H} t \tag{5.4}$$

where $f_0 = \omega_0/2\pi$ is the nominal frequency of the SGBMG. When Δf becomes greater or less than the relay threshold, say β, the relay will be triggered and its timer starts counting. After elapsing the intentional prespecified time delay (Δt), the relay will distinguish the islanding event. The time of islanding detection, including intentional delay Δt, should be less than the required detection time (t_d) specified by the related standard and grid operator. Thus, based on (5.4), considering $\Delta f = \beta$, to successfully detect the islanding condition within t_d, the following condition should be fulfilled:

$$\left| \frac{2H\beta}{f_0 \times \Delta P} \right| \leq (t_d - \Delta t) \tag{5.5}$$

or:

$$\frac{2H|\beta|}{f_0 \times (t_d - \Delta t)} \leq |\Delta P| \tag{5.6}$$

In other words, if the mismatch power becomes less than ΔP determined by (5.6), the islanding condition could not be recognized within t_d, and thus, this point is considered in the NDZ region. In this analysis, the damping coefficient (D) of the SSSG has been neglected.

5.2.1.1.2 *Rate of Change of Frequency (ROCOF) Technique* According to (5.2), the ROCOF, i.e., df/dt, could be calculated by

$$\frac{df}{dt} = \frac{1}{2\pi}\frac{d\omega}{dt} = -\frac{f_0}{2H}\Delta P \tag{5.7}$$

Commonly, in the ROCOF technique, the measured ROCOF, df/dt, is processed through a first-order LPF with a time constant, T_f. In the Laplace domain, the filtered signal, $K(s)$, can be expressed as [21]

$$K(s) = -\frac{f_0 \Delta P}{2H(T_f s + 1)} \tag{5.8}$$

Consequently, in time domain, the processed signal, $K(t)$, could be written as

$$K(t) = -\frac{f_0 \Delta P}{2H}\left(1 - e^{\frac{-t}{T_f}}\right) \tag{5.9}$$

The ROCOF relay identifies an islanding condition when the magnitude of the filtered signal, $K(t)$, becomes greater than or less than a predefined threshold, β, for a specified time delay, Δt. After elapsing the intentional prespecified delay, the relay can confirm the islanding event. The total IDT, including the intentional delay Δt, should be less than the required detection time, t_d, as specified by the relevant standard or the grid operator. Accordingly, based on (5.9), and considering that the threshold of ROCOF function is equal to β, i.e., $df/dt = \beta$, the following condition should be satisfied to successfully detect the islanding condition within the required detection time, t_d:

$$T_f \times \ln\left(1 - \frac{2H|\beta|}{f_0 \times |\Delta P|}\right)^{-1} \leq (t_d - \Delta t) \tag{5.10}$$

or:

$$\frac{2H|\beta|}{f_0}\left(1 - e^{-\frac{(t_d - \Delta t)}{T_f}}\right)^{-1} \leq |\Delta P| \tag{5.11}$$

In other words, if the mismatch power falls below the value determined by (5.11), the islanding condition cannot be detected within the required detection time, t_d, and consequently, this operating point is considered to lie within the NDZ region. The analytical equations for simple calculation of parameters ΔP_1 and ΔP_2 are summarized in Table 5.2.

Table 5.2 Appropriate analytical functions for calculating minimum detectable active power mismatch [21].

Technique	Minimum active power imbalance for reliable operation of frequency-based IDSs
OF/UF	$\lvert\Delta P\rvert = \dfrac{2H \cdot \lvert\beta\rvert}{f_0(t_d - \Delta t)}$
ROCOF	$\lvert\Delta P\rvert = \dfrac{2H \cdot \lvert\beta\rvert}{f_0}\left(1 - e^{-\frac{(t_d-\Delta t)}{T_f}}\right)^{-1}$

5.2.1.1.3 Under/Over-Voltage Technique Up to this point, the value of ΔP has been determined, which influences the frequency-based relays. Since the X/R ratio of the SSSG is very high, it may be assumed that ΔP depends mainly on frequency, while the reactive power variation ΔQ depends mainly on voltage. This assumption is principally made to simplify the problem and could be arguable in a real system, in which the distribution system model including the line X/R ratio and dynamic response of the loads are not taken into account. If the distribution system is modeled, the X/R ratio could approach one, and therefore, the influence of active power P on voltage V, as well as the influence of reactive power Q on frequency f, shall also be considered.

By neglecting the dependence of reactive power on frequency, the reactive power imbalance ΔQ could be identified only by voltage-based relays. Thus, as is a common practice, UV/OV techniques are employed for islanding detection when the imbalance is solely due to reactive power. As such, in this section, the corresponding NDZ of UV/OV relays is derived. In this regard, Fig. 5.3(a) is considered. As shown, after islanding occurrence (i.e., when the CB is opened), the entire reactive load is supplied solely by the SSSG, while before islanding, this load was supplied jointly by the network and the generator. Prior to islanding, the situation can be described as follows:

$$Q_{Load} - Q_{SSSG} = \Delta Q \tag{5.12}$$

where before islanding occurrence, the reactive power mismatch between load and SSSG, i.e., ΔQ, is supplied by the upstream system. After islanding occurrence, the reactive power output of the SSSG will be equal to the load. For simplicity and without loss of generality, the reactive power imbalance, ΔQ, can be represented by a switched inductance X_e, as illustrated in the equivalent circuit in Fig. 5.3(b) [21]. In this figure, a reactance of value X_e is switched in series with the SSSG. To assess the impact of this reactance on the generator terminal voltage and the SGBMG, the SSSG's transient behavior at the instant of switching should be analyzed.

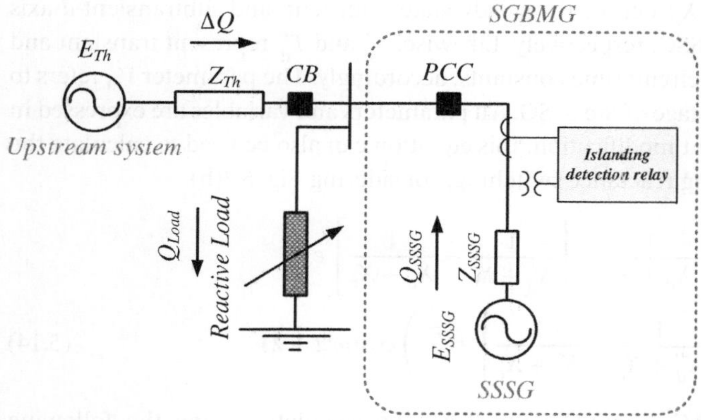

(a)

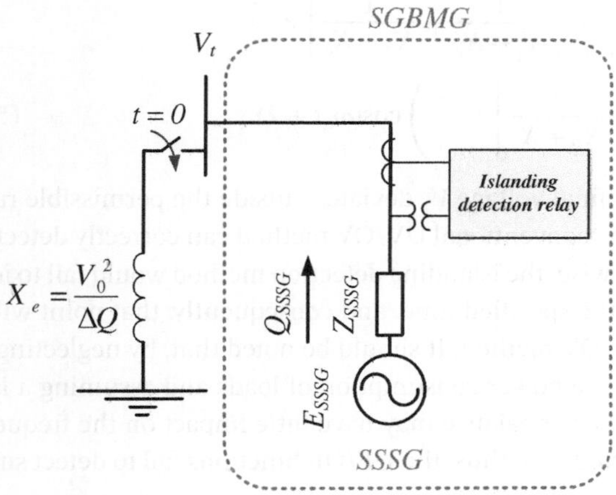

(b)

Fig. 5.3 (a) A typical distribution system containing an SGBMG. (b) An equivalent system used for determining reactive power imbalance ΔQ. Adapted from [21].

When a three-phase fault occurs at the terminals of a SG, neglecting the asymmetrical components, the armature current of SSSG (i_{SSSG}) can be expressed as [21]

$$i_{SSSG} = V_0\left(\frac{1}{X_d} + \left[\frac{1}{X'_d} - \frac{1}{X_d}\right]e^{-\frac{t}{T'_d}} + \left[\frac{1}{X''_d} - \frac{1}{X'_d}\right]e^{-\frac{t}{T''_d}}\right)\cos(\omega_0 t + \lambda) \tag{5.13}$$

where X_d, X'_d, and X''_d denote the steady-state, transient, and subtransient d-axis reactances of the SSSG, respectively. Likewise, T'_d and T''_d represent transient and subtransient short-circuit time constants, accordingly. The parameter V_0 refers to the steady-state voltage of the SSSG. All parameters and variables are expressed in per-unit. With slight modification, this equation can also be used to calculate the SSSG current during reactance switching, considering Fig. 5.3(b):

$$i_{SSSG} = V_0 \left(\frac{1}{X_d + X_e} + \left[\frac{1}{X'_d + X_e} - \frac{1}{X_d + X_e} \right] e^{-\frac{t}{T'_d}} + \left[\frac{1}{X''_d + X_e} - \frac{1}{X'_d + X_e} \right] e^{-\frac{t}{T''_d}} \right) \cos(\omega_0 t + \lambda) \tag{5.14}$$

Therefore, the SSSG terminal voltage can be calculated using the following equation:

$$V_t = V_0 X_e \left(\frac{1}{X_d + X_e} + \left[\frac{1}{X'_d + X_e} - \frac{1}{X_d + X_e} \right] e^{-\frac{t}{T'_d}} + \left[\frac{1}{X''_d + X_e} - \frac{1}{X'_d + X_e} \right] e^{-\frac{t}{T''_d}} \right) \cos(\omega_0 t + \lambda) \tag{5.15}$$

If, for a given ΔQ, the terminal voltage V_t deviates outside the permissible range within a time period, t_d, the conventional UV/OV method can correctly detect the islanding condition. Otherwise, the islanding detection method would fail to identify this condition within the specified time, and consequently, that point will lie within the NDZ of the UV/OV method. It should be noted that, by neglecting the effect of voltage on the active power consumption of loads and assuming a large X/R ratio, the reactive power imbalance may have little impact on the frequency of the SGBMG after islanding, and thus, the UF/OF functions fail to detect such a condition.

5.2.1.1.4 *Typical NDZ shape for SGBMGs* The NDZ shapes for static loads having constant consumed power and constant impedance are illustrated in Fig. 5.4, based on an approach given in [21]. As shown, for both constant power and constant impedance loads, the shape of NDZ is rectangular. This rectangle should be rotated clockwise for impedance loads by an angle of 30°. Meanwhile, parameters ΔP_1 and ΔP_2 are mainly influenced by the setting values of the frequency-based relays. In addition, parameters ΔQ_1 and ΔQ_2 are commonly influenced by the predefined thresholds of voltage-based relays. Moreover, an iterative technique is presented for calculating the amounts of ΔQ_1 and ΔQ_2. This technique employs the analytical equations of the SSSG whose detailed information is given in [21].

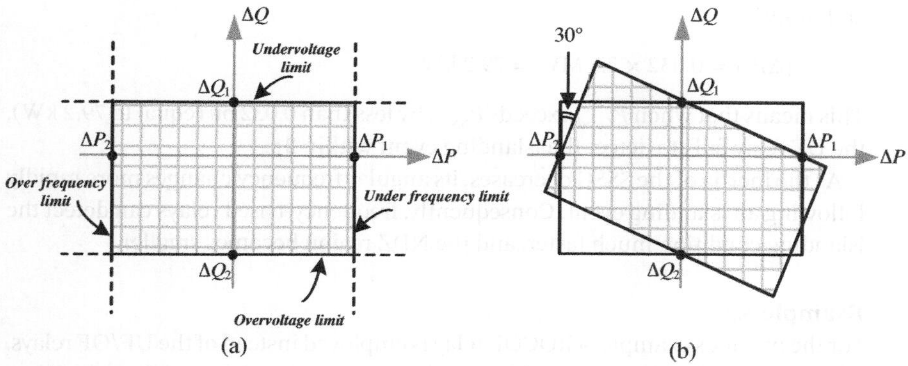

Fig. 5.4 NDZ region for (a) constant power load and (b) constant impedance load [21].

Example 5.1

Consider a synchronous small-scale generator with a nominal frequency of 60 Hz and a rated capacity of 2.5 MVA connected to an medium-voltage distribution network. The thresholds for the UF/OF functions are set to 62 Hz (corresponding to a +2 Hz frequency deviation) and 56.5 Hz (corresponding to a −3.5 Hz frequency deviation), respectively, and both functions incorporate an intentional delay of 160 ms. Determine the range of active power imbalance for which the UF/OF functions are unable to detect the islanding event within 2 s. The inertia constant of the SSSG is assumed to be 0.5 s, and the load is modeled as a constant power load.

Solution:

For the OF function, the threshold $\beta_1 = \Delta f_1$ is considered 2 Hz. Furthermore, t_d and Δt are 2 and 0.16 s, respectively. Based on (5.6),

$$\frac{2H|\beta_1|}{f_0 \times (t_d - \Delta t)} \leq |\Delta P_1| \Rightarrow |\Delta P_1| = \frac{2 \times 0.5 \times 2}{60 \times (2 - 0.16)} = 0.018 \text{ pu}$$

and in kW

$$\Delta P_1 = 0.018 \times 2.5 \text{ MW} = 45 \text{ kW}$$

This indicates that when P_{SSSG} is less than 0.018 pu (equivalent to 45 kW) above the P_{Load}, the OF function would be unable to detect the islanding event.

For the UF function, the threshold $|\beta_2| = |\Delta f_2|$ is considered 3.5 Hz. Furthermore, t_d and Δt are 2 and 0.16 s, respectively. Based on (5.6),

$$\frac{2H|\beta_2|}{f_0 \times (t_d - \Delta t)} \leq |\Delta P_2| \Rightarrow |\Delta P_1| = \frac{2 \times 0.5 \times 3.5}{60 \times (2 - 0.16)} = 0.032 \text{ pu}$$

and in kW

$$|\Delta P_2| = 0.032 \times 2.5\,\text{MW} = 79.2\,\text{kW}$$

This means that when P_{Load} exceeds P_{SSSG} by less than 0.032 pu (equal to 79.2 kW), the UF relay fails to detect the islanding event within 2 s.

As the inertia of the SSSG decreases, its angular frequency changes more rapidly following an islanding event. Consequently, frequency-based relays can detect the islanding condition much faster, and the NDZ region becomes smaller.

Example 5.2

For the previous example, a ROCOF relay is employed instead of the UF/OF relays. The relay threshold is set to ±3 Hz/s, with an intentional time delay of 150 ms. Determine the range of active power imbalance for which the ROCOF function is unable to detect the islanding event within 2 s.

Solution:

For the ROCOF function, the threshold β is set to 3 Hz/s. Furthermore, t_d and Δt are 2 and 0.15 s, respectively. Moreover, parameter T_f is assumed to be 0.1 s. Based on (5.11),

$$\frac{2H|\beta|}{f_0}\left(1 - e^{-\frac{(t_d-\Delta t)}{T_f}}\right)^{-1} \leq |\Delta P| \Rightarrow \frac{2 \times 0.5 \times 3}{60}\left(1 - e^{-\frac{(2-0.15)}{0.1}}\right)^{-1} = 5\%$$

and in MW

$$|\Delta P| = 0.05 \times 2.5\,\text{MW} = 0.125\,\text{MW}$$

This means that for an active power imbalance less than ±0.05 pu (equal to ±0.125 MW), the ROCOF relay fails to detect the islanding event within 2 s.

5.2.1.1.5 Impact of SSSG Governor Response on Islanding Detection So far, the impact of active power imbalance on the frequency deviation following an islanding event has been examined under two simplifying assumptions:

- The damping coefficient in the swing equation is neglected.
- The governor response of the SSSG is ignored by assuming a constant mechanical power input.

In this section, the postislanding frequency behavior of the SSSG is analyzed while accounting for both the damping and governor dynamics.

The swing equation of the SSSG can be refined as follows by considering the damping coefficient, D:

$$\frac{2H}{\omega_0}\frac{d\omega}{dt} = P_{SSSG} - P_{Load} - D\Delta\omega = -\Delta P - D\Delta\omega \tag{5.16}$$

By solving this first-order differential equation and assuming constant value for ΔP after the islanding event, the frequency deviation of the SSSG could be determined by

$$\Delta\omega\,(t) = -\frac{\Delta P}{D}\left(1 - e^{-(\omega_0 D/2H)t}\right) \tag{5.17}$$

Therefore, the frequency would change monotonically following an islanding event if the mechanical power is assumed constant. However, the mechanical power may vary due to both the governor response and the dependency of the load on frequency and voltage. Unlike large-scale SGs, whose governors exhibit significant delays, the governor of an SSSG responds rapidly; hence, assuming a constant mechanical power input may not be entirely accurate. In this section, the impact of the governor response on postislanding frequency is investigated.

Grid-connected SSSGs are commonly operated under either constant power or droop control modes, as discussed in the previous chapters. Therefore, following an islanding event, the mechanical power of the SSSG, i.e., P_{SSSG}, can be expressed in Laplace domain as follows based on the model provided in Chapters 1 and 2:

$$P_{SSSG}(s) = [P^* - P_{output} - R(\omega_0 - \omega(s))] \times C(s) \times T(s) \tag{5.18}$$

where P^* is the active power setpoint of the SSSG, and R is the droop gain, which is set to zero for the constant power control strategy. P_{output} denotes the electrical power output of the SSSG, which equals P_{Load} following an islanding event. Moreover, $T(s)$ represents the turbine transfer function, which is assumed to be unity for simplicity. $C(s)$ is the power controller transfer function, which can be expressed as follows:

$$C(s) = \frac{K_i}{s} + K_p \tag{5.19}$$

where K_i and K_p are the integrator and proportional gains of the SSSG's governor system, respectively. Assuming a constant value for P_{Load} and, consequently, a constant P_{output}, P_{SSSG} can be expressed as

$$\Delta P(s) = -R\Delta\omega(s) \times \left(\frac{K_i}{s} + K_p\right) \tag{5.20}$$

Combining (5.16) and (5.20) results in

$$\frac{2H}{\omega_0} s\Delta\omega(s) = R\Delta\omega(s) \times \left(\frac{K_i}{s} + K_p\right) - D\Delta\omega(s) \tag{5.21}$$

The eigenvalues can be computed by solving

$$\frac{2H}{\omega_0} s^2 + (D + K_p R)s + K_i R = 0 \tag{5.22}$$

whose results are given in (5.23):

$$
\begin{aligned}
s_{1,2} &= \frac{-D' \pm \sqrt{D'^2 - 8k'H/\omega_0}}{4H/\omega_0} \\
D' &= D + K_p R \\
k' &= K_i R
\end{aligned}
\tag{5.23}
$$

Based on (5.23), if $k' > D'^2\omega_0/8H$, the SSSG frequency deviation exhibits an oscillatory behavior. In contrast, for $k' < D'^2\omega_0/8H$, the frequency deviation, and therefore the system frequency, changes exponentially, i.e., without oscillations. A key conclusion is that the response of the SSSG's governor system can produce an oscillatory frequency response following an islanding event, rather than a monotonic increase or decrease. This oscillatory behavior may make the detection of islanding more challenging for UF/OF functions.

5.2.1.2 IBMG

As discussed earlier, the inherent differences between SSSGs and IBDGs lead to fundamentally distinct methodologies for determining the NDZ in IBMGs compared with that of SGBMGs. In this section, a simple procedure for calculating the NDZ in IBMGs is presented.

Fig. 5.5 shows a typical distribution network used for islanding studies and NDZ evaluation. The system includes an RLC load and an IBDG. Once breaker CB is opened, the IBMG becomes islanded, and the entire RLC load is subsequently supplied by the IBDG. According to the IEEE Standard 1547, assessment of the NDZ region for islanding detection methods in IBDGs should be carried out under the following conditions [1, 2]:

- The resonant frequency of the RLC load is assumed to be equal to the grid frequency.

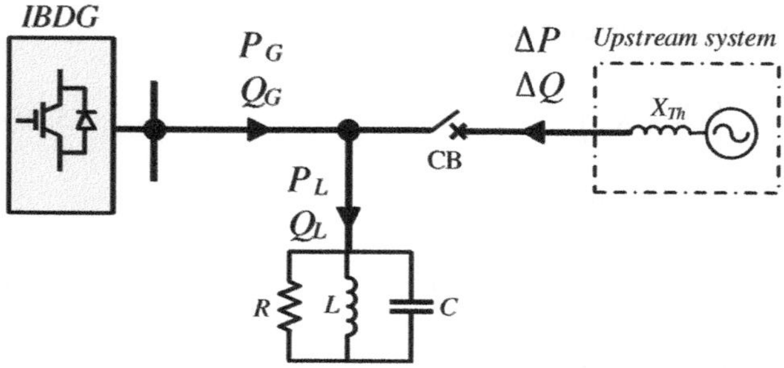

Fig. 5.5 A typical distribution system including an IBDG and an RLC load. Adapted from [2].

- The load quality factor (Q_f) is set to 1.
- Under the grid-connected operation (i.e., with breaker CB closed), the IBDG output power and the load power are equal such that the net power exchanged with the upstream system is zero.

Based on these conditions, and recognizing this scenario as the most challenging case for islanding detection, it is assumed that the IBDG operates at unity power factor before islanding, and therefore, its reactive power output Q_G is zero.

In this section, the NDZ associated with the conventional islanding detection functions, including UF/OF and UV/OV functions, are analyzed [2]. Throughout this analysis, the IBMG voltage and frequency are denoted by V and f, respectively. Moreover, the IBDG's active and reactive output powers are represented by P_G and Q_G, while the load's active and reactive powers are denoted by P_L and Q_L, as shown in Fig. 5.5. In addition, all quantities corresponding to the postislanding condition are marked with the symbol "^."

The active and reactive power imbalances, also known as power mismatches, denoted by ΔP and ΔQ, respectively, are defined as the differences between the power generated by the IBDG and the load power prior to islanding, as expressed by

$$\begin{aligned} \Delta P &= P_L - P_G \\ \Delta Q &= Q_L - Q_G \end{aligned} \tag{5.24}$$

Considering that after islanding the generator's output power and the load power become equal, i.e., $\widehat{P}_G = \widehat{P}_L$ and $\widehat{Q}_G = \widehat{Q}_L$, this equation can be rewritten as follows [2]:

$$\begin{aligned} \Delta P &= P_L - P_G = \Delta P_G - \Delta P_L \\ \Delta Q &= Q_L - Q_G = \Delta Q_G - \Delta Q_L \end{aligned} \tag{5.25}$$

where

$$\begin{aligned} \Delta P_G &= \widehat{P}_G - P_G, \quad \Delta Q_G = \widehat{Q}_G - Q_G \\ \Delta P_L &= \widehat{P}_L - P_L, \quad \Delta Q_L = \widehat{Q}_L - Q_L \end{aligned} \tag{5.26}$$

In addition, it is assumed that the active and reactive powers of the IBDG remain constant after islanding, i.e., $\Delta P_G = 0, \Delta Q_G = 0$. On the other hand, the active and reactive powers consumed by the RLC load, before and after islanding, can be expressed as follows [2]:

$$\begin{aligned} P_L &= \frac{V^2}{R}, \quad Q_L = V^2\left(\frac{1}{\omega L} - \omega C\right) \\ \widehat{P}_L &= \frac{\widehat{V}^2}{R}, \quad \widehat{Q}_L = \widehat{V}^2\left(\frac{1}{\widehat{\omega} L} - \widehat{\omega} C\right) \end{aligned} \tag{5.27}$$

where $\omega = 2\pi f$ and $\hat{\omega} = 2\pi\hat{f}$. Variations in voltage and frequency after islanding lead to changes in the active and reactive powers consumed by the load. On the other hand,

$$P_G = \hat{P}_G = \hat{P}_L \tag{5.28}$$

Using this equation, the normalized amount of power imbalance, i.e., $\Delta P/P_G$, could be expressed as [2]

$$\frac{\Delta P}{P_G} = -\frac{\Delta P_L}{P_G} = -\frac{\Delta P_L}{\hat{P}_L} = \frac{(V^2 - \hat{V}^2)/R}{\hat{V}^2/R} = \frac{V^2}{\hat{V}^2} - 1 \tag{5.29}$$

The permissible voltage range is taken as $[V_{min}, V_{max}]$, where V_{max} is the threshold of the OV function and V_{min} is the threshold of the UV function. If the following condition, given by (5.30), is satisfied, islanding conditions cannot be detected by the UV/OV functions and the associated operating point and power imbalance ΔP should be classified within the NDZ region [2].

$$\frac{V^2}{V_{max}^2} - 1 \leq \frac{\Delta P}{P_G} \leq \frac{V^2}{V_{min}^2} - 1 \tag{5.30}$$

In other words, to detect islanding using a voltage relay, the active power imbalance should fall outside the range specified by (5.30).

To determine the range of reactive power imbalance that lies within the NDZ region, it is assumed that the IBDG was operating at unity power factor prior to the islanding event, representing the most challenging condition for detection. Therefore, $Q_G = 0$, and consequently, $\Delta Q = Q_L$. For $\Delta Q \neq 0$, either the inductance L or capacitance C of the RLC load could be adjusted. Here, it is assumed that the load capacitance changes as:

$$L = L_0, \quad C = C_0 + \Delta C = C_0 - \frac{\Delta Q}{2\pi f V^2} \tag{5.31}$$

Consequently, based on (5.27),

$$\Delta Q = Q_L = V^2 \left(\frac{1}{2\pi f L_0} - 2\pi f (C_0 + \Delta C) \right) \tag{5.32}$$

As a result, by combining relations (5.25)–(5.27) and (5.31)–(5.32) [2]:

$$\frac{\Delta Q}{P_G} = Q_f \left(1 - \frac{f^2}{\hat{f}^2} \right) \tag{5.33}$$

The permissible frequency range is considered within $[f_{min}, f_{max}]$, where the OF function threshold is set to f_{max} and the UF function threshold is set to f_{min}. If the condition expressed in equation (5.34) is satisfied, the frequency relay will fail to

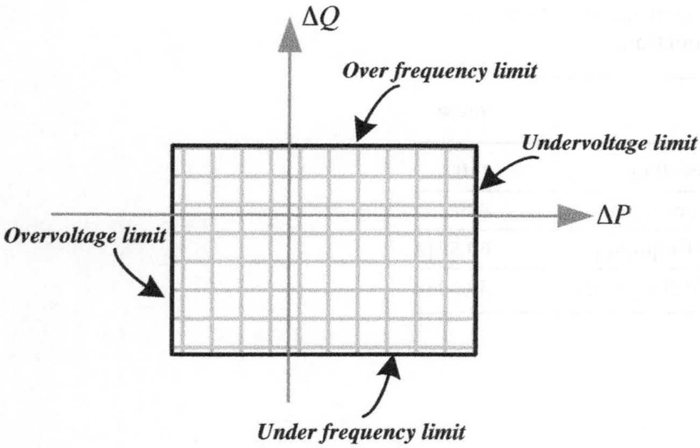

Fig. 5.6 NDZ of traditional islanding detection schemes for identifying IBDG islanding [2].

detect the islanding event, and the associated operating point will lie within the NDZ region [2].

$$Q_f\left(1-\frac{f^2}{f_{min}^2}\right) \leq \frac{\Delta Q}{P_G} \leq Q_f\left(1-\frac{f^2}{f_{max}^2}\right) \tag{5.34}$$

Using equations (5.30) and (5.34), the NDZ characteristic of the conventional voltage- and frequency-based methods is determined by specifying the values of ΔP and ΔQ. An example of the IBDG's NDZ region is illustrated in Fig. 5.6 [2].

Example 5.3

Consider an IBDG unit with a nominal frequency of 60 Hz and a rated capacity of 350 kVA connected to an low-voltage distribution network.

a) The UV/OV and UF/OF protection settings of the IBDG unit are configured according to the Table 5.3. Determine and plot the NDZ region $Q_f = 2.5$.
b) Repeat (a) for $Q_f = 1$.

Solution:

a) Using (5.30), the upper band and lower band of the active power imbalance (ΔP) in the NDZ are calculated from the following equations:

$$\left(\frac{\Delta P}{P_G}\right)_{max} = \frac{V^2}{V_{min}^2} - 1 = \frac{1}{0.88^2} - 1 = 29.1\%$$

$$\left(\frac{\Delta P}{P_G}\right)_{min} = \frac{V^2}{V_{max}^2} - 1 = \frac{1}{1.1^2} - 1 = -17.3\%$$

Table 5.3 Settings of over/undervoltage and under/overfrequency functions.

Function		Value
Voltage	Overvoltage	110%
	Undervoltage	88%
Frequency	Overfrequency	60.5 Hz
	Underfrequency	59.3 Hz

and in kW

$$\Delta P_{max} = 29.1\% \times 350 = 101.8\text{ kW}$$
$$\Delta P_{min} = -17.3\% \times 350 = -60.5\text{ kW}$$

Based on (5.34), the upper band and lower band of the reactive power imbalance (ΔQ) in the NDZ are calculated from the following equations:

$$\left(\frac{\Delta Q}{P_G}\right)_{max} = Q_f\left(1 - \frac{f^2}{f_{max}^2}\right) = 2.5 \times \left(1 - \frac{60^2}{60.5^2}\right) = 4.11\%$$
$$\left(\frac{\Delta Q}{P_G}\right)_{min} = Q_f\left(1 - \frac{f^2}{f_{min}^2}\right) = 2.5 \times \left(1 - \frac{60^2}{59.3^2}\right) = -5.94\%$$

and in kVAr

$$\Delta Q_{max} = 4.11\% \times 350 = 14.38\text{ kVAr}$$
$$\Delta Q_{min} = -5.94\% \times 350 = -20.79\text{ kVAr}$$

The NDZ is also illustrated in Fig. 5.7. As these results indicate, conventional voltage- and frequency-based methods, even with relatively sensitive settings, exhibit a fairly large NDZ and, in many cases, are unable to detect islanding conditions.

b) When Q_f changes, the values of ΔP remain unchanged. However, ΔQ_{min} and ΔQ_{max} decrease proportionally according to (5.34).

$$\left(\frac{\Delta Q}{P_G}\right)_{max} = Q_f\left(1 - \frac{f^2}{f_{max}^2}\right) = \left(1 - \frac{60^2}{60.5^2}\right) = 1.64\%$$
$$\left(\frac{\Delta Q}{P_G}\right)_{min} = Q_f\left(1 - \frac{f^2}{f_{min}^2}\right) = 2.5 \times \left(1 - \frac{60^2}{59.3^2}\right) = -2.38\%$$

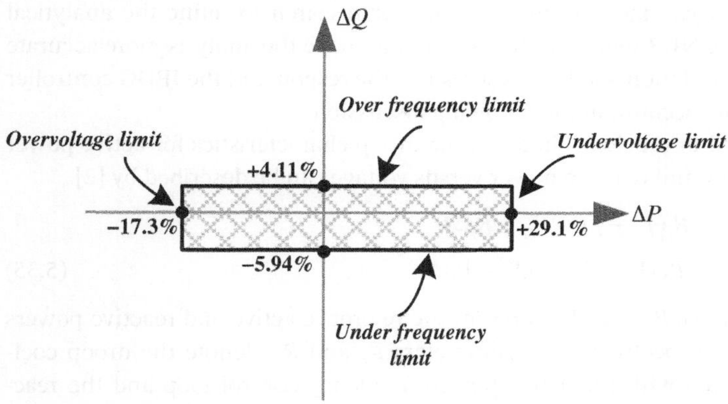

Fig. 5.7 NDZ of conventional islanding detection schemes of Example 5.3.

One of the key distinctions between the NDZ characteristics of IBDGs and SSSGs is the way power imbalances are detected. In IBDGs, the active power imbalance ΔP is governed by the voltage protection settings, whereas the reactive power imbalance ΔQ is governed by the frequency protection settings. In contrast, for SSSGs, ΔP is determined by the frequency relay settings, while ΔQ is determined by the voltage relay settings.

As evident from the above equations, ΔP is determined solely based on voltage value and the settings of the UV/OV relay, while ΔQ is determined solely based on frequency value and the settings of the UF/OF relay. In other words, the dependence of ΔP on frequency and ΔQ on voltage is neglected in this method. This simplification makes the resulting calculations approximate and can lead to discrepancies when compared to simulation-based and field-test results.

A notable consequence of this simplification is that, in this analysis, the NDZ region is represented as a rectangle (as shown in Fig. 5.6), whereas in simulation-based and field-test results, the region is not perfectly rectangular due to the dependence of ΔP on frequency and ΔQ on voltage. One of the main reasons for this behavior is the presence of active power–frequency droop and reactive power–voltage droop control in IBDG units. These droop characteristics adjust the IBDG's active power output in response to frequency changes and the reactive power output in response to voltage changes.

Additionally, in the analysis presented in this section, the load is modeled purely as an impedance. In reality, loads may comprise a combination of constant power, constant current, and constant impedance components. Moreover, dynamic loads, especially motor loads, may have a considerable impact on the analysis. To improve the accuracy of the results, the sensitivity of loads to voltage and frequency, as well as the behavior of dynamic loads, should be considered.

These challenges highlight areas for further research to refine the analytical derivation of the NDZ region. In this section, to make the analysis more accurate and closer to simulation and field-test results, the response of the IBDG controller is also taken into account in the following discussion.

As mentioned in previous chapters, the droop characteristics for active power versus frequency and reactive power versus voltage can be described by [2]

$$\hat{P}_G^* = P_G^* + R_f(f^* - \hat{f}) = P_G^* - R_f\Delta f$$
$$\hat{Q}_G^* = Q_G^* + R_V(V^* - \hat{V}) = Q_G^* - R_V\Delta V \tag{5.35}$$

In these equations, P^* and Q^* represent the reference active and reactive powers of the inverter, respectively. The parameters R_f and R_V denote the droop coefficients associated with the active power–frequency control loop and the reactive power–voltage control loop, respectively. Under normal operating conditions, when the system frequency is equal to its reference value (reference frequency f^*) and the voltage is equal to its reference value (reference voltage V^*), the IBDG operates in constant power mode. However, when the voltage deviates from its reference value, the IBDG adjusts its reactive power output in order to bring the voltage back toward the reference level. Similarly, when the frequency deviates from its reference, the IBDG modifies its active power output so that the frequency is restored to its reference value. After an islanding event, the deviations in voltage and frequency, together with the droop characteristics, cause the inverter to no longer operate in constant power mode, resulting in changes in its output power. The amount of change in the IBDG's active and reactive power outputs (i.e., ΔP_G and ΔQ_G), corresponding to variations in frequency and voltage, is determined by [2]

$$\Delta P_G = -R_f\Delta f$$
$$\Delta Q_G = -R_V\Delta V \tag{5.36}$$

By combining (5.36) with (5.27), the normalized value of the active power variation is obtained as

$$\frac{\Delta P}{P_G} = \frac{V^2}{\hat{V}^2} - 1 - \frac{V^2 R_f(\hat{f} - f)}{\hat{V}^2 P_G} \tag{5.37}$$

In addition, for the reactive power, it is similarly assumed, as in the previous section, that the IBDG operates at a power factor of one under normal conditions and, therefore, delivers zero reactive power. Hence, the reactive power variation is equal to the load reactive power, that is, $\Delta Q = Q_L$. Accordingly, by combining equations (5.27) and (5.35):

$$R_V\Delta V = -\hat{V}^2\left(\frac{1}{2\pi\hat{f}L_0} - 2\pi\hat{f}(C_0 + \Delta C)\right) \tag{5.38}$$

Substituting this into (5.31) yields the following equation:

$$\frac{\Delta Q}{P_G} = \left(1 + \frac{\Delta P}{P_G}\right)\left(Q_f\left(1 - \frac{f^2}{\hat{f}^2}\right) - R_R\frac{f\Delta V}{\hat{f}\hat{V}^2}\right) \tag{5.39}$$

Equations (5.37) and (5.38) can be rewritten using (5.26) and (5.27) as

$$\begin{aligned} R_f\Delta f &= -\frac{\hat{V}^2}{R} + P_G \\ R_V\Delta V &= -\hat{V}^2\left(\frac{1}{2\pi\hat{f}L} - 2\pi\hat{f}C\right) \end{aligned} \tag{5.40}$$

Finally, based on these equations, the voltage and frequency following an islanding event can be computed using the expressions given below. A detailed derivation of these expressions is provided in [2].

$$\begin{aligned} \hat{f}(V,\hat{V},\Delta Q) &= \frac{\beta_V + \sqrt{\beta_V^2 + 8\pi f C_0\left(\frac{\Delta Q}{\hat{V}^2} + 2\pi f C_0\right)}}{4\pi C_0} \\ \hat{V}(f,\hat{f},\Delta P) &= \sqrt{\frac{V^2}{\Delta P + P_G(\min\{P_{nom},\hat{P}_G\})}} \end{aligned} \tag{5.41}$$

Using (5.37) and (5.39), the NDZ region can ultimately be obtained from the following expressions [2]:

$$\begin{aligned} &\frac{V^2}{V_{max}^2} - 1 - \frac{V^2 R_f\Delta f_1}{V_{max}^2 P_G} \le \frac{\Delta P}{P_G} \le \frac{V^2}{V_{min}^2} - 1 - \frac{V^2 R_f \Delta f_2}{V_{min}^2 P_G} \\ &\xi_{min} \le \frac{\Delta Q}{P_G} \le \xi_{max} \end{aligned} \tag{5.42}$$

where

$$\begin{aligned} \xi_{min} &= \left(1 + \frac{\Delta P}{P_G}\right)\left(Q_f\left(1 - \frac{f^2}{f_{min}^2}\right) - R_R\frac{f\Delta V_1}{f_{min}\hat{V}_1^2}\right) \\ \xi_{max} &= \left(1 + \frac{\Delta P}{P_G}\right)\left(Q_f\left(1 - \frac{f^2}{f_{max}^2}\right) - R_R\frac{f\Delta V_2}{f_{max}\hat{V}_2^2}\right) \\ &\hat{f}_1 = \hat{f}(V, V_{max}, \Delta Q), \quad \hat{f}_2 = \hat{f}(V, V_{min}, \Delta Q) \\ &\hat{V}_1 = \hat{V}(f, f_{min}, \Delta P), \quad \hat{V}_2 = \hat{V}(f, f_{max}, \Delta P) \\ &\Delta f_i = \hat{f}_i - f, \quad \Delta V_i = \hat{V}_i - V, \quad i = \{1, 2\} \end{aligned} \tag{5.43}$$

5.2.2 Islanding Detection Time

IDT is defined as the duration between the instant of MG islanding occurrence and the detection moment of IDS. To preserve the MG stability, proper protection and control actions should be performed after the islanding detection [26]. Hence, IDT should be small enough to prevent MG instability. According to the IEEE 1547 recommendations, IDT should be less than 2 s. However, for some MGs, IDT should be considerably smaller than the recommended criteria due to stability issues [24–27]. It should be noted that different islanding scenarios under various operating conditions should be scrutinized to evaluate the detection time of each IDS.

5.2.3 False Detection Ratio

IDSs should remain secure for nonislanding disturbances [28, 29]. To evaluate this feature, the false detection ratio (FDR) is defined as [10]

$$\mathrm{FDR} = N_F / N_T \times 100 \tag{5.44}$$

where N_T is the total number of analyzed islanding or nonislanding case studies and N_F is the number of false decisions. The following scenarios can be considered for evaluating the correctness of the IDS decision [10]:

- **Load switching:** Large loads or large capacitors switching can lead to variations of voltage and frequency [30]. Hence, the IDS performance should be evaluated for the maximum possible loads and capacitors switching.
- **Starting of induction motors:** Large IM starting may lead to a considerable increment of reactive power consumption and also voltage drop in the MG feeders. Therefore, voltage-based IDS may operate incorrectly in this case. Furthermore, during motor starting, the reactance seen from the DG terminal considerably varies, and thus, the impedance-based IDS may issue an incorrect islanding command in such a condition [31].
- **Short-circuit faults:** IDSs should stay secure during short-circuit events due to the LVRT requirements.
- **Cold load pickup:** Reenergization of the MG feeders after a long duration of power interruption, which is commonly known as "cold load pickup," may result in simultaneous overcurrent and UV phenomena. The IDS performance should be evaluated during cold load pickup conditions.
- **Fault-induced delayed voltage recovery (FIDVR):** FIDVR phenomenon is a severe voltage-degraded condition, which commonly occurs after short-circuit fault clearance due to the large slip of IMs. In such a condition, the MG voltage magnitude remains low for a long duration, i.e., about 3–30 s. Meanwhile, this phenomenon commonly occurs at MG feeders with high penetration of IMs. The

IDS should remain secure during both recoverable and unrecoverable FIDVR conditions [32].

- **Power swing:** Both stable and unstable power swings may take place in MGs, especially after short-circuit faults clearance. Thus, it is necessary to evaluate the IDS performance in this condition [33].
- **Fault-initiated islanding:** This scenario takes place when the point of common coupling (PCC) breaker or the breaker of the upstream feeders is tripped by their associated protection relays due to the upstream system faults. In this case, the IDS should identify the islanding condition within a short time duration [34].

5.2.4 Nondetection Index

Nondetection index (NDI) is presented by [35] to scrutinize the effectiveness of an IDS to identify the islanding events for a specific MG. Each IDS can be proper or improper for a specific MG depending on various factors, including the MG specifications, the configuration of the upstream grid, and load behavior. For some MG configurations, low-cost and simple conventional IDS may be suitable for islanding detection. However, for some MGs, more complex IDS may be employed for secure islanding detection. NDI can considerably help system engineers to adopt the appropriate IDS for a specific MG.

To express NDI, a sample test system depicted in Fig. 5.8 is considered [35]. The DG may be isolated from the upstream system by opening different breakers. For the network illustrated in Fig. 5.8, the opening of breakers RC01, RC02, and RC03 builds Island 01, Island 02, and Island 03, respectively.

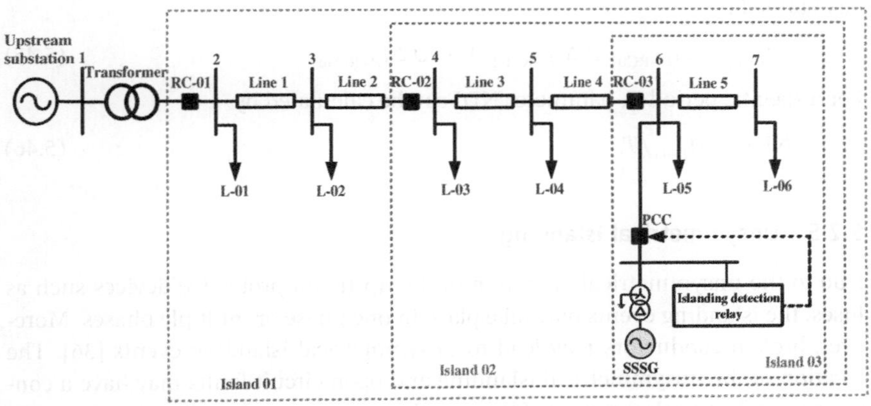

Fig. 5.8 Sample distribution system. Adapted from [35].

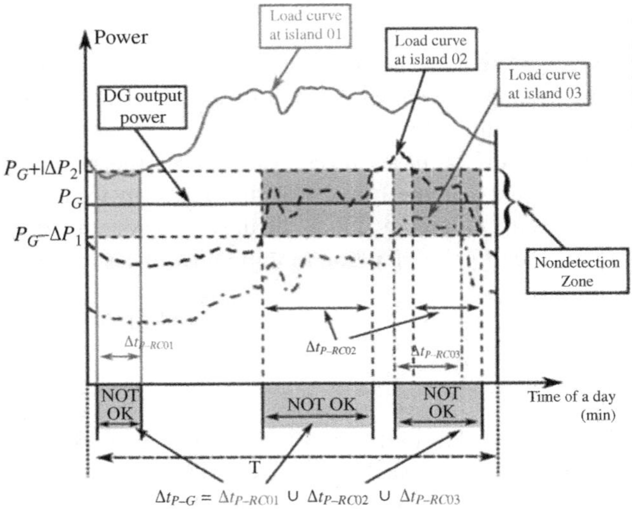

Fig. 5.9 Example of nondetectable periods. [35] / with permission of IEEE.

The power consumption curve of loads of each islanded portion is shown in Fig. 5.9. In this figure, the NDZ boundary of the exploited IDS is presented by the parameters ΔP_1 and ΔP_2. As shown, for some periods of time, the power mismatch ΔP may be located in the NDZ region where the islanding events cannot be identified by the IDS. For example, within the $\Delta t_{P\text{-RC01}}$ period, the power mismatch of Island 01 is sited in the NDZ region of the IDS, and thus, within the mentioned period, the islanding event of Island 01 cannot be identified by the IDS. The summation of time periods in which the IDS fails to identify the islanding events can be obtained by

$$\Delta t_{P\text{-}G} = \Delta t_{P\text{-RC01}} \cup \Delta t_{P\text{-RC02}} \cup \cdots \cup \Delta t_{P\text{-RC0}n} \tag{5.45}$$

For a specific period T, parameter NDI can be calculated by

$$\text{NDI} = \Delta t_{P\text{-}G}/T \tag{5.46}$$

5.2.5 Unsymmetrical Islanding

Due to the unsymmetrical operation of the upstream protective devices such as fuses, the islanding events may take place in one phase or multiple phases. Moreover, broken conductors may lead to unsymmetrical islanding events [36]. The location of the unsymmetrical islanding and open circuit faults may have a considerable effect on the amount of the negative sequence current. To prevent equipment damage, the IDS should identify the unsymmetrical islanding and then the proper control/protection actions have to be performed to preserve the MG stability. Consequently, the unsymmetrical islanding detection capability of the IDS should be evaluated for various cases.

5.3 Islanding Detection Methods

To identify the islanding events of MGs, different techniques have been proposed by scholars and researchers, which can be categorized into local and remote approaches, as presented in Fig. 5.10 [11–17, 37]. Local techniques are based on using local information at the DG or PCC locations. However, the remote approaches identify the islanding events using the transferred measurements from different locations by suitable communication media. Although the remote techniques can completely eliminate the NDZ, the local techniques are more preferred by system engineers and researchers due to financial reasons. In the following subsections 5.3.1 and 5.3.2, the local and remote techniques applicable to MGs are scrutinized.

5.3.1 Remote Methods

The communication link between the DGs and remote equipment is one of the main requirements for implementing remote approaches. This communication link transfers the signals from remote devices, which considerably helps the IDS to identify the islanding events. The remote approaches are applicable to various types of MGs, including SGBMGs, IBMGs, and HMGs. The major benefit of the remote approaches is their zero NDZ [37].

Remote techniques can be divided into four major groups, as illustrated in Fig. 5.10. Power line carrier is exploited by some remote techniques to transfer a signal from the upstream substation [38]. In this approach, a high-frequency signal is continuously injected into the power lines. At the MG location, this signal is identified by a suitable receiver [39]. If the receiver cannot detect the injected

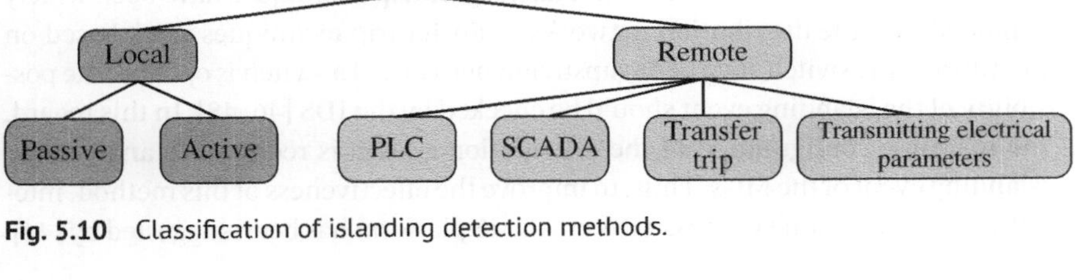

Fig. 5.10 Classification of islanding detection methods.

signal, the islanding event would be detected [39, 40]. However, this technique may not be feasible due to its high capital expenditure. One of the main remote IDSs is based on supervisory control and data acquisition (SCADA), which can be properly exploited to identify MG islanding events [41]. The electrical parameters of the upstream network, including voltage and current at various locations, are gathered in a center through suitable communication media [42]. This approach significantly depends on the communication media and may have large IDT in some cases [43].

Due to the recent advancements in smart systems, SCADA-based algorithms including IDS can be properly implemented through an advanced distribution management system (ADMS) platform [44]. The ADMS is an integrated platform that contains advanced control, monitoring, planning, and protection functions. This high-speed platform can securely be connected to other intelligent electrical devices (IEDs) such as control and protection devices, as well as remote-controlled switches (RCSs). Hence, by collecting the information of the IEDs and RCSs through the ADMS platform, the islanding event can be quickly identified in each portion of the distribution system. Furthermore, the ADMS platform considerably enhances the IDT by exploiting features of IEC-61850 such as "GOOSE messages." However, the practical implementation of the ADMS platform may face various challenges, such as requiring high bandwidth communication media. In addition, implementing this platform for some distribution systems may not be feasible due to the lack of proper infrastructures and high capital expenditure [44].

To decrease the expenditure of IDS implementation, some techniques have been proposed recently, which exploit few electrical parameters for islanding detection and thus require simple communication media. For example, in [37], a synchronized method measuring phase voltages at the upstream grid and MG is suggested to identify the islanding event. In this technique, the synchronized measurement units are mainly exploited to gather and transfer the required information. In this method, a time tag should be also provided by a GPS device. Micro-phasor measurement units (μPMUs) are utilized by [45] to estimate the phase angle and frequency at the SSSG terminals and upstream reference point. Afterward, to identify the islanding event, the angle difference (between DG terminals and the reference point) and the cumulative sum of the estimated frequency are exploited.

To lessen the volume of the transmitted information and decrease the implementation cost of the communication media, transfer trip techniques have been widely exploited in some distribution networks. Transfer trip techniques work based on monitoring the switch state of the upstream network. If a switch is opened, the possibility of the islanding event should be checked by the IDS [46–48]. In this regard, the real-time configuration of the distribution system is required to analyze the islanding event of the MGs. Thus, to improve the effectiveness of this method, integrating this technique with SCADA or ADMS platform has been suggested by [44].

Different communication infrastructures may be utilized to enhance IDT of transfer trip schemes including optical fiber and cellular data such as general packet radio service (GPRS), 2G, 3G, and wireless communications. However, some of these technologies are not secure against cyber-attacks. Hence, for reliable and fast data transfer, a suitable protocol should be adopted. This requirement can be provided through the features of the IEC 61850 protocol such as its GOOSE message capability [44, 47]. However, the implementation of this method would become complex for large number of switches, i.e., breakers and reclosers, between the SGBMG and upstream substation [48].

5.3.2 Local Methods

The local techniques exploit local suitable signals to detect islanding without utilizing communication links. These approaches are conventionally divided into active and passive techniques [13–17]. Active techniques inject intentional perturbations to MG and its associated controllers, which results in the voltage and frequency drift from their permissible ranges after the islanding occurrence. These perturbations can be injected into the DG controllers or a disturbance can be applied to the MG by intentional switching of loads in the feeders [13, 14, 17]. On the other hand, passive techniques only monitor and analyze suitable indices or extract hidden features of locally measured signals to identify islanding events without injecting any perturbation to the MG. Active and passive methods are scrutinized in the following sections 5.4 and 5.5.

5.4 Active Islanding Detection Methods

In this section, several widely used active techniques applicable to different types of MGs, i.e., SGBMGs and IBMGs, are discussed. Figure 5.11 illustrates the typical flowchart of active methods. As shown, intentional disturbances may be applied continuously, during specific time intervals, or only after a potential islanding event has been detected. To trigger the injection of perturbations, and also to determine the level of the injected perturbations, various indices based on local parameters such as voltage, frequency, and current may be utilized. After the perturbations are injected, appropriate passive indices calculated from the measured local signals are analyzed and compared with predefined thresholds to discriminate islanding events from nonislanding ones. These thresholds are typically determined through off-line analysis. By combining passive indices with active methods, the technique benefits from the advantages of both approaches and enhances overall effectiveness. Such combined strategies are generally referred to as hybrid techniques [49–52].

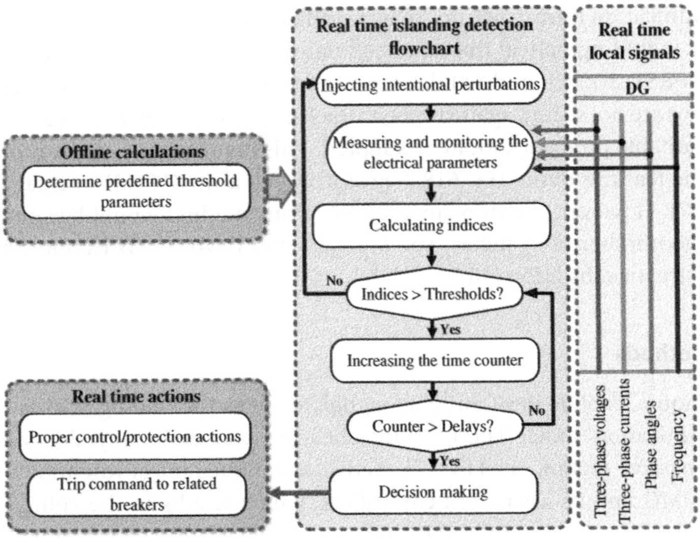

Fig. 5.11 IDS flowchart for active methods.

As previously explained, active islanding detection approaches are primarily applied to IBMGs, while their application in SGBMGs remains quite limited. In fact, for the following reasons, active methods may not be a practical solution for islanding detection in SGBMGs:

- Missing the smooth transition from grid-connected mode to islanded mode
- Increasing the possibility of SSSG damages due to the mechanical stresses during the steady-state operation
- Reduction of the lifetime of the SSSG and SGBMG equipment
- Increasing overhaul and maintenance expenditures
- SSSG stability challenges during nonislanding events
- Monotonically changes of SGBMG frequency even without any intentional disturbance, as shown in Fig. 5.12 [53]
- Adverse impacts on the SGBMG power quality.

The well-known active methods commonly applicable only to IBMGs include (i) active frequency drift (AFD), (ii) slip-mode frequency shift (SMS), (iii) Sandia frequency shift (SFS), and (iv) Sandia voltage shift (SVS) [1, 3–8]. Among the active methods that can be applied to SGBMGs, three main categories can be identified: (i) switching of constant impedance loads, (ii) opening and closing the SGBMG breakers, and (iii) injecting intentional perturbation signals into control loops such as governor and AVR [49–52]. The adverse impacts of the first category, switching constant impedance loads, on SSSG stability and mechanical fatigue are lower

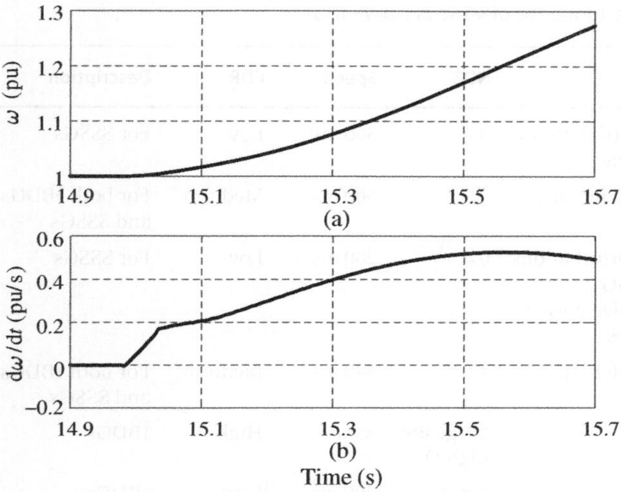

Fig. 5.12 Simulation results for an islanding scenario: (a) SGBMG frequency and (b) rate of change of SGBMG frequency.

than those of the other categories. Conversely, the third category generally provides greater effectiveness in identifying islanding events compared to that of the others. The specifications of various active techniques, which are applicable to MGs, including SGBMGs and IBMGs, are given in Table 5.4. Brief explanations of these techniques are presented in the following paragraphs.

5.4.1 Active Frequency Drift

AFD is one of the well-known active techniques used for detecting islanding conditions. In this method, intentional continuous disturbances are injected into the inverter's output current. The output current waveform of the inverter is illustrated in Fig. 5.13. As shown, during the first half-cycle, the current signal follows a sinusoidal waveform. However, when the signal crosses zero, it is forced to remain at zero for a dead-time t_z. As a result, the frequency of the output current slightly deviates from the grid frequency during the grid-connected operation of the IBDG.

The chopping factor (c_f) is defined as [3]

$$c_f = \frac{2t_z}{T_{grid}} \tag{5.47}$$

where T_{grid} is the period of the grid voltage. As the chopping fraction increases, the PCC voltage frequency drifts away from its rated value, eventually resulting in a shorter or longer voltage zero-crossing interval. Thus, these disturbances cause

Table 5.4 Comparing performance of various active IDSs.

References	Technique	NDZ	Speed	FDR	Description
[49]	Open-close of the SSSG circuit breaker	0%	300 ms	Low	For SSSGs
[50]	Capacitor switching	0%	600 ms	Medium	For both IBDGs and SSSGs
[51]	Injecting perturbations into the SSSG active/reactive power control loops	0%	200 ms	Low	For SSSGs
[52]	RL load switching	1.67%	440 ms	Medium	For both IBDGs and SSSGs
[1, 3]	AFD	Large for high Q_f	<2 s	High	IBDGs
[1]	SMS	Small	400 ms	Low	IBDGs
[1]	SFS	Small	500 ms	Low	IBDGs
[1]	SVS	Small	N/A	Low	IBDGs

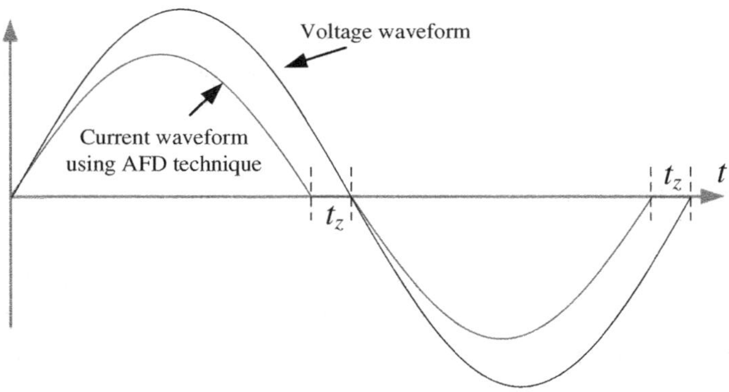

Fig. 5.13 Output current of the IBDG using AFD technique. Adapted from [3].

drift of the frequency, and the IBDG becomes unstable when islanding occurs. During the normal operation, the frequency of the PCC voltage remains stable and within the permissible range, and the impact of perturbation on IBDG stability is negligible. However, after islanding occurrence, the intentional disturbance acts as a destabilizing factor and causes the frequency of the PCC voltage to drift from the nominal frequency and get out from the permissible range, enabling UF/OF functions to detect the islanding condition [1].

The main benefit of the AFD technique is its simplicity of implementation compared to that of other active islanding detection methods. It can also effectively detect islanding events for purely resistive loads. However, when the inverter supplies RLC loads, the method faces significant operational limitations. In such cases, although AFD introduces a positive frequency shift to the inverter output, the resonant nature of the LC load can counteract this shift, resulting in a large NDZ in the presence of RLC loads. Additionally, AFD may introduce high harmonic distortion into both the IBMG and the connected distribution system. Recent research has focused on addressing these challenges, including reducing the NDZ, mitigating harmonic distortion, and improving the security and robustness of the AFD technique [1, 3].

In multi-inverter systems, achieving reliable performance requires that all inverters use similar or coordinated control systems. If different AFDs with the opposite drift biases are implemented in the IBDGs located in the same IBMG, the frequency perturbations could cancel each other out. As a result, AFD method may fail to detect the islanding event. Therefore, it is crucial that all IBDGs employ the same AFD scheme so that the induced frequency drift occurs in a consistent drift bias.

5.4.2 Sandia Frequency Shift

The SFS technique is an enhanced version of the classical AFD technique whose detection speed has been improved. This method has been proposed to mitigate the inherent operational shortcomings of AFD, most notably the reduction of the NDZ and the suppression of harmonic distortion. The principal innovation of SFS lies in replacing the fixed chopping factor of AFD with a variable chopping factor, dynamically adjusted through a positive feedback loop. In this scheme, this factor is continuously updated as a function of the frequency error, defined as the deviation between the frequency measured at the PCC and the nominal grid frequency. This chopping factor (c_f) is determined by [1]

$$c_f = c_{f_0} + K(f_{PCC} - f_0) \tag{5.48}$$

where c_{f0} is the initial chopping factor, K is the acceleration constant, f_{PCC} is the measured frequency at the PCC, and f_0 is the nominal grid frequency.

In the grid-connected operation, the system frequency remains stable and very close to its nominal value. As a result, the term ($f_{PCC} - f_0$) stays small and nearly zero, and the chopping factor remains limited to its small base value cf_0. This leads to a very low total harmonic distortion (THD) during normal operation, which is a key advantage. After the grid is disconnected, even a slight deviation in frequency causes an increase in the frequency error. Through the positive feedback loop, this rapidly increases the value of c_f. A larger chopping factor further amplifies the frequency deviation. This reinforcing cycle produces a very fast frequency drift, significantly reducing the detection time and greatly reducing the NDZ compared to the classical AFD method.

Tuning the parameters of the SFS method is a critical and challenging factor, which could directly affect the islanding detection performance and IBMG's stability. For instance, study of [4] reveals that large values of K may lead to dynamic instability even in grid-connected operation of the IBMG, especially when the upstream system is weak or when the IBDG penetration level is high. On the other hand, small values of K may significantly increase the NDZ of this technique. Moreover, cf_0 could influence THD. Therefore, parameters K and cf_0 should be determined using an optimal tradeoff between the adverse impact of this method on steady-state behavior of the IBDG, including power quality and stability issues, and its performance for islanding detection, including its detection time and NDZ. The interdependence between detection performance and system stability represents a major challenge in designing the SFS method and highlights the need for further research to develop techniques capable of eliminating the NDZ without relying on large gain values.

Various research works propose suitable frameworks to optimally determine the constant coefficient of SFS. A proper approach to parameter tuning involves the use of machine learning and artificial intelligence techniques to determine appropriate values for cf_0 and K. For instance, in [5], SFS parameters are tuned using an adaptive fuzzy logic (FL)-based algorithm. This algorithm estimates local load parameters and determines the minimum value of K required for NDZ elimination. Moreover, in [6], a variable SFS version based on the concept of an oscillatory initial chopping factor is introduced. In this regard, the constant value cf_0 has been replaced with a periodic oscillatory signal that continuously varies over the time.

It should be noted that the controller's loop and its dynamic response could have significant impact on the performance of the SFS method. Thus, optimal adjustment of the parameters of the SFS technique should be performed considering the control strategy. In addition, using suitable control loops and associated gains could be very helpful to enhance the performance of the SFS method. Similar to AFD, one of the challenges of the SFS method is that its NDZ may be affected by the parallel inverters connected to a single PCC, which may have adverse impact on the detection performance. To address this limitation, [7] proposed a new SFS design methodology to ensure optimal tuning even in the presence of parallel IBDGs.

5.4.3 Slip-Mode Frequency Shift

The most widely used well-known active method for detecting islanding in IBDGs is the SMS technique, which is commonly implemented in commercial inverters [7]. In SMS, the perturbation θ_{SMS} is intentionally injected into the inverter's output current (i_{inv}) to modify the phase angle between the inverter's current and voltage, thereby enabling fast and secure islanding detection. In this technique, the inverter's output current could be expressed as [7]

$$i_{inv} = I_m \sin(\omega t + \theta_{SMS}) \tag{5.49}$$

The perturbation θ_{SMS} is determined by

$$\theta_{SMS} = \theta_m \sin\left(\frac{\pi}{2}\frac{f_{PCC} - f_0}{f_m - f_0}\right) \tag{5.50}$$

where θ_m is the maximum phase shift and f_m is the corresponding frequency at which this maximum takes place. Moreover, f_0 represents the nominal grid frequency and f_{PCC} is the frequency of the PCC voltage. When the PCC voltage frequency slightly deviates from f_0 after the islanding event, the current phase angle also intentionally changes by the perturbation injected by the controller (θ_{SMS}). The controller recognizes this behavior, and thus, in the islanding condition, the frequency starts to drift from its nominal value and enables UF/OF relays to operate and detect the islanding condition. Although the NDZ of this method is relatively small, it still may fail to detect islanding condition in certain conditions. For instance, when the islanded portion of the system operates at the nominal frequency. Moreover, the characteristics of IBMG's control loops, including Volt–VAr, frequency–Watt, etc., and also dynamic grid support functions, such as voltage-ride through and frequency-ride through, have considerable impacts on the performance of this technique. These control loops and grid supporting functions may has adverse impacts on the operation of the active techniques, especially SMS, and thus, they may fail to detect islanding condition in some cases.

5.4.4 Sandia Voltage Shift

SVS is an active islanding detection method that slightly alters the PCC voltage magnitude by applying positive feedback and adjusting the output current of the IBDG. If the IBDG becomes islanded, this positive feedback amplifies the perturbation and pushes the voltage magnitude outside the permissible range. However, when the IBDG remains connected to the grid, the grid's stiffness limits the disturbance, resulting in only very small voltage variations. To implement this technique, a perturbation current signal I_p is added to the current controller reference. The signal I_p is determined by the following linear equation [8]:

$$I_p = k_e V_e \tag{5.51}$$

where V_e is the voltage error between the nominal voltage and the measured PCC voltage. The constant gain k_e determines the overall performance of the

SVS method, including its detection time, NDZ, and its impact on IBDG stability during steady-state operation. The main challenges associated with this technique include its adverse effects on IBMG power quality and the maximum power point tracking (MPPT) performance of the inverter. Several studies have proposed improvements to the SVS method. For example, in [8], it is suggested to use an exponential function instead of a linear one, expressed as:

$$I_p = \left[e^{|k_e V_e|} - 1\right] \text{sign}(k_e V_e) \tag{5.52}$$

Using this exponential function instead of the linear one can increase the perturbation after IBDG disconnection and, therefore, reduces both the detection time and the NDZ. Furthermore, it has been proposed to employ a self-adaptive selection of the parameter k_e rather than using a constant value. In this approach, k_e is adjusted to minimize the adverse impact on the system during the steady-state operation and to ensure that the THD remains within the permissible range.

5.4.5 Open-Close of the SSSG Breaker

The open-close incident of the SSSG breaker is suggested by [49] to identify the SGBMG islanding events even for small amounts of power mismatch. In this technique, one phase of the SSSG breaker is opened and closed immediately, which results in considerable variations in the SSSG exciter voltage and also MG voltage. Afterward, the rate of change of exciter voltage is exploited as a novel index to identify the islanding events. The value of this index would become greater than a predefined threshold for islanding events. The NDZ of this technique is zero, and its IDT is about 300 ms. This open–close incident may take place even for nonislanding disturbances, and thus, it may lead to excessive mechanical stresses on the SSSG and its associated breaker. Furthermore, the single-pole opening and closing leads to a large magnitude negative sequence current, which has considerable adverse impacts on the MG and its equipment. Moreover, this technique may endanger the SSSG transient stability.

An IDS is suggested by [50] to identify the SGBMG islanding events through capacitor switching. The rated power of this capacitor is specified based on a systematic approach. In this regard, at first, the maximum values of reactive power mismatches for which the conventional IDS fails to identify the islanding event are calculated. Based on this, the minimum reactive power of this capacitor bank will be analytically determined. For increasing the power mismatch, this capacitor is switched after opening the SGBMG breaker. Then, the SGBMG islanding event can be identified by voltage- or frequency-based IDSs. The IDT of this approach is less than 600 ms, and its NDZ is about zero. However, if the SGBMG islanding takes place due to the opening of a breaker in the upstream grid, this technique may fail to switch the capacitor bank and detect the islanding event.

5.4.6 Injecting Perturbations into the SSSG Active/Reactive Power Control Loops

Zamani *et al.* [51] suggests injecting an intentional small perturbation signal into the SSSG controllers. This perturbation may lead to the SSSG instability and also considerable undamped variations of the SGBMG frequency and voltage signals after islanding events. Afterward, to detect these undamped variations, the SGBMG voltage/frequency signals are processed via a signal processing block. Then, after detecting undamped variations of voltage/frequency, the islanding event would be realized. If the islanded operation of the SGBMG is required, the parameters of the SGBMG controller will be automatically adjusted by this technique at the islanding detection instant. However, the stable operation of the islanded SGBMG is highly affected by the injected perturbation signals, and thus, the SSSG may lose its stability after the islanding occurrence. In such a condition, the chance of creating a stable islanded SGBMG may be lost. The IDT of this technique considerably depends on the SSSG controllers. For a droop control system, the islanding event at the zero-power mismatch can be effectively identified within 200 ms. However, the IDT of this technique is about 1.5–2 s for a constant power control system.

5.4.7 RL Load Switching

In [52], RL load switching is proposed to identify islanding events. To initiate the RL load switching, the rate of change of reactive power (*dq/dt*) is exploited in this technique. The value of *dq/dt* is determined again after the RL load switching, to identify the possible islanding events. A suitable breaker or solenoid contactor should be exploited in this technique for intentional switching of the RL load. The capital expenditure for implementing this technique is relatively high, and it may not be beneficial for MGs. Furthermore, this technique cannot discriminate some types of short-circuit faults, such as three-phase ones, from islanding conditions. In addition, this method may insert the RL load during nonislanding events, such as short circuits, which may cause SGBMG instability. It cannot completely eliminate the NDZ, and its IDT is about 440 ms. Hence, the effectiveness of this technique is not suitable in comparison to the other active approaches.

5.5 Passive Islanding Detection Methods

Passive approaches are the most viable techniques for identifying islanding events in SGBMGs, although they may exhibit weaker performance in IBMGs. In this section, passive islanding detection techniques, including OV/UV and OF/UF methods, both conventional and newly developed, are discussed. Conventional passive techniques typically suffer from a high NDZ value, long IDT, and high false detection rate. To improve their effectiveness, various enhanced algorithms have been proposed in the literature, which are presented in the following subsections 5.5.1, 5.5.2, 5.5.3, and 5.5.4.

5.5.1 Setting Rules for Adjusting Conventional Relays

The widely used passive IDSs that are conventionally employed include (i) OV/UV, (ii) OF/UF, (iii) ROCOF, and (iv) vector shift (VS). The main shortcomings of these approaches are their high amounts of NDZ and FDR. Moreover, the effectiveness of these approaches considerably depends on the values of the predefined thresholds and settings [54, 55]. Thus, selecting suitable values for these thresholds and settings can effectively enhance the effectiveness of the conventional relays. Some research works have studied the impact of setting values on the efficacy of the conventional relays, and based on these studies, some new rules are recommended for adjusting these settings. A brief summary of these rules is presented in the following.

In [56], a new data-mining technique is recommended for determining the optimal values of conventional IDS thresholds. In this technique, the NDZ region and the amount of FDR are minimized by selecting the optimal settings. For a specific SSSG, the optimal values of settings for UF, ROCOF, and UV approaches are equal to 0.9975 pu, 0.035 pu/s, and 0.915 pu, respectively.

A novel graphical technique is suggested by [57] to adjust the setting values of frequency-based IDSs. In this regard, to minimize the relay NDZ and its false tripping, a graphical power mismatch region is introduced for selecting the proper setting values of frequency-based IDS. The setting values of ROCOF for a sample SSSG are proposed to be 0.0083 pu/s with a 330 ms operation delay.

5.5.2 Monitoring Proper Indices

Fig. 5.14 shows the typical decision-making flowchart of the passive techniques, which can be divided into different parts. As shown, the locally measured signals of DG and MG such as voltage, current, frequency, and phase angles are exploited as the input of the flowchart. The first step of this flowchart is utilized for initializing the parameters. Afterward, the system disturbances are identified through the second step by monitoring the real-time measured signals. The algorithm will be triggered if any disturbance is detected in the second step. It is worth mentioning that the second step is overlooked for some IDSs. The proper indices are calculated by the third step based on the predefined procedure. In the final step, these indices are compared to the associated thresholds, and the islanding events are

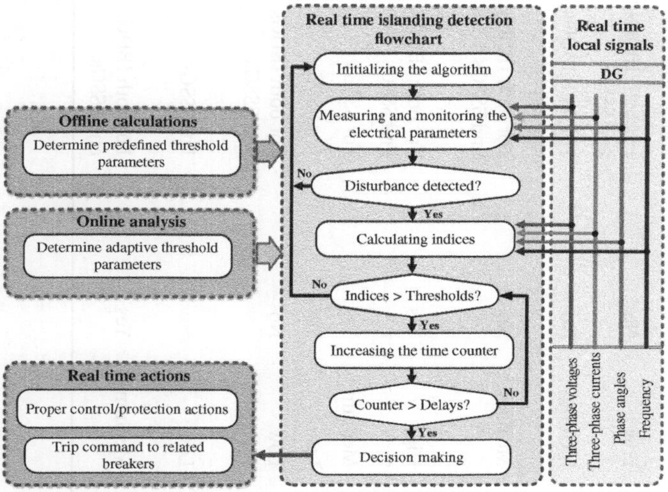

Fig. 5.14 IDS flowchart based on monitoring proper indices.

realized after elapsing the intentional operation time delay. It should be mentioned that the operation time delay is zero for some IDSs. The predefined thresholds are calculated through the off-line calculations part. In some IDSs, the threshold values are not constant, and they are adaptively calculated through the online analysis part. Suitable real-time actions, such as issuing trip commands and appropriate control/protection actions, will be performed after realizing the islanding condition.

The effectiveness of various types of these islanding detection approaches is presented in Table 5.5 [58–66]. These approaches are mainly suggested to resolve the issues of conventional techniques. In [60] and [67], the variations of different indices, tabulated in Table 5.6, during various islanding and nonislanding events are investigated through simulation-based studies. Figure 5.15 depicts the changes of these indices during various islanding scenarios and nonislanding events, such as load switching and short-circuit faults. Furthermore, the MG power mismatch value for each scenario is shown in this figure. As shown in Fig. 5.15(a), only dp/dq, dq/dp, dq/df, dq/dv, dv/df, df/dq, and dp/dv have considerable changes during islanding events, and the other indices do not change significantly. These indices, except dq/df, have large oscillations during nonislanding disturbances, as illustrated in Figs. 5.15(b) and (c). Thus, the dq/df index, which has small variations during nonislanding events and significant changes during islanding events, may be more suitable than the other ones to discriminate the islanding events from the nonislanding disturbances.

Table 5.5 Comparing performance of various IDSs based on monitoring proper indices.

References	Technique	Threshold	NDZ (%)	Speed (ms)	FDR	Unsymmetrical detection	Description
[58]	Equivalent resistance with respect to time	Constant	1	173	Medium	No	For both IBDGs and SSSGs
[59]	Rate of change of power factor angle	Adaptive	0.5	<100	Low	No	For both IBDGs and SSSGs
[60]	Rate of change of frequency over reactive power	Constant	1.65	200	Medium	No	For both IBDGs and SSSGs
[61]	Estimating the parameters of a linear model of the distribution system	Constant	NR	NR	Medium	No	For both IBDGs and SSSGs
[62]	Sequence superimposed component of impedance	Constant	0	20	High	Yes	For both IBDGs and SSSGs
[63]	Reactive power rate of change	Constant	NR	100	Medium	Yes	For SSSGs
[64]	Rate of change of impedance	Constant	3.4	70	Low	Yes	For SSSGs
[65]	Phase angle between the positive-sequence components of voltage and current	Constant	<0.1	20	Low	No	For both IBDGs and SSSGs
[25]	Equivalent resistance with respect to angular frequency	Adaptive	0.2	<250	Low	No	For SSSGs
[66]	Rate of change of sequence components of currents	Constant	0	NR	Medium	Yes	For both IBDGs and SSSGs

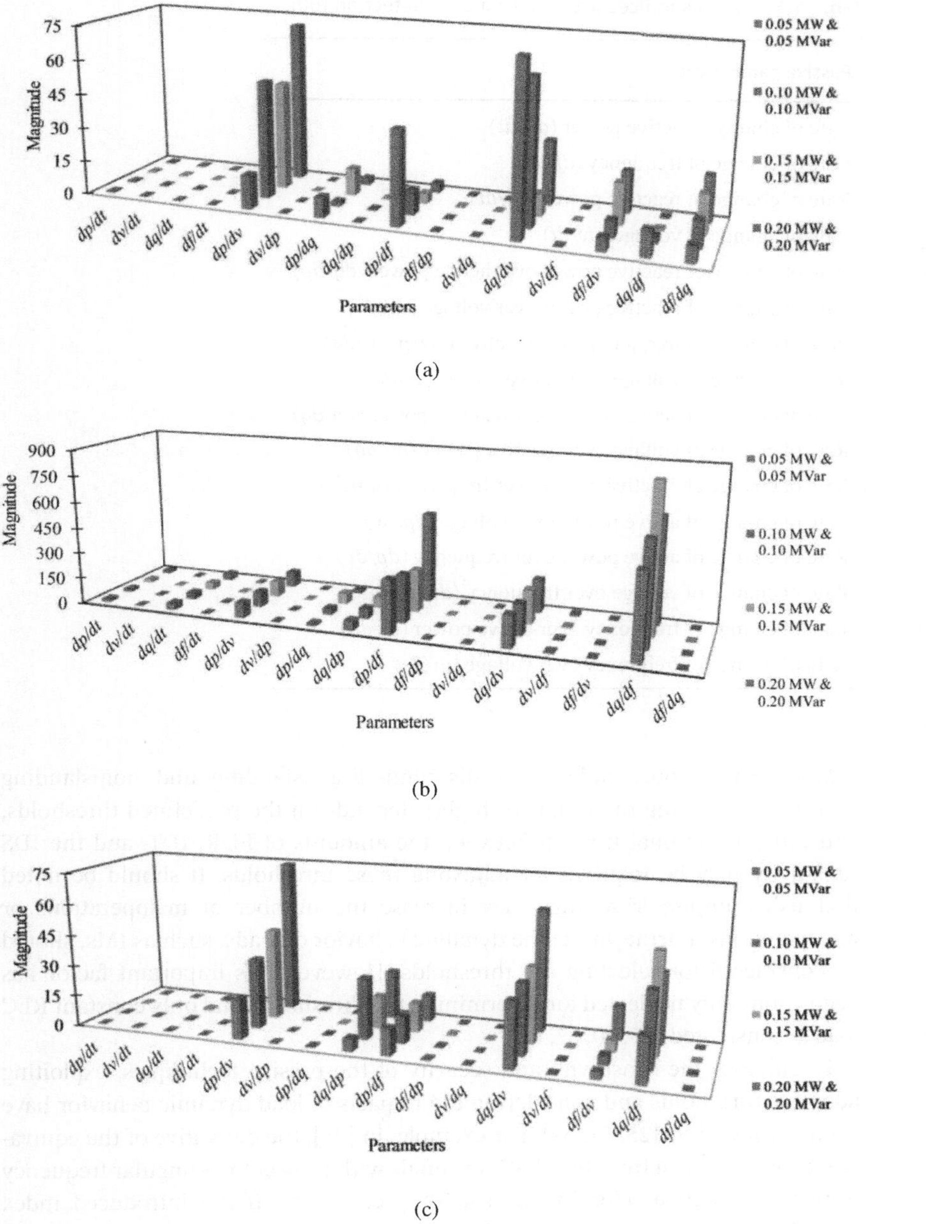

Fig. 5.15 Variations of passive indices during (a) islanding event, (b) nonislanding short-circuit event, and (c) nonislanding load increment event. [67] / with permission of John Wiley & Sons.

Table 5.6 Various indices used for islanding detection [60].

Passive parameters
Rate of change of active power (*dp/dt*)
Rate of change of frequency (*df/dt*)
Rate of change of reactive power (*dq/dt*)
Rate of change of voltage (*dv/dt*)
Rate of change of reactive power over active power (*dq/dp*)
Rate of change of reactive power over voltage (*dq/dv*)
Rate of change of frequency over reactive power (*df/dq*)
Rate of change of voltage over active power (*dv/dp*)
Rate of change of active power over reactive power (*dp/dq*)
Rate of change of voltage over reactive power (*dv/dq*)
Rate of change of reactive power over frequency (*dq/df*)
Rate of change of active power over voltage (*dp/dv*)
Rate of change of active power over frequency (*dp/df*)
Rate of change of voltage over frequency (*dv/df*)
Rate of change of frequency over active power (*df/dp*)
Rate of change of frequency over voltage (*df/dv*)

Monitoring proper indices for discriminating islanding and nonislanding events by exploiting these indices highly depends on the predefined thresholds, and thus, an optimal tradeoff between the amounts of FDR, IDT, and the IDS sensitivity may be required for adjusting these thresholds. It should be noted that using improper settings may increase the number of maloperations or misoperations. Furthermore, the dynamic behavior of loads, such as IMs, should be considered for selecting the thresholds. However, this important factor has been commonly neglected for determining the thresholds, and only constant RLC load is considered [55, 60, 67, 68].

To enhance the sensitivity and security of the passive techniques, exploiting adaptive thresholds and considering the impacts of load dynamic behavior have been suggested by [25, 58, 59]. For example, in [25], the derivative of the equivalent resistance seen from the SSSG terminals with respect to its angular frequency is recommended to identify the islanding condition. If the introduced index value becomes greater than a threshold, the islanding event would be concluded. To improve the sensitivity and reduce the detection time, the amount of this threshold changes adaptively. In fact, this amount is analytically determined in

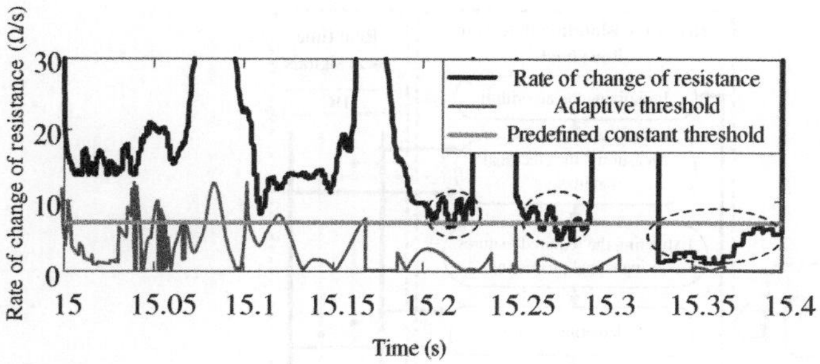

Fig. 5.16 Simulation results for nonislanding short-circuit scenarios for both adaptive and constant threshold values. Adapted from [58].

real time considering the dynamic behavior of the SGBMG and the local loads. For nonislanding events, the amount of this adaptive threshold is very large, whereas its value becomes small for the islanding events.

An adaptive threshold has been proposed by [59] based on the relation between the resonant frequency of local loads and reactive power consumption. This can significantly improve the algorithm security, especially during motor starting conditions. In addition, the IDSs with adaptive thresholds have lower FDRs, as presented in Table 5.5. For example, the equivalent resistance is exploited by both the approaches proposed by [25] and [58]. However, adopting the adaptive thresholds by [25] avoids its false operation during nonislanding disturbances, such as three-phase short circuits. As shown in Fig. 5.16, the approach proposed by [58] maloperates during a three-phase short-circuit fault, while the technique suggested by Bekhradian *et al.* [25] remains stable.

5.5.3 Feature Extraction and Signal Processing Methods

To decrease the amounts of NDZ and IDT, hidden features of the measured signals such as voltage, current, and frequency can be extracted through signal processing techniques [69, 70]. Figure 5.17 shows the typical flowchart of the feature extraction approaches. Local measured signals of MG, such as voltage, current, phase angles, and frequency, are exploited as the real-time inputs of the flowchart, as depicted in Fig. 5.17. Then, the hidden features of these real-time signals will be extracted through signal processing methods. Afterward, based on the extracted features, proper indices can be proposed to detect the islanding event. These proper indices are compared to the predefined thresholds, and the islanding condition would be realized after elapsing the operation time delay. It is worth mentioning that these predefined thresholds are specified through off-line calculations.

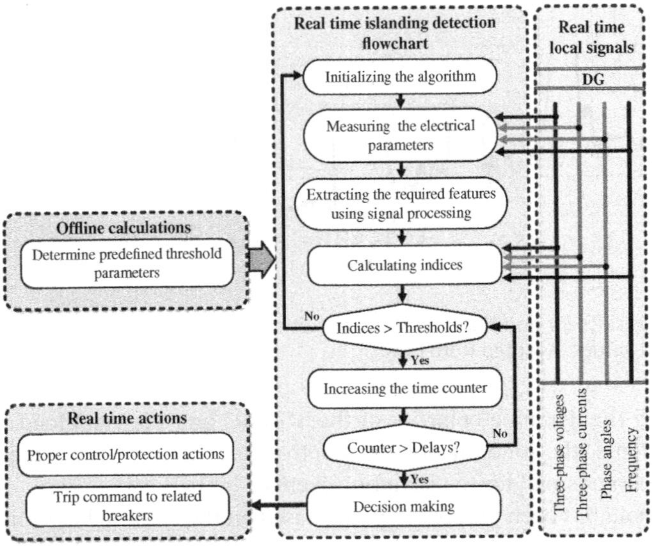

Fig. 5.17 Flowchart of feature extracting-based IDS methods.

Different signal processing techniques have been proposed in research works for islanding detection purposes. In this section, various IDSs, which operate based on signal processing techniques, have been described. The specifications of these IDSs are given in Table 5.7 [71–79]. A thorough summary of these signal processing methods is expressed in the following.

One of the widely used frequency-domain techniques is the Fourier transform (FT), which is commonly utilized to capture the harmonics of the input signals [80]. Harmonic extraction techniques are mainly effective for islanding detection of IBMGs, and they may not be suitable for SGBMGs due to the lack of high-magnitude harmonics. Thus, transient information of the signals should be extracted for detecting the islanding events of SGBMGs. However, FT-based techniques are commonly exploited for stationary signals, and they may not suitable for extracting the high-frequency transient information of the signals [80]. Therefore, to resolve this issue and to extract the transient information of the input signals, other signal processing techniques such as modified FT and wavelet transform (WT) are suggested [81]. These techniques can decompose the signal into different frequency bands [82]. WT extracts the signal features in both frequency and time domains [82, 83]. However, FT only provides the frequency-domain contents of the signal [82–85]. Furthermore, WT-based techniques exploit small moving windows to extract high-frequency information and large moving windows for low-frequency contents [85]. For example, the WT

Table 5.7 Comparing various feature extraction-based IDSs.

References	Extraction method	Input signal	NDZ (%)	Speed	FDR	Multi-phase detection	Description
[71]	TVF-EMD	Voltage	0	15 ms	Medium	No	Both SSSGs and IBDGs
[72]	HHT	Angular frequency	<0.05	<450 ms	Low	No	SSSGs
[73]	Tufts-Kumaresan	Angular frequency	0.5	611.5 ms	Low	No	SSSGs
[74]	WT	DG voltage	0	Half cycle	Medium	Yes	Both SSSGs and IBDGs
[75]	TMF	Voltage	1	55 ms	Medium	Yes	Both SSSGs and IBDGs
[76]	MM	Negative sequence voltage	0	15 ms	Low	Yes	Hybrid
[77]	TMF	DG voltage and current	0	<1 cycle	Medium	No	Both SSSGs and IBDGs
[78]	WT	Voltage and frequency	—	<1 s	Medium	No	SSSGs
[79]	HHT	PCC voltage and current	5	<2 cycles	Medium	No	SSSGs

method is used by [78] to extract the high-frequency contents of the voltage and frequency signals after MG islanding occurrence. Afterward, if the amount of these high-frequency contents becomes greater than the predefined thresholds, the islanding event would be identified within 1 s. It is worth mentioning that this approach may maloperate for some nonislanding disturbances.

WT technique may have nuisance incorrect tripping due to the high-frequency transients and noises of the measurement devices. To diminish the shortcomings of WT-based techniques and enhance their security against measurement transients, the wavelet singular entropy (WSE) method has been suggested [74]. In this technique, three-phase voltage signals at the DG terminals are measured. Then, WT is used to process the measured signals and calculate the associated coefficient matrix. Afterward, the probability array and singular value matrix will be generated. Then, for each phase, the WSE value will be computed. Finally, the WSE linear summation of each phase is exploited as a suitable index to identify islanding events. This technique can considerably minimize the NDZ boundary and reduce the IDT. However, the main shortcomings of this technique include heavy computational burden and false operation for nonislanding events.

A modified version of WT, known as "S-transform" (ST), has been recommended by [86–89] to improve the WT effectiveness. In fact, the ST technique can be assumed as a combination of short-time FT (STFT) and WT. ST provides frequency-dependent resolution and extracts both amplitude and phase information. However, it may fail to localize some events in the time domain [83]. Thus, the modified version of ST, called "Hyperbolic ST (HST), has been developed to resolve the mentioned shortcoming [84]. HST exploits a pseudo-Gaussian hyperbolic window, which gives better resolution for different frequency bands. In [90], the HST method is exploited to analyze the voltage signal measured at the PCC and extract its components. Afterward, based on the extracted components, the signal energy is computed. Finally, the islanding event will be identified if the calculated energy becomes greater than a predefined threshold. The investigations of [82] properly reveal that the HST has a better resolution for identifying islanding events and also better performance against noisy signals compared to the ST and WT techniques. In addition, the HST-based methods have lower FDR and are more secure against nonislanding disturbances.

The empirical mode decomposition (EMD) technique is also used by various research works to identify islanding events [71, 91]. This technique extracts the various mono-components of the input signal, which are known as meaningful intrinsic mode functions (IMFs) [92]. However, the main drawbacks of this technique include mode mixing impact and computational complexity [71, 91, 92]. To resolve these issues, a time-varying filter-based EMD (TVF-EMD) is suggested by [71, 92] whose cutoff frequency may change and is not a constant value. This variable cutoff frequency improves the decomposition effectiveness, especially for nonstationary signals. To calculate the energy of IMFs, TVF-EMD technique is exploited by [71]. When the computed energy becomes greater than a predefined threshold, the islanding condition would be realized. The FDR of this technique for islanding and nonislanding disturbances are 0.056% and 1.96%, respectively. In addition, this technique can effectively detect islanding events in less than 15 ms under the zero-power mismatch condition. Although the computation of TVF-EMD has some complexities, reference [71] claims that TVF-EMD techniques can be practically implemented through the new digital relays.

A novel EMD-based technique, which is known as "Hilbert–Huang transform" (HHT), has been proposed to process nonstationary and nonlinear signals. Indeed, the HHT technique exploits frequency and amplitude modulation to decompose the signal through EMD. To analyze the dynamic response of the SSSG, HHT is exploited by [72]. Afterward, four features, including identity participation matrix, damping factor, zero mode, and frequency of oscillations, will be computed to discriminate the islanding events from nonislanding ones. This method can identify the islanding event within 450 ms, and its NDZ is less than 0.05%.

To improve the time localization feature, especially for high-frequency components, the time–time transform (TTT) method has been recommended by some research works [93–95], which is computed by inverse FT of the ST filter [82]. However, similar to EMD, its computation has some complexities. Mathematical morphology (MM) is another powerful signal processing method that is adopted based on integral geometry and set theory. This technique is a time-domain nonlinear transformation that has a lower computational burden compared to TTT [96]. MM exploits some simple operators, such as Dilation and Erosion, for processing the signals [97, 98]. Dilation and Erosion operate based on the set theory and simple addition or subtraction; thus, their computational burden is low. In [76], an islanding detection technique is suggested utilizing MM to realize the islanding condition of each DG in an HMG. In this regard, for each DG, the three-phase voltage and current signals are acquired for one cycle. Afterward, MM filters are exploited to analyze these signals through simple operators, including Erosion, Dilation, Opening, and Closing. Then, utilizing the MM output, proper islanding detection indices will be calculated. Finally, the islanding event will be realized by comparing the amount of these indices with the prespecified thresholds. The IDT of this method is less than 15 ms, and its NDZ value is almost zero.

Another signal processing technique is the transient monitoring function (TMF), which can be exploited to discriminate islanding events from nonislanding disturbances [75]. This technique computes the difference between the real measured signals and estimated signals considering the normal grid-connected operation of the MG. Large computed differences indicate the MG islanding event. The main challenge of this technique is the accurate computing of the estimated signal. In addition, this method may maloperate during nonislanding events, such as three-phase short circuits. The NDZ and IDT of this technique are about 1% and 55 ms, respectively.

5.5.4 Intelligent Techniques

To improve the performance of islanding detection techniques, intelligent-based methods have been adopted using the indices introduced in Section 5.5.2 and signal processing methods presented in Section 5.5.3. Figure 5.18 shows the typical flowchart of intelligent techniques. As shown, signal processing methods are used to extract suitable features and compute proper indices. Afterward, these computed indices and extracted features would be exploited for training and testing intelligent classifiers. Finally, these trained classifiers are used in real time to identify islanding events. Choosing suitable features, which can effectively discriminate islanding events from nonislanding ones, is the most important

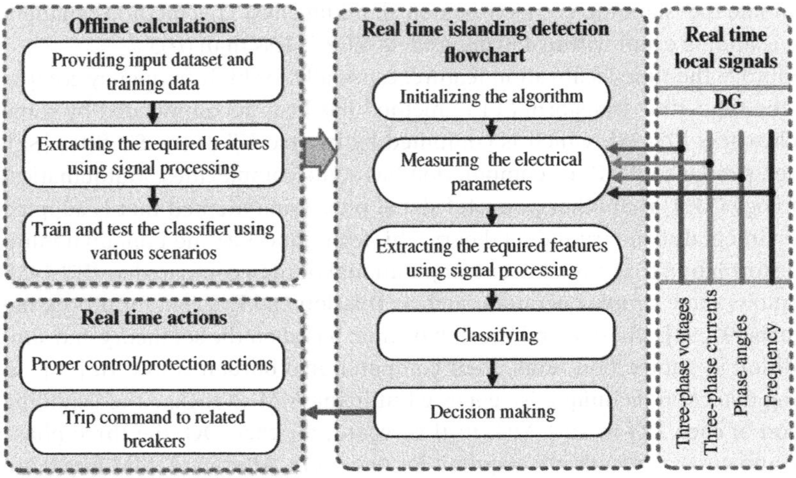

Fig. 5.18 Flowchart of intelligent IDS methods.

part of intelligent techniques. Thus, sensitive features to islanding events should be selected to improve the IDT and reduce FDR. Furthermore, to decrease the computational burden, the correlation between the selected features should be minimized [99].

The classifier type has a considerable effect on the islanding detection effectiveness. Various classifiers are suggested by the previous research works for islanding detection. The effectiveness of some of the intelligent IDSs and their exploited classifiers is compared in Table 5.8. These techniques are briefly explained in the following [100–107].

The most common data-mining method is decision tree (DT), which provides a hierarchical decision-making model [100]. Indeed, DT decomposes complex decisions into simple decisions exploiting "if-then" rules. Classification by DT starts at the "Root node," which contains the initial classification. This node is splattered into some "Child nodes" by utilizing the suitable classifications [107]. These conversions will be continued till the final decision or "Leaf node" is achieved. A DT-based IDS is suggested by [99], which exploits 27 features. However, due to employing low sensitive features, the NDZ of this technique is relatively large. In [107], the harmonic content of equivalent reactance is exploited as a new feature to enhance the IDS effectiveness. Afterward, a new relay logic is adopted utilizing a classification and regression tree. The NDZ value of this algorithm is zero, and its IDT is about 300 ms.

FL is a computational technique that is exploited by many research works to identify islanding events. In [101], a novel algorithm is proposed based on FL, which exploits 11 different features, such as ROCOF, to identify the SGBMG islanding events. Furthermore, to specify the initial classification boundaries, DT-based approaches are commonly exploited. This approach can identify the islanding events only when the SGBMG power mismatch is larger than 40%.

Artificial neural network (ANN) is one of the main intelligent approaches, which is mainly adopted for machine learning purposes. Different components, including various layers and analytical functions, are employed by ANN-based techniques to learn a pattern from the input data. After the learning procedure, ANN can recognize the pattern of the new input data. Various MG features and conditions should be exploited for training or testing the ANN-based IDSs. If the ANN-based IDS has not been trained for some system conditions, it may operate incorrectly during the unexperienced islanding or nonislanding events. A four-layer ANN is employed in [108] for islanding detection. For the training and testing stages of such an ANN-based IDS, the measured voltage signals at DG terminals are used. For considering different operating conditions of MG, about 2000 simulations have been performed. Four major islanding scenarios have been considered in the training and testing stages, including positive/negative active/reactive power mismatches. In addition, load switching and frequency variations are considered for nonislanding events. However, for nonislanding events, fault-initiated islanding and the dynamic behavior of IMs have been neglected. The IDT of this approach is about 2.75 s, and its NDZ is 10% [108]. To improve the accuracy of islanding detection approaches, an adaptive combination of FL and ANN methods, which is commonly known as "ANFIS," has been used by various research works [109].

Extreme learning machine (ELM) is adopted to improve the speed of conventional classifiers [104]. Indeed, ELM is a "randomized single-hidden layer feed-forward neural network," which can be properly used for detecting islanding events. For example, ELM is utilized by [104] in which the values of ELM optimal parameters are specified through "evolutionary computation" [104]. In this technique, the features would be specified by the phase-space approach. The FDR of this technique is quite low, and its IDT is less than six cycles. In addition, an ensemble learning method called "random forest (RF) classifier" is exploited by [102] to identify islanding events. This classifier works based on various DTs, and its effectiveness is presented in Table 5.8.

Support-vector machine (SVM) is another machine learning tool that works based on separating hyperplanes. SVM can be exploited for both classification and regression [110]. To classify the input data with the maximum accuracy, the suitable hyperplane should be specified using different linear or nonlinear kernel functions. In [111], an intelligent technique is adopted employing SVM to enhance the effectiveness of VS method. To do so, some novel features are extracted using

Table 5.8 Comparing various intelligent-based IDSs.

References	Method	Inputs	NDZ (%)	Speed	FDR	Multi-phase detection	Description
[100]	DT	27 features	<30	1.5 cycles	High	No	Both SSSGs and IBDGs
[101]	DT & FL	11 features	<40	—	Medium	No	SSSGs
[102]	RF	—	<30	—	Medium	No	Both SSSGs and IBDGs
[103]	EkNN	—	0	70 ms	Low	No	Both SSSGs and IBDGs
[104]	ELM	Phase space method	—	6 cycles	Low	No	Both SSSGs and IBDGs
[105]	DT	11 indices	0	45 ms	Medium	No	SSSGs
[106]	DT	11 indices	—	163.7 ms	Medium	No	SSSGs
[107]	DT	Harmonic content df/dt	0	300 ms	Low	No	SSSGs

five MG variables and then the features would be injected into the SVM classifier. The effectiveness of the proposed approach is analyzed by different kernels including Gaussian RBF, polynomial, and linear. Based on these studies, the NDZ of this approach is almost zero, and its detection accuracy for 0.5% power mismatch is equal to 89.6%, 92.6%, and 82.2% employing Gaussian RBF, polynomial, and linear functions, respectively. These investigations indicate that the polynomial function has a better performance compared to the other functions.

5.5.5 Comparison of Different Methods

Different features of various types of IDSs are compared with each other as given in Table 5.9. In this table, "very fast," "fast," and "medium" refer to "less than 40 ms," "less than 200 ms," and "less than 1 s," respectively. In addition, high implementation cost means that those IDS methods require additional devices to be installed. Moderate cost means they do not require new devices while they may require hardware improvements and minor changes in the islanding detection relays. Moreover, low-cost techniques can be implemented using the existing equipment without requiring additional expenditures.

As shown in this table, remote approach and active methods have appropriate performance, and they can quickly identify islanding events even with zero-power

Table 5.9 Comparison of various islanding detection algorithms.

Method		Adverse impact on steady-state operation	NDZ	Speed	FDR	Cost	Network configuration dependency	Threshold dependency	Training data dependency	Computation burden
Communication-based (remote)		No	Zero	Very fast	Low	High	High	Low	None	Low
Active (local)		Yes	Zero	Fast	Moderate	High	Moderate	Moderate	None	Low
Passive (local)	Conventional	No	High	Medium	High	Low	Moderate	High	None	Low
	Using proper setting rules	No	Moderate	Medium	Moderate	Low	Moderate	High	None	Low
	Monitoring proper indices	No	Low	Fast	Moderate	Low	Moderate	Moderate	None	Low
	Monitoring proper indices and using adaptive thresholds	No	Low	Fast	Low	Low	Low	Low	None	Moderate
	Feature extraction and signal processing	No	Almost zero	Very fast	Low	Medium	Low	Moderate	None	High
	Intelligent techniques	No	Almost zero	Very fast	Low	Medium	Moderate	Low	High	High

mismatch. However, the capital expenditure of remote techniques is relatively high, and using these approaches may not be practical due to their implementation complexities. Furthermore, active techniques are suitable for islanding detection in IBMGs, and therefore, IBDGs commonly employ active methods that inject perturbations into their control systems. However, these active approaches may have adverse effects on the steady-state operation and stability of SGBMGs. Since local passive techniques are simple and do not impose negative impacts on SGBMG performance, they are generally recommended for islanding detection in SGBMGs.

Conventional passive methods have a large NDZ and low IDT. In addition, the performance of these techniques highly depends on threshold values, which is a negative feature. Thus, suitable adjustment of these threshold parameters improves the effectiveness of the conventional methods. Employing appropriate monitoring indices can enhance the IDT and decrease the NDZ. However, using these indices cannot completely eliminate the NDZ region. It is highly recommended to exploit adaptive thresholds to enhance the security of these indices against nonislanding events and reduce their dependency on the predefined thresholds. Feature extraction using signal processing techniques and intelligent methods can considerably reduce the amount of NDZ. In addition, these approaches enhance both IDT and FDR. The main shortcoming of data-mining techniques is their high dependency on the training data and system configuration. Since the feeder configuration may change frequently in distribution systems, these techniques are not suggested for real applications. Furthermore, both the signal processing-based approaches and intelligent methods suffer from high computation burden, and they require more powerful hardware and processors compared to the other passive methods.

5.6 Ideas for the Future Works

In this section, the following recommendations are presented for further research works and to enhance the effectiveness of existing IDSs [10].

5.6.1 Integrated Central Islanding Detection

An MG may include multiple DGs, while each of them uses different control systems and different technologies. In such a condition, it is suggested to adopt a central islanding detection unit for the whole MG instead of using a separate IDS for each DG. In this regard, to provide a suitable performance, the measured signals of the whole DGs should be exploited.

5.6.2 Using Adaptive Thresholds

Utilizing adaptive thresholds, instead of constant predefined ones, for both passive and active methods, has been recently proposed by a few research works that can considerably enhance the IDS security against nonislanding events and diminish the NDZ [25]. To improve the IDS effectiveness, it is proposed to consider the following factors for determining the adaptive thresholds:

- Control strategy of MG including control modes of DGs (e.g., constant power and droop modes)
- Real-time variations of MG power mismatch
- MG configuration (e.g., feeder configuration, the status of DGs, the status of the large loads)
- MG total inertia
- Dynamic behavior of the local loads.

5.6.3 Using Negative-Sequence Current

As discussed earlier, the islanding event can occur unsymmetrically due to the open-circuit condition of one or two phases. This condition may lead to MG equipment damage, especially to SSSGs. Thus, it should be quickly identified and then proper control/protection actions should be performed. To detect such a condition, the ratio of the negative-sequence to the positive-sequence current (I_2/I_1) and also its rate of change can be exploited.

5.6.4 Combined Application of Signal Processing Techniques

Each individual signal processing technique may include some benefits and shortcomings concurrently. Thus, it is recommended to adopt a combination of the signal processing techniques and fuse their output decisions to improve the IDS detection effectiveness for various transient types and minimize the FDR and NDZ values.

5.6.5 Utilizing the Neutral Current

A very high percentage of the short-circuit faults include the ground, i.e., single-phase and double-phase to ground faults for which, large zero-sequence currents pass through the feeder. However, the zero-sequence current is not usually present during the islanding condition. Therefore, this current component and the associated harmonics can be effectively exploited to discriminate islanding events from the short-circuit faults.

5.6.6 Using Off-line Studies to Adjust the Time Delay Setting

The maximum permissible value of IDS intentional time delay has been provided by various standards [9, 15]. Using the maximum value may unnecessarily lead to various challenges, including (i) stability issues, (ii) equipment damages, and (iii) large IDTs. Therefore, instead of using the maximum value, the exact required amount of IDS time delay should be determined based on the real configuration of the MG and its protection/control system. To do so, proper off-line studies should be performed by the system engineers to determine the suitable amount of time delay. These off-line studies should include the following investigations:

- Investigating various types of MG stability (as discussed in Chapter 2) after islanding and fault-initiated islanding events
- Investigating possible thermal/mechanical damages of MG equipment, especially DG damages due to deviations of voltage and frequency
- Investigating FDR and NDI values for various amounts of the intentional delay.

5.6.7 Using Proper Control Strategy

The control strategies of the MG, including DG control modes, the controller parameters, and droop coefficients values, have considerable impacts on the deviations of the MG parameters after the islanding event and IDS operation. In fact, the features of the MG measured signals, such as the oscillations cycle, rate of variations, and damping ratio, may be highly affected by the MG control strategy. For instance, when the grid-connected SSSG operates under the constant active/reactive power control mode, the SGBMG frequency changes monotonically after the islanding event. However, for the droop control mode, the frequency may experience considerable oscillations after the islanding event depending on the droop coefficient. Moreover, the SGBMG voltage may have larger deviations when the SSSG operates under the constant reactive power control mode.

Different dynamic features of MG parameters can result in various decisions by using different IDS methods during the islanding event. For example, when the frequency/voltage deviations of the MG are large, the conventional IDSs, such as UF/OF and UV/OV, can identify the islanding event fast. Therefore, the control strategy can have significant impacts on the successful IDS operation. On the other hand, the stability of MG during the steady-state operation and transient disturbances may be affected by the abovementioned controlling factors. Therefore, a proper control strategy for MG operation should be exploited to achieve a proper tradeoff between preserving the stability of MG after transient events and preventing thermal/mechanical damages of MG equipment, versus a quick and secure islanding detection.

5.6.8 Probabilistic Approach for Adjustment of IDS Settings

For a given MG, the probability of occurrence of each specific islanding event may be different from that of the other ones. For instance, for a particular MG, the likelihood of occurrence of a harmful islanding event under the zero-power imbalance may be negligible, while the chance of a destructive islanding event occurrence due to the single-phase short-circuit faults in the upstream grid may be much larger. Consequently, although using sensitive adjustment of IDS settings improves the IDS performance for low-probability islanding events, the security of IDS will be endangered during the nonislanding disturbances. Therefore, it is evident that the philosophy of considering the worst scenarios, such as zero-power imbalance, may have adverse effects on IDS security. Accordingly, to deal with the related uncertainties and consider the probability of occurrence of different islanding events, an appropriate probabilistic method should be utilized for optimal adjustment of IDS setting values.

5.7 Summary

The MG islanding events should be quickly identified to perform proper control actions and prevent possible adverse impacts on the upstream network. In this chapter, a thorough technical review of islanding detection approaches of MGs has been presented. At first, some appropriate indices have been introduced that can be practically exploited to analyze the effectiveness of the islanding detection techniques and select the most suitable one. Afterward, different IDSs applicable to MGs are scrutinized and carefully compared with each other. Each of the presented IDSs may be proper for some MGs; hence, the most compatible IDS for a specific MG should be chosen considering the obtained comparison results. Based on the previous studies, islanding events in SGBMGs can be reliably detected using passive techniques, even under near zero-power mismatch conditions. In contrast, for IBMGs, active techniques are necessary to identify such conditions. In addition, active methods are not recommended for SGBMGs due to their adverse impacts on the SGBMG steady-state operation, stability, and power quality. The investigations provided in this chapter properly indicate that the most viable passive technique is monitoring the proper islanding detection indices along with utilizing feature-extracting methods, which can considerably reduce the NDZ. In addition, the effectiveness of these passive techniques can be further enhanced by exploiting intelligent techniques. Moreover, using adaptive thresholds improves the IDS security against nonislanding events and also reduces the IDS dependency on the system configuration. Additionally, for future research works and further enhancements of the existing passive techniques, some practical and useful suggestions have been provided in this chapter of the book.

Problems

5.1 A sample SGBMG is illustrated in Fig. 5.19, which contains an SSSG, some static load, and an IM whose parameters are given in Table 5.10. A ROCOF relay is located at the PCC to detect islanding events whose threshold and operation delay is adjusted as 0.05 pu/s and 50 ms, respectively.

a) Calculate the maximum active power mismatch for which the relay cannot identify the islanding event in 1 s. To do so, the analytical functions given in Table 5.2 can be exploited.

b) Implement this SGBMG and ROCOF relay in an appropriate software considering the guidelines provided in Problem 1.1. Then, determine the NDZ region through simulation studies for which the relay cannot identify islanding events within 1 s. In this regard, disconnect the local loads of SGBMG, at first. Afterward, provide various amounts of SGBMG power mismatches exploiting constant power loads to test the relay performance

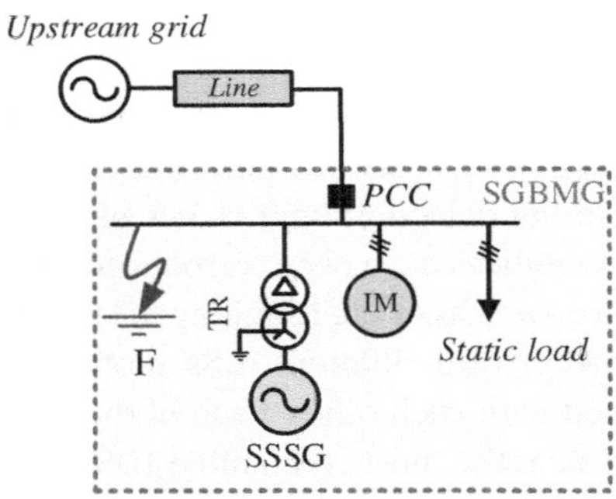

Fig. 5.19 Sample SGBMG and distribution system.

Table 5.10 System parameters.

Equipment name	Parameters values
Line	Length = 5 km, impedance = $0.35 + j0.4$ (Ω/km)
Static loads	$P = 4$ MW, PF = 0.95
Induction motor (IM)	Nominal power = 5 MW, inertia = 0.4 s
SSSG	Voltage = 11 kV, $P = 10$ MW, inertia = 0.57 s
	$X_d = 1.9$ pu, $X'_d = 0.28$ pu, $X''_d = 0.18$ pu,
	$X_q = 0.99$ pu, $X''_q = 0.18$ pu
Transformer (TR)	Ratio = 20/11 kV, $S = 15$ MVA, $U_k = 10\%$
Upstream grid	Voltage = 20 kV, short-circuit capacity = 170 MVA, $X/R = 5$

and extract its NDZ. Compare the results with analytical calculations of part (a).

c) Repeat part (b) by exploiting constant impedance loads for extracting NDZ through simulation studies. Compare the results with those of part (b).

d) Simulate a three-phase short-circuit fault at point F that is cleared after 200 ms. Analyze the islanding relay decision in this case. Does the relay maloperate in this example?

e) Analyze the islanding relay decision during motor starting conditions using simulation studies. Does the relay maloperate in this case?

f) Revise the settings of the relay to reduce its NDZ by 30%. How does the relay FDR amount change in this case?

5.2 In Problem 5.1, UF/OF functions are exploited instead of the ROCOF function.

a) Calculate proper settings, i.e., threshold and operation time delay, for these functions, assuming that the maximum active power mismatch, which is not identifiable by the relay, is equal to 10%. To do so, the analytical functions given in Table 5.2 can be exploited.

b) Determine the NDZ through an appropriate simulation software considering the calculated settings of part (a). In this regard, disconnect the local loads of SGBMG, at first. Afterward, provide various amounts of SGBMG power mismatches exploiting constant power loads.

c) Simulate a three-phase short-circuit fault at point F, which is cleared after 200 ms. Analyze the islanding relay decision in this example. Does the relay maloperate in this case?

5.3 Resolve Problem 5.2 using VS islanding relays.
Note: To implement the VS relay, at first, the first harmonic of the voltage signal should be extracted using the FT algorithm. Then, the vector shift amount can be estimated by calculating the first harmonic component. Meanwhile, to determine this period, various techniques can be exploited, such as the zero-crossing method. More details about the implementation of VS relay can be found in [112].

5.4 In Problem 5.1, for various amounts of SSSG inertias presented below, find proper setting values of the ROCOF islanding relay to identify the islanding within 1 s while the NDZ value be less than 20%.

(a) $H = 0.4$ s, (b) $H = 0.75$ s, (c) $H = 1.3$ s.

References

1 F. Aslani and S. Li, "Islanding detection methods and challenges for distribution generation: A technological review," *IEEE Access*, vol. 13, pp. 133568–133595, Aug. 2025.

2 U. Markovic, D. Chrysostomou, P. Aristidou, and G. Hug, "Impact of inverter-based generation on islanding detection schemes in distribution networks," *Electr. Power Syst. Res.*, vol. 190, Jan. 2021, Art. no. 106610.

3 A. Yafaoui, B. Wu, and S. Kouro, "Improved active frequency drift anti-islanding detection method for grid connected photovoltaic systems," *IEEE Trans. Power Electron.*, vol. 27, no. 5, pp. 2367–2375, May 2011.

4 X. Wang and W. Freitas, "Impact of positive-feedback anti-islanding methods on small-signal stability of inverter-based distributed generation," *IEEE Trans. Energy Convers.*, vol. 23, no. 3, pp. 923–931, Sep. 2008.

5 H. Vahedi and M. Karrari, "Adaptive fuzzy Sandia frequency-shift method for islanding protection of inverter-based distributed generation," *IEEE Trans. Power Del.*, vol. 28, no. 1, pp. 84–92, 2012.

6 M. Al Hosani, Z. Qu, and H. H. Zeineldin, "Scheduled perturbation to reduce nondetection zone for low gain Sandia frequency shift method," *IEEE Trans. Smart Grid*, vol. 6, no. 6, pp. 3095–3103, Nov. 2015.

7 D. Lin *et al.*, "Performance evaluation of islanding detection for distributed energy resources with grid support functions," *IEEE Trans. Ind. Appl.*, vol. 61, no. 6, pp. 9602–9614, Nov.-Dec. 2025.

8 E. Vazquez, N. Vazquez, and R. Femat, "Modified Sandia voltage shift anti-islanding scheme for distributed power generator systems," *IET Power Electron.*, vol. 13, no. 18, pp. 4226–4234, Dec. 2020.

9 IEEE Std 1547-2018, "IEEE standard for interconnection and interoperability of distributed energy resources with associated electric power systems interfaces," Revision of IEEE Std 1547-2003, 2018.

10 R. Bekhradian and M. Sanaye-Pasand, "A comprehensive survey on islanding detection methods of synchronous generator-based microgrids: Issues, solutions and future works," *IEEE Access*, vol. 10, pp. 76202–76219, Jan. 2022.

11 A. L. Finkler, J. M. Lenz, A. Sausen, M. de Campos, and P. S. Sausen, "Challenges on the use of vector shift relays in distributed generation: A Brazilian case study," *Electr. Power Syst. Res.*, vol. 204, Mar. 2022. Art. no. 107688.

12 K. Naraghipour, I. Abdelsalam, K. H. Ahmed, and C. D. Booth, "Modified Q–f droop curve method for islanding detection with zero non-detection zone," *IEEE Access*, vol. 9, pp. 158027–158040, Nov. 2021.

13 N. K. Swarnkar, O. P. Mahela, B. Khan, and M. Lalwani, "Identification of islanding events in utility grid with renewable energy penetration using

current based passive method," *IEEE Access*, vol. 9, pp. 93781–93794, Jul. 2021.

14 K. Jia, H. Wei, T. Bi, D. W. P. Thomas, and M. Sumner, "An islanding detection method for multi-DG systems based on high-frequency impedance estimation," *IEEE Trans. Sustain. Energy*, vol. 8, no. 1, pp. 74–83, Jan. 2017.

15 K. N. E. K. Ahmad, J. Selvaraj, and N. A. Rahim, "A review of the islanding detection methods in grid-connected PV inverters," *Renew. Sustain. Energy Rev.*, vol. 21, pp. 756–766, 2013.

16 M. R. Alam, M. T. A. Begum, and K. M. Muttaqi, "Assessing the performance of ROCOF relay for anti-islanding protection of distributed generation under subcritical region of power imbalance," *IEEE Trans. Ind. Appl.*, vol. 55, no. 5, pp. 5395–5405, 2019.

17 M. Mishra, S. Chandak, and P. K. Rout, "Taxonomy of islanding detection techniques for distributed generation in microgrid," *Renew. Energy Focus*, vol. 31, pp. 9–30, Dec. 2019.

18 J. C. M. Vieira, W. Freitas, W. Xu, and A. Morelato, "An investigation on the nondetection zones of synchronous distributed generation anti-islanding protection," *IEEE Trans. Power Del.*, vol. 23, no. 2, pp. 593–600, Apr. 2008.

19 H. Abdi, A. Rostami, and N. Rezaei, "A novel passive islanding detection scheme for synchronous-type DG using load angle and mechanical power parameters," *Electr. Power Syst. Res.*, vol. 192, 2021.

20 A. K. Özcanlı and M. Baysal, "A novel multi-LSTM based deep learning method for islanding detection in the microgrid," *Electr. Power Syst. Res.*, vol. 202, 2022.

21 D. Salles, W. Freitas, J. C. M. Vieira, and B. Venkatesh, "A practical method for nondetection zone estimation of passive anti-islanding schemes applied to synchronous distributed generators," *IEEE Trans. Power Del.*, vol. 30, no. 5, pp. 2066–2076, Oct. 2015.

22 Y. Li, *et al.*, "A generic method for the determination of non-detection zones in DER-dominated distribution grids," 2018 IEEE Power & Energy Society General Meeting (PESGM), pp. 1–5, 2018.

23 Y. Li *et al.*, "Nondetection zone analytics for unintentional islanding in a distribution grid integrated with distributed energy resources," *IEEE Trans. Sustain Energy*, vol. 10, no. 1, pp. 214–225, Jan. 2019.

24 Z. Ye, A. Kolwalkar, Y. Zhang, P. Du, and R. Walling, "Evaluation of anti-islanding schemes based on nondetection zone concept," *IEEE Trans. Power Electron.*, vol. 19, no. 5, pp. 1171–1176, Sep. 2004.

25 R. Bekhradian, M. Davarpanah, and M. Sanaye-Pasand, "Novel approach for secure islanding detection in synchronous generator based microgrids," *IEEE Trans. Power Del.*, vol. 34, no. 2, pp. 457–466, Apr. 2019.

26 W. Zheng, P. Crossley, B. Xu, and H. Qi, "Transient stability of a distribution subsystem during fault-initiated switching to islanded operation," *Int. J. Electr. Power Energy Syst.*, vol. 97, pp. 418–427, 2018.

27 A. Pouryekta, V. K. Ramachandaramurthy, N. Mithulananthan, and A. Arulampalam, "Islanding detection and enhancement of microgrid performance," *IEEE Syst. J.*, vol. 12, no. 4, pp. 3131–3141, Dec. 2018.

28 R. Zamani, M. P. Moghaddam, H. Panahi, and M. Sanaye-Pasand, "Fast islanding detection of nested grids including multiple resources based on phase criteria," *IEEE Trans. Smart Grid*, vol. 12, no. 6, pp. 4962–4970, Nov. 2021.

29 M. W. Altaf, M. T. Arif, S. Saha, S. N. Islam, M. E. Haque, and A. M. T. Oo, "Effective ROCOF-based islanding detection technique for different types of microgrid," *IEEE Trans. Ind. Appl.*, vol. 58, no. 2, pp. 1809–1821, Mar.–Apr. 2022.

30 A. Rostami *et al.*, "Islanding detection based on ROCOV and ROCORP parameters in the presence of synchronous DG applying the capacitor connection strategy," *Electr. Power Comp. Syst.*, vol. 45, no. 3, pp. 315–330, 2017.

31 S. Reza *et al.*, "Minimum-features-based ANN-PSO approach for islanding detection in distribution system," *IET Renew. Power Gener.*, vol. 10, no. 9, pp. 1255–1263, 2016.

32 R. Bekhradian, M. Davarpanah, and M. Sanaye-Pasand, "Current-based blocking scheme to stabilize distribution network relays against FIDVR," *Int. J. Electr. Power Energy Syst.*, vol. 132, 2021, Art. no. 107205.

33 M. A. Ebrahim, F. Wadie, and M. A. Abd-Allah, "An algorithm for detection of fault, islanding, and power swings in DG-equipped radial distribution networks," *IEEE Syst. J.*, vol. 14, no. 3, pp. 3893–3903, 2019.

34 A. H. K. Alaboudy, H. H. Zeineldin, and J. Kirtley, "Microgrid stability characterization subsequent to fault-triggered islanding incidents," *IEEE Trans. Power Del.*, vol. 27, no. 2, pp. 658–669, 2012.

35 D. Salles, W. Freitas, J. C. M. Vieira, and W. Xu, "Nondetection index of anti-islanding passive protection of synchronous distributed generators," *IEEE Trans. Power Delivery*, vol. 27, no. 3, pp. 1509–1518, Jul. 2012.

36 A. Mazloomzadeh, M. H. Cintuglu, and O. A. Mohammed, "Islanding detection using synchronized measurement in smart microgrids," in *2013 IEEE PES Conference on Innovative Smart Grid Technologies (ISGT Latin America)*, pp. 1–7, 2013.

37 M. M. Ostojić and M. B. Djurić, "The algorithm with synchronized voltage inputs for islanding detection of synchronous generators," *Int. J. Electr. Power Energy Syst.*, vol. 103, pp. 431–439, Dec. 2018.

38 W. Xu, G. Zhang, C. Li, W. Wang, G. Wang, and J. Kliber, "A power line signaling based technique for anti-islanding protection of distributed generators—Part I: scheme and analysis," *IEEE Trans. Power Del.*, vol. 22, no. 3, pp. 1758–1766, Jul. 2007.

39 W. Wang, J. Kliber, G. Zhang, W. Xu, B. Howell, and T. Palladino, "A power line signaling based scheme for anti-islanding protection of distributed generators—Part II: field test results," *IEEE Trans. Power Del.*, vol. 22, no. 3, pp. 1767–1772, Jul. 2007.

40 W. Wang, J. Kliber, and W. Xu, "A scalable power-line-signaling-based scheme for islanding detection of distributed generators," *IEEE Trans. Power Del.*, vol. 24, no. 2, pp. 903–909, 2009.

41 A. Khamis, H. Shareef, E. Bizkevelci, and T. Khatib, "A review of islanding detection techniques for renewable distributed generation systems," *Renew. Sustain. Energy Rev.*, vol. 28, pp. 483–493, 2013.

42 IEA, "Evaluation of islanding detection methods for photovoltaic utility-interactive power systems," Report IEA PVPS, T5-09, 2002.

43 C. Li, C. Cao, Y. Cao, Y. Kuang, L. Zeng, and B. Fang, "A review of islanding detection methods for microgrid," *Renew. Sustain. Energy Rev.*, vol. 35, pp. 211–220, 2014.

44 H. Laaksonen, "Advanced islanding detection functionality for future electricity distribution networks," *IEEE Trans. Power Del.*, vol. 28, no. 4, pp. 2056–2064, Oct. 2013.

45 G. P. Kumar and P. Jena, "Pearson's correlation coefficient for islanding detection using micro-PMU measurements," *IEEE Syst. J.*, vol. 15, no. 4, pp. 5078–5089, 2020.

46 B. Dob and C. Palmer, "Communications assisted islanding detection: contrasting direct transfer trip and phase comparison methods," in *71st Annual Conference for Protective Relay Engineers (CPRE)*, pp. 1–6, 2018.

47 F. Coffele *et al.*, "Centralized loss of mains protection using IEC-61850," in *10th IET International Conference on Developments in Power System Protection (DPSP 2010), Managing the Change, 2010*, pp. 1–5, 2010.

48 A. Timbus, A. Oudalov, and C. N. M. Ho, "Islanding detection in smart grids," in *2010 IEEE Energy Conversion Congress and Exposition*, pp. 3631-3637, 2010.

49 A. Rostami, A. Jalilian, M. T. Hagh, K. M. Muttaqi, and J. Olamaei, "Islanding detection of distributed generation based on rate of change of exciter voltage with circuit breaker switching strategy," *IEEE Trans. Ind. Appl.*, vol. 55, no. 1, pp. 954–963, 2019.

50 D. Bejmert and T. S. Sidhu, "Investigation into islanding detection with capacitor insertion-based method," *IEEE Trans. Power Del.*, vol. 29, no. 6, pp. 2485–2492, 2014.

51 R. Zamani, M. E. H. Golshan, H. H. Alhelou, and N. Hatziargyriou, "A novel hybrid islanding detection method using dynamic characteristics of synchronous generator and signal processing technique," *Electr. Power Syst. Res.*, vol. 175, 2019, Art. no. 105911.

52 J. A. Laghari, H. Mokhlis, A. H. A. Bakar, and M. Karimi, "A new islanding detection technique for multiple mini hydro based on rate of change of reactive power and load connecting strategy," *Energy Convers. Manage.*, vol. 76, pp. 215–224, 2013.

53 G. Marchesan, M. R. Muraro, G. Cardoso, L. Mariotto, and A. P. de Morais, "Passive method for distributed-generation island detection based on oscillation frequency," *IEEE Trans. Power Del.*, vol. 31, no. 1, pp. 138–146, Feb. 2016.

54 J. C. M. Vieira, D. S. Correa, W. Freitas, and W. Xu, "Performance curves of voltage relays for islanding detection of distributed generators," *IEEE Trans. Power Syst.*, vol. 20, no. 3, pp. 1660–1662, Aug. 2005.

55 B. Liu, D. Thomas, Ke Jia, and M. Woolfson, "Advanced ROCOF protection of synchronous generator," in *IEEE PES Innovative Smart Grid Technologies Conference*, pp. 1–6, 2011.

56 K. El-Arroudi and G. Joos, "Data mining approach to threshold settings of islanding relays in distributed generation," *IEEE Trans. Power Syst.*, vol. 22, no. 3, pp. 1112–1119, Aug. 2007.

57 J. C. M. Vieira, D. Salles, and W. Freitas, "Power imbalance application region method for distributed synchronous generator anti-islanding protection design and evaluation," *Electr. Power Syst. Res.*, vol. 81, no. 10, pp. 1952–1960, 2011.

58 X. Xie, C. Huang, and D. Li, "A new passive islanding detection approach considering the dynamic behavior of load in microgrid," *Int. J. Electr. Power Energy Syst.*, vol. 117, 2020, Art. no. 105619.

59 X. Xie, W. Xu, C. Huang, and X. Fan, "New islanding detection method with adaptively threshold for microgrid," *Electr. Power Syst. Res.*, vol. 195, 2021, Art. no. 107167.

60 S. Raza, H. Mokhlis, H. Arof, J. A. Laghari, and H. Mohamad, "A sensitivity analysis of different power system parameters on islanding detection," *IEEE Trans. Sustain. Energy*, vol. 7, no. 2, pp. 461–470, Apr. 2016.

61 A. R. Di Fazio *et al.*, "A smart device for islanding detection in distribution system operation," *Electr. Power Syst. Res.*, vol. 120, pp. 87–95, 2015.

62 P. Kumar, V. Kumar, and B. Tyagi, "A novel islanding detection technique based on event index value for reconfigurable microgrid," *IEEE Trans. Ind. Appl.*, vol. 57, no. 4, pp. 3451–3462, May 2021.

63 S. Nikolovski, H. R. Baghaee, and D. Mlakic, "Islanding detection of synchronous generator-based DGs using reactive power rate of change," *IEEE Syst. J.*, pp. 1–11, 2019.

64 G. Marchesan, K. Maresch, G. Cardoso, A. P. D. Morais, and M. R. Muraro, "Distributed synchronous generation ride-through enhancement by anti-islanding protection blocking," *Electr. Power Syst. Res.*, vol. 196, p. 55, 2021.

65 H. Muda and P. Jena, "Phase angle-based PC technique for islanding detection of distributed generations," *IET Renew. Power Gener.*, vol. 12, no. 6, pp. 735–746, 2018.

66 K. Sareen, B. R. Bhalja, and R. P. Maheshwari, "Universal islanding detection technique based on rate of change of sequence components of currents for distributed generations," *IET Renew. Power Gener.*, vol. 10, no. 2, pp. 228–237, 2016.

67 S. Raza, H. Arof, H. Mokhlis, H. Mohamad, and H. A. Illias, "Passive islanding detection technique for synchronous generators based on performance ranking of different passive parameters," *IET Gener. Transm. Distrib.*, vol. 11, no. 17, pp. 4175–4183, 2017.

68 J. A. Laghari, H. Mokhlis, M. Karimi, A. H. A. Bakar, and H. Mohamad, "An islanding detection strategy for distribution network connected with hybrid DG resources," *Renew. Sustain. Energy Rev.*, vol. 45, pp. 662–676, 2015.

69 S. Raza *et al.*, "Application of signal processing techniques for islanding detection of distributed generation in distribution network: A review," *Energy Convers. Manage.*, vol. 96, pp. 613–624, 2015.

70 Y. M. Makwana and B. R. Bhalja, "Experimental performance of an islanding detection scheme based on modal components," *IEEE Trans. Smart Grid*, vol. 10, no. 1, pp. 1025–1035, 2017.

71 B. K. Chaitanya, A. Yadav, and M. Pazoki, "An advanced signal decomposition technique for islanding detection in DG system," *IEEE Syst. J.*, vol. 15, no. 3, pp. 3220–3229, 2021.

72 R. Zamani, M. E. H. Golshan, H. H. Alhelou, and N. Hatziargyriou, "A novel synchronous DGs islanding detection method based on online dynamic features extraction," *Electr. Power Syst. Res.*, vol. 195, 2021.

73 M. Bakhshi, R. Noroozian, and G. B. Gharehpetian, "Anti-islanding scheme for synchronous DG units based on Tufts–Kumaresan signal estimation method," *IEEE Trans. Power Del.*, vol. 28, no. 4, pp. 2185–2193, Oct. 2013.

74 A. Samui and S. R. Samantaray, "Wavelet singular entropy-based islanding detection in distributed generation," *IEEE Trans. Power Del.*, vol. 28, no. 1, pp. 411–418, 2012.

75 R. Dubey, M. Popov, and S. R. Samantaray, "Transient monitoring function-based islanding detection in power distribution network," *IET Gener. Transm. Distrib.*, vol. 13, no. 6, pp. 805–813, 2019.

76 M. A. Farhan, "Mathematical morphology-based islanding detection for distributed generation," *IET Gener. Transm. Distrib.*, vol. 11, no. 14, pp. 3449–3457, 2017.

77 R. Nale, M. Biswal, and N. Kishor, "A transient component based approach for islanding detection in distributed generation," *IEEE Trans. Sustain. Energy*, vol. 10, no. 3, pp. 1129–1138, Aug. 2018.

78 C. T. Hsieh, J. M. Lin, and S. J. Huang, "Enhancement of islanding-detection of distributed generation systems via wavelet transform-based approaches," *Int. J. Electr. Power Energy Syst.*, vol. 30, no. 10, pp. 575–580, 2008.

79 B. K. Chaitanya and A. Yadav, "Hilbert–Huang transform based islanding detection scheme for distributed generation," in *2018 IEEE 8th Power India International Conference (PIICON)*, pp. 1–5, 2018.

80 A. Ezzat, B. E. Elnaghi, and A. A. Abdelsalam, "Microgrids islanding detection using Fourier transform and machine learning algorithm," *Electr. Power Syst. Res.*, vol. 196, 2021.

81 C. Zhao, M. He, and X. Zhao, "Analysis of transient waveform based on combined short time Fourier transform and wavelet transform," in *2004 International Conference on Power System Technology, vol. 2*, 2004.

82 S. R. Mohanty, N. Kishor, P. K. Ray, and J. P. S. Catalo, "Comparative study of advanced signal processing techniques for islanding detection in a hybrid distributed generation system," *IEEE Trans. Sustain. Energy*, vol. 6, no. 1, pp. 122–131, Jan. 2015.

83 P. K. Ray, N. Kishor, and S. R. Mohanty, "Islanding and power quality disturbance detection in grid-connected hybrid power system using wavelet and *S*-transform," *IEEE Trans. Smart Grid*, vol. 3, no. 3, pp. 1082–1094, Sep. 2012.

84 P. K. Ray, S. R. Mohanty, and N. Kishor, "Disturbance detection in grid-connected distributed generation system using wavelet and S-transform," *Electr. Power Syst. Res.*, vol. 81, no. 3, pp. 805–819, 2011.

85 M. Karimi, H. Mokhtari, and M. R. Iravani, "Wavelet based on-line disturbance detection for power quality applications," *IEEE Trans. Power Del.*, vol. 15, no. 4, pp. 1212–1220, Nov. 2000.

86 P. K. Dash, B. K. Panigrahi, and G. Panda, "Power quality analysis using S-transform," *IEEE Trans. Power Delivery*, vol. 18, no. 2, pp. 406–411, Apr. 2003.

87 C. N. Bhende, S. Mishra, and B. K. Panigrahi, "Detection and classification of power quality disturbances using S-transform and modular neural network," *Electr. Power Syst. Res.*, vol. 78, no. 1, pp. 122–128, 2008.

88 H. S. Behera, P. K. Dash, and B. Biswal, "Power quality time series data mining using s-transform and fuzzy expert system," *Appl. Soft Comput.*, vol. 10, no. 3, pp. 945–955, 2010.

89 S. Ventosa, C. Simon, M. Schimmel, J. J. Danobeitia, and A. Mànuel, "The S-transform from a wavelet point of view," *IEEE Trans. Signal Process.*, vol. 56, no. 7, pp. 2771–2780, Jul. 2008.

90 S. R. Mohanty, N. Kishor, and P. K. Ray, "Islanding detection in a distributed generation based hybrid system using intelligent pattern recognition techniques," in *Third IEEE PES Innovative Smart Grid Technologies Europe (ISGT Europe)*, Germany, 2012.

91 S. Agrawal, S. Patra, S. R. Mohanty, V. Agarwal, and M. Basu, "Use of matrix-pencil method for efficient islanding detection in static DG and a parallel comparison with DWT method," *IEEE Trans. Ind. Electron.*, vol. 66, no. 11, pp. 8937–8946, Nov. 2019.

92 H. Li, Z. Li, and W. Mo, "A time varying filter approach for empirical mode decomposition," *Signal Process.*, vol. 138, pp. 146–158, 2017.

93 C. Simon, M. Schimmel, and J. J. Danobeitia, "On the TT-transform and its diagonal elements," *IEEE Trans. Signal Process.*, vol. 56, no. 11, pp. 5709–5713, Dec. 2008.

94 S. Suja and J. Jerome, "Pattern recognition of power signal disturbances using *S*-transform and TT-transform," *Int. J. Electr. Power Energy Syst.*, vol. 32, no. 1, pp. 37–53, 2010.

95 C. R. Pinnegar, "Time–frequency and time–time filtering with the S-transform and TT-transform," *Digital Signal Process.*, vol. 15, no. 6, pp. 604–620, 2005.

96 M. Jing, W. Zengping, X. Yan, and M. Lei, "Single-ended transient positional protection of transmission lines using mathematical morphology," in *2005 International Power Engineering Conference*, 2005.

97 P. Li, J. Liu, X. Li, C. Tian, and J. Jie, "Detection of power quality disturbances in microgrid based on generalized morphological filter and backward difference," in *2011 4th International Conference on Electric Utility Deregulation and Restructuring and Power Technologies (DRPT)*, pp. 1663–1666, 2011.

98 S. Gautam and S. M. Brahma, "Overview of mathematical morphology in power systems — A tutorial approach," in *2009 IEEE Power & Energy Society General Meeting*, pp. 1-7, 2009.

99 A. Hussain, C. -H. Kim, and A. Mehdi, "A comprehensive review of intelligent islanding schemes and feature selection techniques for distributed generation system," *IEEE Access*, vol. 9, pp. 146603–146624, 2021.

100 S. Kar and S. R. Samantaray, "Data-mining-based intelligent anti-islanding protection relay for distributed generations," *IET Gener. Transmiss. Distrib.*, vol. 8, no. 4, pp. 629–639, 2014.

101 S. R. Samantaray, K. El-Arroudi, G. Joós, and I. Kamwa, "A fuzzy rule-based approach for islanding detection in distributed generation," *IEEE Trans. Power Del.*, vol. 25, no. 3, pp. 1427–1433, 2010.

102 O. N. Faqhruldin, E. F. El-Saadany, and H. H. Zeineldin, "A universal islanding detection technique for distributed generation using pattern recognition," *IEEE Trans. Smart Grid*, vol. 5, no. 4, pp. 1985–1992, Jul. 2014.

103 B. K. Chaitanya, A. Yadav, and M. Pazoki, "Reliable islanding detection scheme for distributed generation based on pattern-recognition," *IEEE Trans. Ind. Inform.*, vol. 17, no. 8, pp. 5230–5238, Aug. 2021.

104 A. Khamis, Y. Xu, Z. Y. Dong, and R. Zhang, "Faster detection of microgrid islanding events using an adaptive ensemble classifier," *IEEE Trans. Smart Grid*, vol. 9, no. 3, pp. 1889–1899, May 2018.

105 K. El-Arroudi, G. Joos, I. Kamwa, and D. T. McGillis, "Intelligent-based approach to islanding detection in distributed generation," *IEEE Trans. Power Del.*, vol. 22, no. 2, pp. 828–835, Apr. 2007.

106 S. Li, A. J. Rodolakis, K. El-Arroudi, and G. Joós, "Islanding protection of multiple distributed resources under adverse islanding conditions," *IET Gener. Transm. Distrib.*, vol. 10, no. 8, pp. 1901–1912, 2016.

107 M. Vatani, T. Amraee, A. M. Ranjbar, and B. Mozafari, "Relay logic for islanding detection in active distribution systems," *IET Gener. Transm. Distrib.*, vol. 9, no. 12, pp. 1254–1263, 2015.

108 V. L. Merlin *et al.*, "A new artificial neural network based method for islanding detection of distributed generators," *Int. J. Electr. Power Energy Syst.*, vol. 75, pp. 139–151, 2016.

109 H. Shayeghi, B. Sobhani, E. Shahryari, and A. Akbarimajd, "Optimal neuro-fuzzy based islanding detection method for Distributed Generation," *Neurocomputing*, vol. 177, pp. 478–488, 2016.

110 S. R. Mohanty, P. K. Ray, N. Kishor, and B. K. Panigrahi, "Classification of disturbances in hybrid DG system using modular PNN and SVM," *Int. J. Electr. Power Energy Syst.*, vol. 44, no. 1, pp. 764–777, 2013.

111 M. R. Alam, K. M. Muttaqi, and A. Bouzerdoum, "A multifeature-based approach for islanding detection of DG in the subcritical region of vector surge relays," *IEEE Trans. Power Del.*, vol. 29, no. 5, pp. 2349–2358, 2014.

112 K. Swierczynski, M. Habrych, and B. Brusilowicz, "A novel method of vector shift criterion utilization in power system automation," *IEEE Access*, vol. 11, pp. 38301–38308, 2023.

6
Solutions for Microgrid Protection and Stability Enhancement

6.1 Introduction

The shortcomings of microgrid (MG) protection system have been thoroughly discussed in Chapter 4. Various schemes and configurations have been presented in the literature to resolve these shortcomings and enhance the reliability, especially during the transition between the grid-connected and islanded modes. In this chapter, proper schemes for improving the security and dependability of protection system of distribution systems/MGs have been carefully reviewed and discussed. In this regard, at first, appropriate rules for proper adjustment of overcurrent relays (OCRs) are given, which could be useful to improve the reliability of these relays and reduce their miscoordinations. However, proper adjustment of the relays could not resolve all of the challenges caused by the connection of distributed generations (DGs) and MGs to the distribution systems. To cope with this issue, some research works suggest application of adaptive schemes that can effectively enhance protection security without losing sensitivity and dependability. These adaptive methods have been investigated in this chapter as well. In this chapter, solutions to the protection challenges are first discussed and then some practical methods are presented to improve the stability of MGs, especially during severe disturbances.

6.2 Feeder Protection Solutions

In this section, some practical approaches are presented to overcome the challenges of feeder protection discussed in Chapter 4, especially in adjacent to DGs. These solutions include (i) simple rules for proper setting adjustment and (ii) novel protection schemes. The first solution could be applied to the existing relays without requiring modification in the hardware and protection schemes, and thus, it could be simply implemented. However, for the second category, the relay schemes should be modified and updated by the manufacturer.

The main features that can be exploited to compare various protection schemes applicable to MG protection are as follows:

- **Compatibility with various DG's technologies (feature 1):** One of the most important expected features of the protection scheme is its robust performance in the presence of various DGs based on different technologies, e.g., small-scale synchronous generator (SSSG) and inverter-based DG (IBDG). Many of the methods proposed in the literature perform well in the presence of SSSGs but lose their sensitivity for IBDGs. Moreover, many protection functions that are effective with synchronous sources may not provide satisfactory responses in the presence of inverter-based sources.
- **Adaptability to both grid-connected and islanded modes (feature 2):** After the islanding of the MG, the short-circuit level may significantly decrease, and as discussed in Chapter 4, conventional OCRs lose their sensitivity under such conditions. Therefore, an essential requirement for any protection scheme is its compatibility with both grid-connected and islanded modes of operation.
- **Security against dynamic load behavior (feature 3):** Dynamic loads, such as induction motors (IMs) and thermostatically controlled loads, can adversely affect the performance of conventional protection systems and may lead to maloperations, as discussed in detail in Chapter 4. Therefore, the protection system must be properly designed and configured to remain secure against the dynamic behavior of these types of loads.
- **Compatibility with various configurations (feature 4):** The configuration of distribution systems/MGs may change frequently due to the planned or unplanned events, such as line or DG outages. Therefore, the protection system should be capable of adapting to different configurations and maintain reliable performance even when the system topology changes.
- **Protection scheme cost considerations (feature 5):** Cost is a critical factor in the selection of a protection scheme. While certain methods, such as those utilizing conventional OCRs, offer a low-cost solution, other schemes incur significantly higher expenses due to the need for auxiliary components, such as communication infrastructure, voltage transformers, central processing units, controllers, and other hardware. These additional requirements may cause such schemes become economically impractical, particularly for widespread deployment in distribution systems and MGs.
- **Setting-free operation (feature 6):** A key advantage of an effective protection system is its ability to maintain a reliable and accurate operation under all operating conditions without requiring changes to the settings. This characteristic

enhances the system's adaptability and reduces the need for continuous supervision or reconfiguration.

- **Considering transient stability of DGs (feature 7):** One of the major challenges in distribution systems/MGs that threatens the transient stability of DGs is the prolonged fault-clearing time (FCT). Therefore, taking the critical clearing time (CCT) of DGs into account throughout the relays setting and protection schemes design procedures can significantly enhance the MG stability.
- **Fault direction identification capability (feature 8):** To prevent relay maloperation due to the DG back-feed, fault direction identification could be very helpful. Conventionally, to identify the fault direction, voltage signals are utilized by the relay, which increases the cost of the protection system due to the use of voltage transformers. Recently, various schemes have been presented by the research works, which identify the fault direction only using the current signal.

Table 6.1 presents various methods proposed in different research studies and standards to enhance the feeder protection performance. The key supported features of each method are also included to facilitate a clearer comparison.

Although applications of simple overcurrent protection schemes present several challenges for MGs and active distribution networks, using appropriate relay settings can enhance the protection performance and somewhat reduce these issues. Therefore, at first, this chapter introduces straightforward setting guidelines for the adjustment of OCRs.

6.2.1 Rules for Proper Setting Adjustment

The main protection functions employed in distribution feeders, including MG feeders, are phase overcurrent and earth fault overcurrent functions whose principles are fully explained in Chapter 4. Due to the lack of voltage signals in the protection system of distribution feeders, nondirectional overcurrent functions are more common than the directional overcurrent ones [1]–[5]. Based on the guidelines provided in [1], the following subsections 6.2.1.1 and 6.2.1.2, present the setting rules for configuring these nondirectional protection functions, which are commonly applied in practical distribution systems/MGs (Fig. 6.1).

6.2.1.1 Phase Overcurrent Protection

6.2.1.1.1 MV-Side of Distribution Transformer (Relay R1) A distribution transformer is utilized to supply the low-voltage (LV) loads or exploited to connect IBDGs/SSSGs to the distribution system. The phase overcurrent function of the related relay located at the medium-voltage (MV) side of this transformer commonly includes at least two stages, (i) instantaneous stage, i.e., ANSI code 50,

Table 6.1 Various overcurrent-based protection schemes used for feeder protection.

References	Core concept	Supported features								Additional elements
		1	2	3	4	5	6	7	8	
[1]	Simple one-stage overcurrent using conventional setting rules	×	×	×	×	✓	×	×	×	No additional elements
[1]	Simple multi-stage overcurrent using proper setting rules	×	×	✓	×	✓	×	×	×	No additional elements
[2, 3]	Simple overcurrent using dual stage curves	×	×	✓	×	✓	×	✓	×	No additional elements
[4]	Adaptive directional overcurrent	×	✓	×	✓	×	×	×	✓	Voltage transformer
[5] [6]	Adaptive online adjustment of simple overcurrent	✓ ×	✓ ×	× ×	✓ ✓	× ×	× ×	× ×	× ×	Central processing unit and communication link
[7]	Adaptive setting group	×	✓	×	✓	✓	×	×	×	No additional hardware
[8]	Voltage-supervised overcurrent	✓	✓	×	✓	×	×	×	✓	Voltage transformer
[9]	Current-based directional overcurrent	×	×	×	×	✓	×	×	✓	No additional hardware
[10]	Time–current–voltage curve	✓	✓	×	×	×	×	×	✓	Voltage transformer
[11]	Dual settings with islanding detection	×	✓	×	×	×	×	×	×	Islanding detection scheme

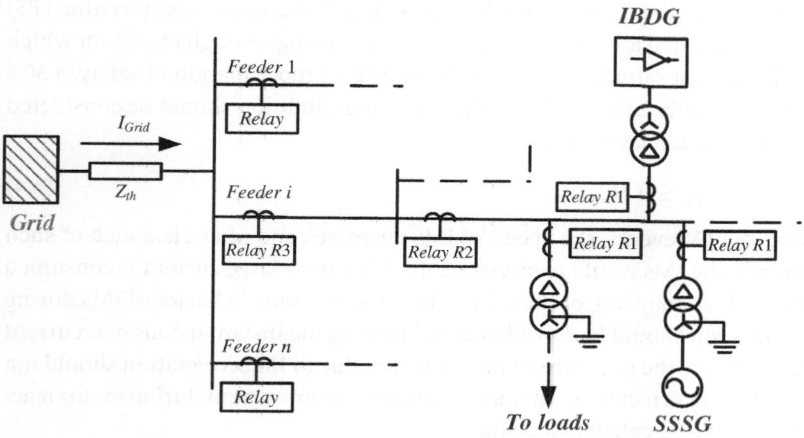

Fig. 6.1 Sample distribution system containing DGs.

and (ii) inverse stage, i.e., ANSI code 51. The proper setting values for these relays are explained in the following, mostly based on the instructions provided in [1].

To adjust the pickup value for the instantaneous stage, the following points should be considered:

- The pickup value of the instantaneous stage should be greater than the maximum three-phase short-circuit current (I_{fmax}) on the secondary side of the distribution transformer. The impedance of the upstream network could be neglected in comparison to the transformer impedance, and thus, I_{fmax} is calculated as

$$I_{fmax} = \frac{I_{nTr}}{U_k} \tag{6.1}$$

where I_{nTr} and U_k are the nominal current and per-unit impedance of the transformer, respectively. Accordingly, the following condition should be considered:

$$I^{50\text{-}(1)}_{Pickup\text{-}R1} \geq 1.2 I_{fmax} \tag{6.2}$$

- If the relay does not support an inrush-blocking unit, the pickup current of the instantaneous stage should be greater than the transformer inrush current. The rms value of the inrush current is about 50% of its peak value. If the peak of the inrush current is not provided by the manufacturer, as a rule of thumb, to prevent maloperation of the relay due to this current, the following condition should be considered:

$$I^{50\text{-}(2)}_{Pickup\text{-}R1} \geq 0.5 \frac{I_{nTr}}{U_k} \tag{6.3}$$

- In cases where the CT has already been selected, the relay pickup setting (PS) should be adjusted less than the maximum AC passing through the CT for which the CT does not saturate ($I_{saturation}$). To ensure a proper margin of safety, a 30% margin is usually adopted. Thus, the following condition should be considered in adjusting the relay pickup:

$$I^{50\text{-}(3)}_{Pickup\text{-}R1} \leq 0.7 I_{Saturation} \tag{6.4}$$

- During the LV events, the speed of IMs decreases and after clearance of such LV events, the IMs would reaccelerate. In this case, a large current is consumed by IMs, which may trigger the OCR. Thus, the dynamic behavior of IMs during reacceleration should be considered in adjusting the instantaneous overcurrent pickup. In fact, the overcurrent phenomenon due to IM acceleration should not result in an incorrect relay operation. The maximum current during motor reacceleration can be calculated using

$$I_{MR} = [(1-k) \times I_{feeder} + k \times n \times I_{feeder}] \times K_{VD} \tag{6.5}$$

where I_{MR} and I_{feeder} are the feeder maximum current under the motor reacceleration condition and the steady-state current of feeder when all static and dynamic loads are connected, respectively. Moreover, per-unit parameters k, n, and K_{VD} are IM penetration factor, locked rotor current of IMs, and distribution substation bus voltage during motor reacceleration, respectively. Therefore, for adjusting the instantaneous stage, the pickup current should be greater than the maximum current consumed by IMs during the dynamic reacceleration to avoid incorrect relay operation:

$$I^{50\text{-}(4)}_{Pickup\text{-}R1} \geq I_{MR} \tag{6.6}$$

Based on the abovementioned explanations, the PS of instantaneous stage ($I^{50}_{Pickup\text{-}R1}$) is adjusted as

$$I^{50}_{Pickup\text{-}R1} = \text{Min}\left\{\text{Max}\left(I^{50\text{-}(1)}_{Pickup\text{-}R1}, I^{50\text{-}(2)}_{Pickup\text{-}R1}, I^{50\text{-}(4)}_{Pickup\text{-}R1}\right), I^{50\text{-}(3)}_{Pickup\text{-}R1}\right\} \tag{6.7}$$

Meanwhile, to adjust the pickup value for the inverse stage, the following points should be considered [1]:

- When an overload function is applied for transformer protection, the overcurrent function is only adopted to protect the transformer against the short-circuit condition. However, in the absence of a dedicated overload protection function, the overcurrent function may be used out of necessity to protect the transformer against both overload and short-circuit conditions, although this approach would not be an ideal solution.
- The pickup current of the inverse stage should be less than the maximum tolerable short-time overload current.

- The above pickup current should be less than the thermal limit of other series equipment, such as CT.

Therefore, the pickup current is determined by the following equation:

$$I^{51}_{Pickup\text{-}R1} = \mathrm{Min}(K_{OverLoad} \times I_{nTr}, K_{OverLoad} \times I_{nCT}) \tag{6.8}$$

where I_{nCT} is the nominal current of the CT. The parameter $K_{OverLoad}$ indicates the presence of an overload function in the transformer protection relay. If the overload function is enabled, this coefficient is considered 1.2; otherwise, the value of this parameter is adjusted at 1.0 to protect the transformer against both overload and short-circuit conditions.

The operation time of the inverse stage (t_{op}) is determined by

$$t_{op} = \mathrm{TMS} \times \frac{k}{\left(\frac{I}{I^{51}_{pickup}}\right)^{\alpha} - 1} \tag{6.9}$$

where parameter I is the measured current and I^{51}_{pickup} is the pickup current. Furthermore, k and α are constant parameters, whose values are given in Chapter 4.

To determine the proper value for the time multiplier setting (TMS) parameter, the following criteria should be considered [1]:

- As a common practice, for determining the *TMS* value, the maximum short-circuit current (i.e., the fault current at the point of the downstream protective equipment) is considered. To do so, the maximum short-circuit current on the secondary side of the distribution transformer I_{fmax} is determined by (6.1). However, the PS of the instantaneous stage of the downstream relay is normally less than I_{fmax}, and thus, the downstream LV-side protective device operates instantaneously when the maximum short-circuit current passes through this relay. Therefore, the operation time of the MV protection relay *R*1 ($t_{op\text{-}51(R1)}$) is calculated by

$$t_{op\text{-}51(R1)} = \min\left(t_{op\text{-}50(LV)}, t_{op\text{-}50(R1)}\right) + \mathrm{CTI} \approx \mathrm{CTI} \tag{6.10}$$

where $t_{op\text{-}50(LV)}$ and $t_{op\text{-}50(R1)}$ are the operation times of the instantaneous stages of the downstream LV-side protective device and relay *R*1 for I_{fmax}, respectively. Meanwhile, CTI is the coordination time interval between the backup relay, i.e., relay *R*1, and the main one, i.e., the downstream LV-side protective device, whose proper value can be considered between 150 and 200 ms. Therefore, TMS_1 is calculated as

$$\mathrm{TMS}_1 \geq \frac{\mathrm{CTI} \times \left[\left(\frac{I_{fmax}}{I^{51}_{Pickup\text{-}R1}}\right)^{\alpha} - 1\right]}{k} \tag{6.11}$$

- The operation time of unit 51 of relay $R1$ should be greater than the operation time of unit 51 of its downstream relay:

$$t_{op\text{-}51(R1)} = t_{op\text{-}51(LV)} + \text{CTI} \quad \text{for } I^{50}_{Pickup\text{-}R1} \tag{6.12}$$

where $t_{op\text{-}51(LV)}$ is the operation time of the downstream LV-side protective device at the pickup current of the instantaneous stage of relay $R1$ (i.e., $I^{50}_{Pickup\text{-}R1}$). Thus, as the second condition:

$$\text{TMS}_2 \geq \frac{t_{op\text{-}51(R1)} \times \left[\left(\dfrac{I^{50}_{Pickup\text{-}R1}}{I^{51}_{Pickup\text{-}R1}}\right)^{\alpha} - 1\right]}{k} \tag{6.13}$$

- The operation time of unit 51 of relay $R1$ should be greater than the operation time of unit 51 of its downstream relay:

$$t_{op\text{-}51(R1)} = t_{op\text{-}51(LV)} + \text{CTI} \quad \text{for } I^{50}_{Pickup\text{-}LV} \tag{6.14}$$

where $t_{op\text{-}51(LV)}$ is the operation time of the downstream LV-side protective device at the pickup current of its instantaneous stage (i.e., $I^{50}_{Pickup\text{-}LV}$). Consequently, as the third condition:

$$\text{TMS}_3 \geq \frac{t_{op\text{-}51(R1)} \times \left[\left(\dfrac{I^{50}_{Pickup\text{-}LV}}{I^{51}_{Pickup\text{-}R1}}\right)^{\alpha} - 1\right]}{k} \tag{6.15}$$

Finally, the TMS value can be calculated as

$$\text{TMS} \geq \max\left(\text{TMS}_1, \text{TMS}_2, \text{TMS}_3\right) \tag{6.16}$$

Example 6.1

Single-line diagram of a grid-connected PV plant is depicted in Fig. 6.2. Calculate the proper settings for phase overcurrent functions, (a) instantaneous stage and (b) inverse stage, at the point of connection. The CT saturation current is considered 25 times its rated current. Meanwhile, the setting values of LV relay are given in Table 6.2.

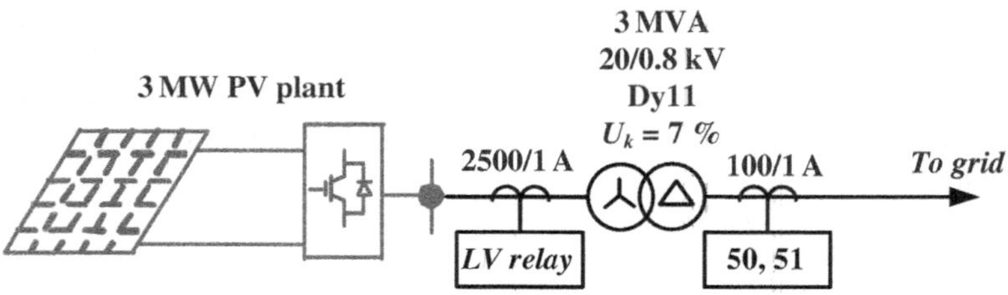

Fig. 6.2 Single-line diagram for Example 6.1.

Table 6.2 Setting values of the LV relay for Example 6.1.

Function	Pickup (secondary)	Delay	Curve
Instantaneous	6 A	0	—
Inverse	0.9 A	TMS = 0.05	IEC-very inverse

Solution:

a) **Instantaneous stage:** The rated current at the MV side of the transformer is calculated by

$$I_{nTr} = \frac{S_{nTr}}{\sqrt{3}V_{nom}} = \frac{3\text{MVA}}{\sqrt{3} \times 20\text{ kV}} = 86.6\text{ A}$$

Moreover, the MV-side current for a three-phase fault at the LV side of the transformer I_{fmax} is calculated by

$$I_{fmax} = \frac{I_{nTr}}{U_k} = \frac{86.6\text{ A}}{7\%} = 1237\text{ A}$$

Thus, based on (6.2):

$$I^{50\text{-}(1)}_{Pickup\text{-}R1} \geq 1.2 I_{fmax} = 1.2 \times 1237\text{ A} = 1484\text{ A}$$

Moreover, based on (6.3):

$$I^{50\text{-}(2)}_{Pickup\text{-}R1} \geq 0.5 \frac{I_{nTr}}{U_k} = 0.5 \times 1237\text{ A} = 618.5\text{ A}$$

Meanwhile, based on (6.4), to consider the CT saturation:

$$I^{50\text{-}(3)}_{Pickup\text{-}R1} \leq 0.7 I_{Saturation} = 0.7 \times 25 \times 100\text{ A} = 1750\text{ A}$$

On the other hand, since no dynamic loads such as IMs are connected to the power plant, I_{MR} could be considered equal to the rated current of the transformer. Thus, according to (6.6):

$$I^{50\text{-}(4)}_{Pickup\text{-}R1} \geq I_{MR} = 86.6\text{ A}$$

Finally, based on (6.7):

$$\begin{aligned} I^{50}_{Pickup\text{-}R1} &= \text{Min}\left\{ \text{Max}\left(I^{50\text{-}(1)}_{Pickup\text{-}R1}, I^{50\text{-}(2)}_{Pickup\text{-}R1}, I^{50\text{-}(4)}_{Pickup\text{-}R1} \right), I^{50\text{-}(3)}_{Pickup\text{-}R1} \right\} \\ &= \text{Min}\{\text{Max}(1484, 618.5, 86.6), 1750\} \\ &= 1484\text{ A} \end{aligned}$$

Thus, the secondary pickup current is calculated as

$$I^{50}_{Pickup\text{-}R1}(secondary) = \frac{1484\ \text{A}}{100} = 14.84\ \text{A}$$

b) **Inverse stage:** Based on (6.8), the pickup current of the inverse stage is determined by the following equation. It should be noted that since the relay does not support the overload thermal protection unit, $K_{OverLoad}$ is considered 1.

$$I^{51}_{Pickup\text{-}R1} = \text{Min}(K_{OverLoad} \times I_{nTr}, K_{OverLoad} \times I_{nCT})$$
$$= \text{Min}(1 \times 86.6\ \text{A}, 1 \times 100\ \text{A}) = 86.6\ \text{A}$$

Thus, the secondary pickup current is

$$I^{51}_{Pickup\text{-}R1}(secondary) = \frac{86.6\ \text{A}}{100} = 0.866\ \text{A}$$

For a better coordination with the LV-side relay, the operation curve of the MV relay is chosen as "IEC-very inverse," and thus, $k = 13.5$ and $\alpha = 1$. To calculate the TMS value, based on (6.11), parameter TMS_1 is calculated by considering $\text{CTI} = 150$ ms:

$$\text{TMS}_1 \geq \frac{\text{CTI} \times \left[\left(\frac{I_{fmax}}{I^{51}_{Pickup\text{-}R1}}\right)^{\alpha} - 1\right]}{k} = \frac{0.15 \times \left[\left(\frac{1237}{86.6}\right)^{1} - 1\right]}{13.5} = 0.147\ \text{s}$$

Moreover, $t_{op\text{-}51(LV)}$ for $I^{50}_{Pickup\text{-}R1}$ is calculated as

$$t_{op\text{-}51(LV)} = \text{TMS} \times \frac{k}{\left(\frac{I}{I^{51}_{pickup}}\right)^{\alpha} - 1} = 0.05 \times \frac{13.5}{\left(\frac{1484 \times \frac{20}{0.8}}{0.9 \times 2500}\right)^{1} - 1} = 43.5\ \text{ms}$$

Therefore, based on (6.12) and (6.13), TMS_2 is calculated as

$$t_{op\text{-}51(R1)} = t_{op\text{-}51(LV)} + \text{CTI} = 43.5 + 150 = 193.5\ \text{ms}$$

$$\text{TMS}_2 \geq \frac{t_{op\text{-}51(R1)} \times \left[\left(\frac{I^{50}_{Pickup\text{-}R1}}{I^{51}_{Pickup\text{-}R1}}\right)^{\alpha} - 1\right]}{k} = \frac{0.1935 \times \left[\left(\frac{1484}{86.6}\right)^{1} - 1\right]}{13.5}$$
$$= 0.23\ \text{s}$$

Moreover, $t_{op\text{-}51(LV)}$ for $I^{50}_{Pickup\text{-}LV}$ is calculated as

$$t_{op\text{-}51(LV)} = \text{TMS} \times \frac{k}{\left(\frac{I}{I^{51}_{pickup}}\right)^{\alpha} - 1} = 0.05 \times \frac{13.5}{\left(\frac{6 \times 2500}{0.9 \times 2500}\right)^{1} - 1} = 119\ \text{ms}$$

Therefore, based on (6.14) and (6.15), TMS_3 is calculated as

$$t_{op\text{-}51(R1)} = t_{op\text{-}51(LV)} + \text{CTI} = 119 + 150 = 269\ \text{ms}$$

$$\mathrm{TMS}_3 \geq \frac{t_{op\text{-}51(R1)} \times \left[\left(\frac{I^{50}_{Pickup\text{-}LV}}{I^{51}_{Pickup\text{-}R1}}\right)^{\alpha} - 1\right]}{k}$$

$$= \frac{0.269 \times \left[\left(\frac{6\times2500\times\frac{0.8}{20}}{86.6}\right)^{1} - 1\right]}{13.5}$$

$$= 0.12\ \mathrm{s}$$

Finally, based on (6.16), the TMS value can be calculated as

$$\mathrm{TMS} \geq \max\left(\mathrm{TMS}_1, \mathrm{TMS}_2, \mathrm{TMS}_3\right) = 0.23\ \mathrm{s}$$

6.2.1.1.2 MV Feeders Relay (Relay R2) This relay is located at the upstream zone of relay $R1$, as shown in Fig. 6.1. To adjust the instantaneous operating current threshold of relay $R2$, the following points should be considered [1]:

- The instantaneous stage of relay $R2$ should be adjusted to only cover the short circuits within the zone between $R2$ and the downstream relays (i.e., $R1$). Therefore, the pickup current of this stage should be greater than the maximum fault current at the location of relay $R1$, i.e., $I_{fmax\text{-}R1}$ (considering the effect of DC component of the fault current):

$$I^{50\text{-}(1)}_{Pickup\text{-}R2} \geq K_{SF} \times I_{fmax\text{-}R1} \quad (6.17)$$

where K_{SF} is a safety margin to cover the impact of the DC component and CT error. Typically, this parameter can be assumed 1.2. If relay $R2$ measures the first harmonic of the current, due to the reduced effect of the DC component, the value of K_{SF} can be reduced to 1.1. On the other hand, if relay $R2$ exploits the rms value, K_{SF} can be increased up to 1.3 to enhance the relay reliability.

- The pickup current of instantaneous stage of relay R2 should be set greater than the instantaneous stage of relay $R1$. In other words,

$$I^{50\text{-}(1)}_{Pickup\text{-}R2} \geq K_{SF} \times I^{50\text{-}(1)}_{Pickup\text{-}R1} \quad (6.18)$$

- The instantaneous pickup current setting should be adjusted less than the maximum AC passing through the CT and for which the CT does not saturate ($I_{saturation}$). To ensure a proper margin of safety, a 30% margin is usually considered. Thus,

$$I^{50\text{-}(3)}_{Pickup\text{-}R2} \leq 0.7 I_{Saturation} \quad (6.19)$$

- The pickup current of the instantaneous stage should be greater than the motor reacceleration current, as described in this chapter and calculated by (6.5)

$$I^{50-(4)}_{Pickup-R2} \geq I_{MR} \quad (6.20)$$

Finally, the instantaneous stage could be adjusted as

$$I^{50}_{Pickup\text{-}R2} = \text{Min}\left\{\text{Max}\left(I^{50\text{-}(1)}_{Pickup\text{-}R2}, I^{50\text{-}(2)}_{Pickup\text{-}R2}, I^{50\text{-}(4)}_{Pickup\text{-}R2}\right), I^{50\text{-}(3)}_{Pickup\text{-}R2}\right\} \tag{6.21}$$

If only one distribution transformer is installed at the downstream area and relay *R*2 is used solely for its protection, and the distance between relays *R*1 and *R*2 is insignificant, the maximum fault current at the relay *R*2 and *R*1 locations is close to each other and thus it is not possible to adjust the instantaneous stage of *R*2. In such a condition, to retain selectivity of the protection system, the instantaneous unit of relay *R*2 should be disabled.

Moreover, to adjust the pickup value for the inverse stage, the following points should be considered:

- When an overload function is applied for feeder protection, the overcurrent function is only adopted to protect the transformer against the short-circuit condition. In the absence of a dedicated overload protection function, the feeder is often protected using the overcurrent function against both overload and short-circuit conditions, despite its inherent limitations.
- The pickup current of the inverse stage should be less than the maximum tolerable short-time overload current of the feeder.
- This pickup current should be less than the thermal limit of other series equipment, such as CT.

Accordingly, the pickup current of the inverse stage, i.e., $I^{51}_{Pickup\text{-}R2}$ is adjusted as

$$I^{51}_{Pickup\text{-}R2} = \text{Min}(K_{OverLoad} \times I_{nFeeder}, K_{OverLoad} \times I_{nCT}) \tag{6.22}$$

where $I_{nFeeder}$ is the rated current of the feeder and $K_{OverLoad}$ is a constant factor. If a separate overload function is exploited for thermal protection of the feeder, $K_{OverLoad}$ is considered 1.2; otherwise, this parameter should be considered 1.

To determine the proper value for the TMS parameter, the following criteria should be considered:

- As a first condition, for determining *TMS*, the maximum short-circuit current (i.e., the fault current at the point of the downstream protective equipment) should be considered. In this regard, the maximum short-circuit fault (three-phase short circuit), at the location of the downstream protective devices, i.e., relay *R*1, should be calculated ($I_{fmax\text{-}R1}$). Therefore, the operation time of the MV protection relay *R*2 ($t_{op\text{-}51(R2)}$) is calculated by

$$t_{op\text{-}51(R2)} = \min\left(t_{op\text{-}50(R1)}, t_{op\text{-}50(R2)}\right) + \text{CTI} \tag{6.23}$$

where $t_{op\text{-}50(R1)}$ and $t_{op\text{-}50(R2)}$ are the operation times of the instantaneous stages of the downstream protective device, i.e., relays *R*1 and *R*2 for $I_{fmax\text{-}R1}$, respectively.

Therefore, TMS_1 is calculated as

$$TMS_1 \geq \frac{t_{op\text{-}51(R2)} \times \left[\left(\frac{I_{fmax\text{-}R1}}{I^{51}_{Pickup\text{-}R2}}\right)^{\alpha} - 1\right]}{k} \quad (6.24)$$

- The operation time of unit 51 of relay $R2$ should be greater than the operation time of unit 51 of its downstream relay at $I^{50}_{Pickup\text{-}R1}$:

$$t^1_{op\text{-}51(R2)} = t_{op\text{-}51(R1)} + \text{CTI} \quad \text{for } I^{50}_{Pickup\text{-}R1} \quad (6.25)$$

where $t_{op\text{-}51(R1)}$ is the operation time of the downstream relay, i.e., relay $R1$, at the pickup current of its instantaneous stage (i.e., $I^{50}_{Pickup\text{-}R1}$). Thus, as the second condition:

$$TMS_2 \geq \frac{t^1_{op\text{-}51(R2)} \times \left[\left(\frac{I^{50}_{Pickup\text{-}R1}}{I^{51}_{Pickup\text{-}R2}}\right)^{\alpha} - 1\right]}{k} \quad (6.26)$$

- The operation time of unit 51 of relay $R2$ should be greater than the operation time of unit 51 of its downstream relay at $I^{50}_{Pickup\text{-}R2}$:

$$t^2_{op\text{-}51(R2)} = t_{op\text{-}51(R1)} + \text{CTI} \quad \text{for } I^{50}_{Pickup\text{-}R2} \quad (6.27)$$

Consequently, as the third condition:

$$TMS_3 \geq \frac{t^2_{op\text{-}51(R2)} \times \left[\left(\frac{I^{50}_{Pickup\text{-}R2}}{I^{51}_{Pickup\text{-}R2}}\right)^{\alpha} - 1\right]}{k} \quad (6.28)$$

Finally, the TMS can be calculated as

$$TMS \geq \max(TMS_1, TMS_2, TMS_3) \quad (6.29)$$

6.2.1.2 Earth Fault Protection

6.2.1.2.1 MV-Side of Distribution Transformer (Relay R1) Commonly for relay $R1$, a two-stage overcurrent earth fault protection is used, which includes instantaneous and inverse stages. For the instantaneous stage, the pickup current ($I^{50N}_{Pickup\text{-}R1}$) is determined by

$$I^{50N}_{Pickup\text{-}R1} = \begin{cases} 0.4 \times I_{nZCT} & \text{for direct measurement of } I_0 \\ 0.1 \times I^{50}_{Pickup\text{-}R1} & \text{for indirect measurement of } I_0 \\ 0.17 \times I^{50}_{Pickup\text{-}R1} & \text{for analytical calculation of } I_0 \end{cases} \quad (6.30)$$

where I_{nZCT} is the nominal current of core-balance/neutral CT utilized for direct measurement of zero-sequence current (I_0). In addition, the pickup current of inverse stage is determined by

$$I^{51N}_{Pickup\text{-}R1} = \begin{cases} \max(0.05 \times I_{nZCT}, 10\,A) & \text{for direct measurement of } I_0 \\ \max(0.1 \times K_{CLPU} \times I_{nTr}, 0.05 \times I_{nCT}) & \text{for indirect measurement of } I_0 \\ \max(0.17 \times K_{CLPU} \times I_{nTr}, 0.05 \times I_{nCT}) & \text{for analytical calculation of } I_0 \end{cases} \tag{6.31}$$

where K_{CLPU} is a safety factor to consider the impact of cold load pickup (CLPU) event on the error of the indirect measurement or analytical calculation of the zero-sequence current. Typically, this factor could be assumed equal to 1.2.

The zero-sequence circuit at the two sides of the transformer are isolated from each other. In fact, when a single-phase fault occurs at the LV side of the transformer, no zero-sequence current passes through the MV side, and vice versa. Therefore, earth fault protection of the MV side could not provide backup protection for the LV-side earth faults. Moreover, passing zero-sequence currents through the MV side of the transformer indicates the occurrence of the earth fault at the MV-side relay protection region, and thus, no intentional delay is required to retain the protection system selectivity. As a result, coordination between the MV and LV sides earth fault units is not required. On the other hand, to avoid maloperation of the earth fault protection due to the partial or asymmetrical saturation of the CTs, a small intentional time delay is recommended to be considered for the inverse stage, i.e., TMS = 0.05 s.

Example 6.2

Calculate the settings of earth fault protection for the system presented in Example 6.1. Meanwhile, the indirect measurement method is exploited to determine the zero-sequence current.

Solution:

Since indirect measurement is exploited, based on (6.30), the pickup of earth fault instantaneous stage is determined as

$$I^{50N}_{Pickup\text{-}R1} = 0.1 \times I^{50}_{Pickup\text{-}R1} = 0.1 \times 1484 = 148.4\text{ A}$$

Therefore, the pickup current of the instantaneous stage at the secondary side of CT is calculated as

$$I^{50N}_{Pickup\text{-}R1}(secondary) = \frac{148.4}{100} = 1.48\text{ A}$$

Moreover, based on (6.31), the pickup of earth fault inverse stage is determined as

$$\begin{aligned} I^{51N}_{Pickup\text{-}R1} &= \max(0.1 \times K_{CLPU} \times I_{nTr}, 0.05 \times I_{nCT}) \\ &= \max(0.1 \times 1.2 \times 86.6, 0.05 \times 100) = 10.4\text{ A} \end{aligned}$$

Consequently, the pickup current of the inverse stage at the secondary side of CT is calculated as

$$I^{50N}_{Pickup\text{-}R1}(secondary) = \frac{10.4}{100} = 0.1\ \text{A}$$

The operation curve is considered IEC-very inverse, and its TMS is assumed 0.05 s.

6.2.1.2.2 MV Feeders Relay (Relay R2) The pickup current of the instantaneous stage of the earth fault unit of relay $R2$, i.e., $I^{50N}_{Pickup\text{-}R2}$, is determined as

$$I^{50N}_{Pickup\text{-}R2} = 1.2 \times I^{50N}_{Pickup\text{-}R1} \tag{6.32}$$

Moreover, the pickup current of the inverse stage of the earth fault unit of relay $R2$ is calculated as

$$I^{51N}_{Pickup\text{-}R1} = \begin{cases} \max(0.1 \times I_{nZCT}, 5\,A) & \text{for direct measurement of } I_0 \\ \max(0.1 \times K_{CLPU} \times I_{LoadMax}, 0.1 \times I_{nCT}) & \text{for indirect measurement of } I_0 \\ \max(0.17 \times K_{CLPU} \times I_{LoadMax}, 0.1 \times I_{nCT}) & \text{for analytical calculation of } I_0 \end{cases} \tag{6.33}$$

6.2.2 Solutions for Sympathetic Tripping

6.2.2.1 DG Back-Feed

As discussed in Chapter 4, one of the main challenges of distribution system's protection is the DG back-feed phenomenon in which the direction of power flow may be reversed and thus the coordination between the relays may get lost. Since the connection of the MV-side winding of the DG transformer is delta, during the single-phase to ground short circuits, the DG does not inject any current to the fault. In fact, the coordination between the earth fault protective functions of the system is not affected for this common winding connection of the DG transformer. However, for phase faults, such as two-phase and three-phase short circuits, the DG participates in the fault current and changes the current direction during the short-circuit event. To take into account this condition and retain the protection system coordination, directional units should be utilized. However, using directional elements for all of the system relays could increase the protection system expenditure and complexity and may not be cost-effective. Meanwhile, it may not be required to utilize directional elements for all of the system relays.

To scrutinize this matter, a simple distribution system is illustrated in Fig. 6.3. In this system, a synchronous generator-based MG (SGBMG) is connected to Bus 3. Initially, before connection of the SGBMG, all of the system relays are nondirectional and are adjusted, considering the setting rules presented in the previous

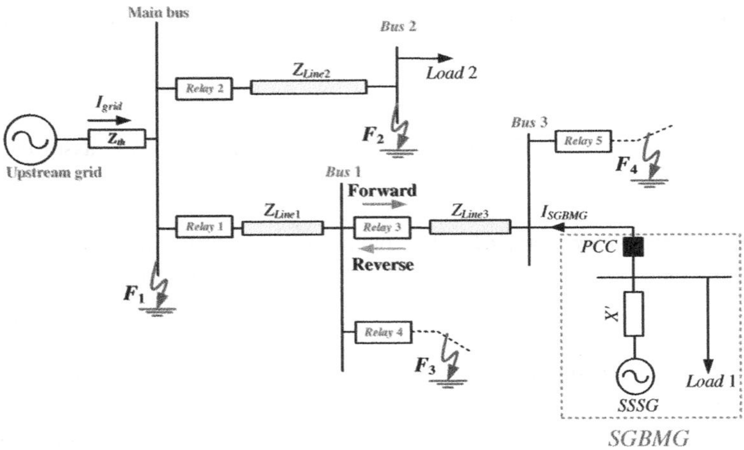

Fig. 6.3 Simple distribution system containing an SGBMG.

section 6.2.1 of this chapter. After the connection of the SGBMG, for the fault occurring at the F_3 location, only *Relay* 3 is prompted by the SSSG back-feed current. Similarly, for faults at F_1 and F_2, *Relay* 3 is picked-up again by the back-feed current, while this back-feed current may not trigger *Relay* 1 due to its higher pickup current. Therefore, in this sample system, if *Relay* 1 is not activated by the DG back-feed current, *Relay* 3 becomes the only relay that may potentially maloperate due to the SGBMG connection. If only this relay is equipped with the directional unit, the challenge of SGBMG back-feed could be resolved in this system. The forward direction of *Relay* 3 should be coordinated with its downstream protective device, i.e., *Relay* 5, as well as the forward direction of the PCC relay. Furthermore, the reverse direction of *Relay* 3 should be coordinated with the upstream protective device, such as *Relay* 2.

To determine the direction of fault, the voltage signal at the relay point is usually exploited. To do so, a potential transformer (PT) should be installed at the relay location. However, using PTs in distribution systems may not be practical and cost-effective. Thus, various schemes have been presented in the literature to determine the fault direction only using current signal and without benefitting voltage signals. In the rest of this section, one of the simplest and practical methods proposed by Jalilian *et al.* [9] is explained for revealing the fault direction. This method works only based on current signal without requiring any additional equipment, and thus, it can be applied to all of the distribution system relays. It is described using a simple distribution system shown in Fig. 6.4. First, it is assumed that the fault occurs at the forward direction of the relay, i.e., at the F_1 location. In

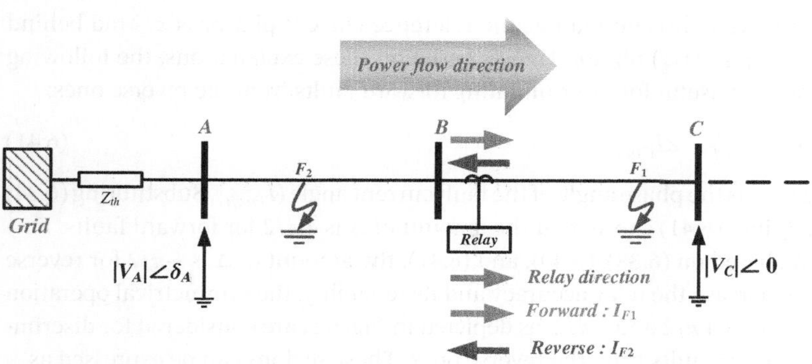

Fig. 6.4 Simple distribution system for analytical analysis of the fault direction.

this case, the fault current can be determined by

$$I_{F1} = \frac{|V_A|\angle\delta_A}{|Z_{AF1}|\angle\theta_1} = |I_{F1}|\angle(\delta_A - \theta_1) \tag{6.34}$$

where $|Z_{AF1}|\angle\theta_1$ is the impedance between bus A and fault location F_1. Moreover, the negative of I_{F1} is described as

$$-I_{F1} = |I_{F1}|\angle(\delta_A - \theta_1 + \pi) \tag{6.35}$$

A new signal, named reference current (I_{Ref}), is defined as follows to determine the fault direction:

$$I_{Ref} = |I_{F1}|\angle\left(\frac{\angle I_{F1} + \angle - I_{F1}}{2}\right) \tag{6.36}$$

For forward faults, the reference current is calculated as

$$I_{Ref} = |I_{F1}|\angle(\delta_A - \theta_1 + \pi/2) \tag{6.37}$$

Comparing (6.37) and (6.34) properly indicates that the reference current phasor is $\pi/2$ rad ahead of the fault current (I_{F1}) phasor for forward direction faults.

Additionally, the fault current for reverse direction faults is computed as

$$I_{F2} = -\frac{|V_C|\angle 0}{|Z_{CF2}|\angle\theta_2} = |I_{F2}|\angle(-\theta_2 + \pi) \tag{6.38}$$

where $|Z_{CF2}|\angle\theta_2$ is the impedance between Bus C and fault location F_2. Furthermore, the negative of I_{F2} is given by

$$-I_{F2} = |I_{F2}|\angle(-\theta_2) \tag{6.39}$$

Similar to (6.36), the reference current is determined as follows for reverse faults:

$$I_{Ref} = |I_{F2}|\angle\left(\frac{\angle I_{F2} + \angle - I_{F2}}{2}\right) = |I_{F2}|\angle(-\theta_2 + \pi/2) \tag{6.40}$$

Thus, for reverse direction faults, the reference current phasor is $\pi/2$ rad behind the fault current (I_{F2}) phasor. Hence, based on these explanations, the following index could be useful for discriminating forward faults from the reverse ones:

$$\Delta = \angle I_{Ref} - \angle I_{Fault} \tag{6.41}$$

where $\angle I_{Fault}$ is the phase angle of the fault current angle (I_{Fault}). Substituting (6.34) and (6.37) into (6.41) shows that the amount of Δ is $+\pi/2$ for forward faults. Furthermore, based on (6.38), (6.40), and (6.41), the amount of Δ is $-\pi/2$ for reverse faults. To increase the relay accuracy and its reliability, the symmetrical operation regions around $+\pi/2$ and $-\pi/2$, as depicted in Fig. 6.5, are considered for discriminating forward faults from the reverse ones. These regions can be expressed as

$$\begin{cases} +\pi/4 < \Delta < +3\pi/4 \Rightarrow \text{Forward} \\ -3\pi/4 < \Delta < -\pi/4 \Rightarrow \text{Reverse} \end{cases} \tag{6.42}$$

Based on the described concept, the algorithm of this current-based method is depicted in Fig. 6.6 to discriminate forward faults from reverse faults without using the voltage signal. In this algorithm, after the fault detection, the fault direction is determined using the I_{Ref} phasor and index Δ. This method works correctly only when the phase angle of the system voltage does not change considerably during the fault. However, in MGs, the voltage phase angle may noticeably alter especially in inverter-based MGs (IBMGs) and after occurrence of fault-initiated islanding events. Therefore, the effectiveness of this method may be questionable in some conditions.

6.2.2.2 Protection Sensitivity and Loss of Coordination

As discussed in Chapter 4, the back-feed current effect of DGs may result in maloperation of the healthy feeders' relays. It can also decrease the sensitivity of faulty

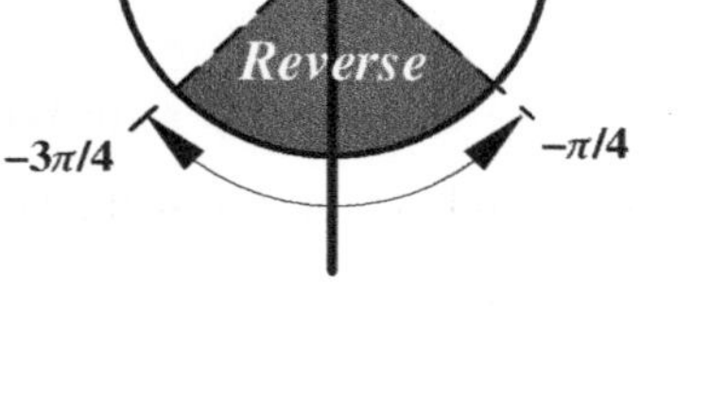

Fig. 6.5 Operation regions for discriminating forward faults from reverse ones [9].

Fig. 6.6 Flowchart for detection of the fault direction.

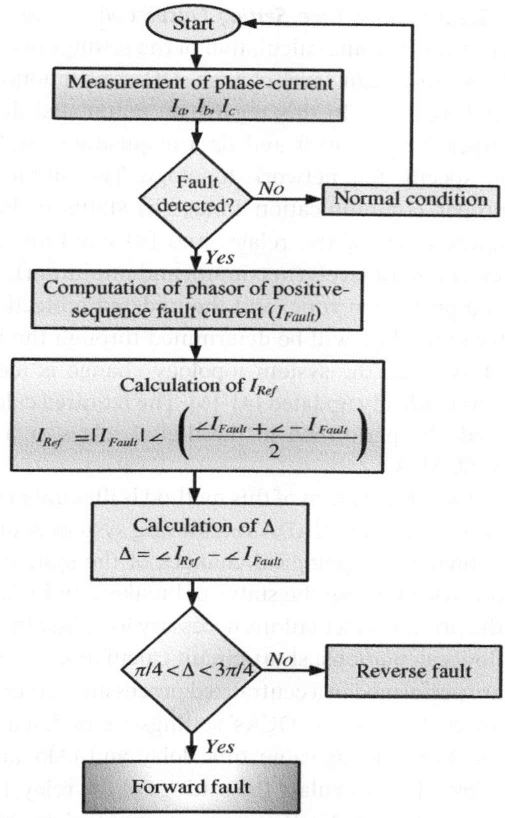

feeders' protection system. Furthermore, connecting DGs in the distribution system may have adverse impact on the coordination between the protection relays due to change of the system short-circuit level.

IBDGs and SSSGs have different impacts on protection relays performance during the fault occurrence. The IBDG current during short-circuit faults is not sizable and usually limited to at most 1.5 pu due to the limitation of power-electronic switches, while the short-circuit level of SSSGs is about 5–6 pu. Thus, connecting SSSGs to the distribution system would result in more adverse impact on the protection sensitivity and loss of coordination between the relays. On the other hand, the short-circuit level may considerably decrease during the islanded operation of the MGs, especially for IBMGs, and thus, the MG relays may not be capable to detect and clear the faults. To resolve these serious issues, some techniques are presented in the rest of this section.

6.2.2.2.1 *Real-Time Setting Calculation* Various literature propose several algorithms for online calculation of the settings of OCRs to mitigate the issues caused by short-circuit level change, DG connection/disconnection, and changing the grid topology. In this regard, an automated distribution system using a proper supervisory control and data acquisition (SCADA) system should be adopted to specify the network topology. This automated system should provide (i) proper communication links, (ii) status of breakers, (iii) status of DGs, (iv) information of the relays, and (v) real-time monitoring capability. Then, the short-circuit levels (maximum and minimum), maximum loading of each section and protection zone, and the updated protection zones including the main and backup relays will be determined through the information provided by SCADA. Thus, once the system topology change is identified, the relay setting will be automatically updated [4]–[6]. The required calculations for updating the settings could be performed in the digital relays or a centralized control system such as SCADA.

A sample pattern of this method is illustrated in Fig. 6.7 using directional OCRs. As shown, the SCADA monitoring system is utilized to detect the possible operational and topological changes of the system. Then, the new configuration of the system using the status of breakers and DGs should be identified. Afterward, the primary calculations necessary for adjusting the relays' settings including load flow calculations, short-circuit calculations, contingency and sensitivity analysis are performed in a centralized processing server (CPS) [5]. Once these calculations are performed, the OCRs' settings are updated in CPS. Then, these new settings, including pickup value, time delay, and TMS, are transmitted to be adjusted in the relays. To recalculate the settings of the relay, the rules presented in this chapter can be utilized. Furthermore, an appropriate objective function is adopted by various literature to find the optimum solution for the relay setting problem. The main purpose of defining this objective function is to minimize the relays' operation times and enhance the security and selectivity of the protection system. In this regard, the following objective function could be utilized [5]:

$$\text{Objective function} = \left(\frac{NV}{NCP}\right) + h_1 \times \left(\frac{\sum_{a=1}^{NCP} t_{pa}}{NCP}\right) + h_2 \times \left(\frac{\sum_{b=1}^{NCP} t_{bb}}{NCP}\right) \tag{6.43}$$

where *NCP* is the number of main–backup coordination pairs, *NV* is the number of miscoordinations or violating constraints, t_{pa} is the operation time of the main relay *a* for the maximum fault current due to the three-phase fault occurring at near-end of the main protection region, and t_{bb} is the operation time of the backup relay *b* for the phase-phase fault occurring at the remote-end ($I_{sc}^{2\varphi}$) of the main

Fig. 6.7 Flowchart for real-time setting calculation [5].

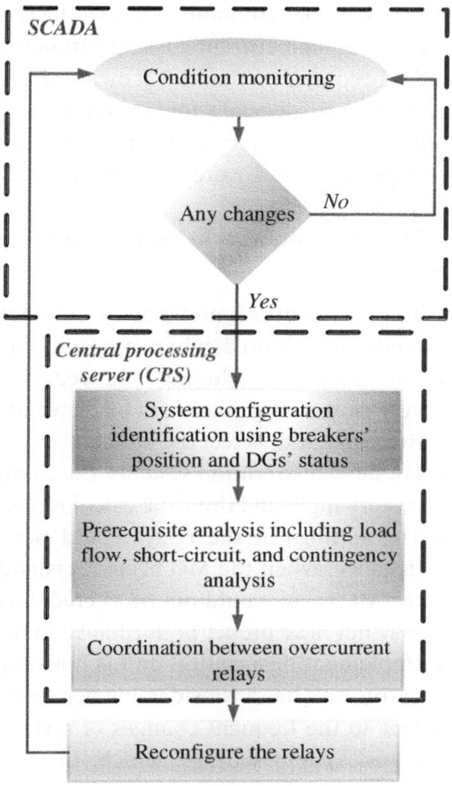

protection relay zone. Moreover, parameters h_1 and h_2 are the constant weighting factors. The main constraints of coordination problem can be given as

$$t_b - t_p \geq \text{CTI}$$

$$\text{TMS}_{min} \leq \text{TMS} \leq \text{TMS}_{max} \tag{6.44}$$

$$I_{pickupmin} \leq I_{pickup} \leq I_{pickupmax}$$

where TMS_{min} and TMS_{max} indicate the minimum and maximum adjustable settings of TMS, respectively. Moreover, parameters $I_{pickupmax}$ and $I_{pickup\,\min}$ are the maximum and minimum values of I_{pickup}, respectively.

Furthermore, to investigate the sensitivity of backup protection relays, a sensitivity analysis index is defined as [5]

$$\text{sensitivity} = \frac{I_{sc}^{2\varphi}}{k \times I_{loadmax}} \tag{6.45}$$

where k is the temporary overload factor of the backup relay and $I_{loadmax}$ is the maximum load current passing through the backup relay. If the amount of this index becomes less than 1.5, it means that the backup relay may not provide proper performance especially for far-end or high-impedance faults. To resolve this issue, other complementary protection functions such as broken conductor protection, voltage-controlled overcurrent protection, and unbalance current protection could be utilized.

The main drawbacks of the studied real-time setting calculations method include:

- To implement this method, the automated system and reliable communication media are required, which may not be cost-effective, especially for MGs.
- For conditions including long feeders, the minimum short-circuit fault can be very close to the maximum load current, and thus, the relay could not be sensitive enough to detect the faults.
- For large distribution systems, the computation burden of the relay or CPS may be very high, and thus, the calculations could not be performed quickly.
- The settings of the relays could not be changed during disturbances. For instance, when the MG becomes islanded for an internal fault, the settings of the MG relays could not be altered during this event, and thus, the MG relays may not have proper performance in this condition.
- Implementing a central unit is not simple and straightforward, and the failure of this unit has a considerable negative impact on the system protection.
- Due to the frequent changes of system topology, there is a continual need to adjust system settings, which, in turn, increases the likelihood of applying improper values.

6.2.2.2.2 Utilizing Predefined Setting Groups To resolve the challenges regarding the high computational burden due to the real-time calculation of the settings, some predefined setting groups can be utilized. As a common practice, for the grid-connected and islanded modes of the MG, two different setting groups are utilized, which is referred as "dual-setting." The setting values are determined using off-line calculations. Dual-setting can only cover the issues related to the variations of the short-circuit level due to the changing of the MG operation mode. To consider various topologies of the distribution system and also connection/disconnection of DGs, clustering technique is suggested by various studies. This technique is adopted to reduce the scenarios of network topology changes and thus it considerably decreases the computational burden compared to the online calculation of the setting values. A cluster includes various topologies, which are similar to each other and not similar to the other topologies [12]. For each cluster, specific setting groups could be assigned. It should be noted that the setting values of each cluster are calculated using off-line calculations.

Fig. 6.8 *k*-means algorithm [12].

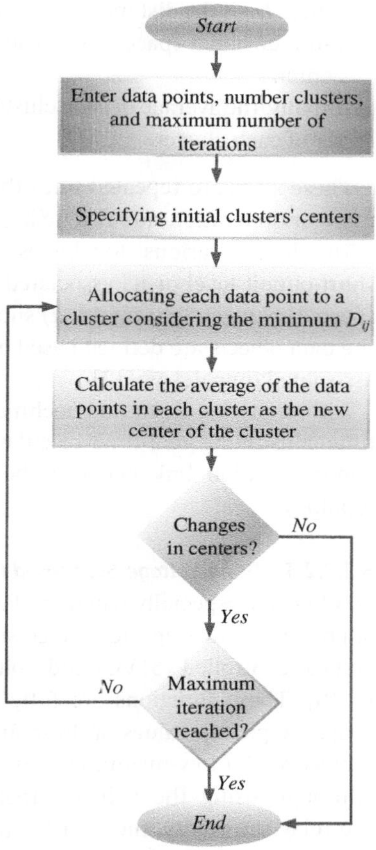

A well-known technique for clustering is *k*-means algorithm, which is widely utilized in various applications including fault classification, fault detection, and, in general, pattern recognition. The flowchart of *k*-means algorithm is illustrated in Fig. 6.8, which includes the following steps [12]:

1) The inputs of the *k*-means algorithm include the number of clusters, maximum number of iterations, and data points, which should be specified at first.
2) Then, *k* data points should be selected as the initial centers.
3) The distance between each data point and the centers is calculated by

$$D_{ij} = \sqrt{\sum_{m=1}^{d} (x_{im} - c_{jm})^2} \tag{6.46}$$

where D_{ij} is the distance between the center point cj and the data point x_i in the d-dimensional space. Then, each data point is assigned to the nearest cluster center.

4) Finally, the average of each cluster is calculated and considered as the new center of each cluster.

These steps are repeated until the clusters remain changing or the maximum number of iterations are reached.

To cluster various topologies of the distribution systems, the near-end short-circuit level or its associated time delays could be utilized. Therefore, the space dimension (parameter d) should be considered equal to 1. In this context, the data objects are defined based on either the near-end short-circuit level or its associated time delays [12].

Although the clustering technique reduces the computational burden and time-consuming real-time calculations, this technique still requires a reliable communication link between the relays and the CPS to determine the grid topology.

6.2.2.2.3 Using Voltage-Supervised Overcurrent Functions To improve the sensitivity of OCRs, especially when the fault current considerably decreases, one viable approach is utilizing the voltage-supervised overcurrent (VS-OC) function, e.g., voltage-controlled (51VC) and voltage-restrained (51VR) overcurrent functions [8, 10]. These functions are fully explained in Chapter 4. During the voltage drop, the pickup values of these functions reduce and the relay could detect the faults even for low-magnitude fault currents, especially in adjacent to the IBDGs. Although using the voltage-supervised function could considerably improve the relay sensitivity, the coordination of these compound relays during various operation modes and topologies would become a challenge.

To diminish the coordination challenges of this function, some methods have been proposed in the research works. A general characteristic of a VS-OC function is depicted in Fig. 6.9. As shown, for $V > V_t$, the PS remains constant and equal

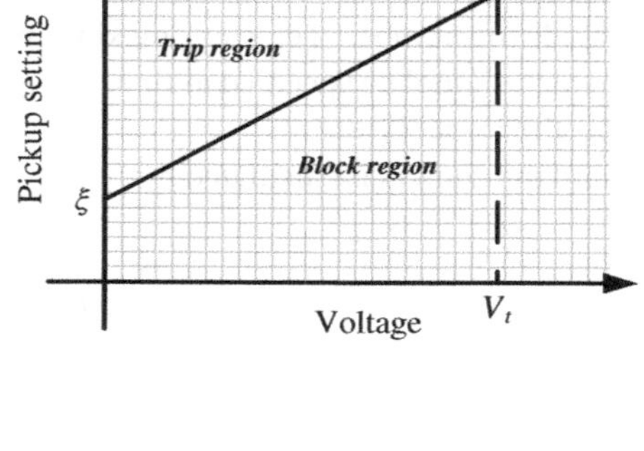

Fig. 6.9 Pickup setting of voltage-supervised overcurrent (VS-OC) function. [8] / with permission of IEEE.

to PS^{max}. Whereas for $V < V_t$, the PS linearly decreases depending on the voltage magnitude (V). The analytical expression of this function is given by [8]

$$PS = \min(\lambda V + \xi, PS^{max}) \tag{6.47}$$

where λ and ξ are the slope and bias of the VS-OC function, respectively. In addition, V_t is the pick value of voltage and V is the positive-sequence value of the measured voltage. Meanwhile, PS^{max} is determined by

$$PS^{max} = \lambda V_t + \xi \tag{6.48}$$

If the passing current (I) would be less than PS, the relay is not triggered. However, if this current is greater than PS, the relay is triggered and operates after elapsing its time delay (t). The time delay of the relay could be determined by

$$t = TMS\left(\frac{a}{\left(\frac{I}{PS}\right)^b - 1}\right) + c \tag{6.49}$$

where a, b, c, and TMS are constant parameters. Since PS is a function of voltage, the time delay would become a function of both voltage and current. A sample curve of operation delay considering TMS = 0.1 is depicted in Fig. 6.10. As shown, by decreasing the voltage, the amount of t also decreases. Furthermore, t reduces with increment of the passing current [8].

The coordination of VS-OC functions with each other may become complex due to the variation of pickup current with the voltage signal. In other words, the voltage magnitude at the main and backup protection relays is different and also the voltage magnitude considerably depends on the fault location. Thus, various

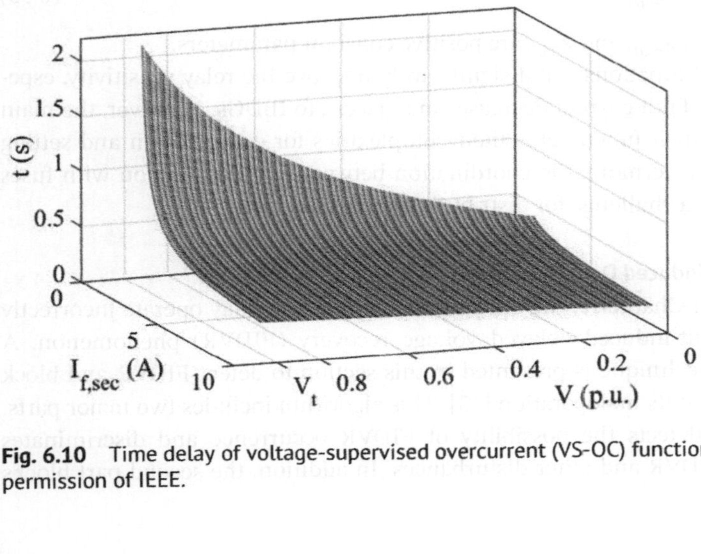

Fig. 6.10 Time delay of voltage-supervised overcurrent (VS-OC) function. [8] / with permission of IEEE.

off-line studies should be performed for coordination of these functions with each other. In the following, an overall procedure for adjustment of this function is explained.

- PS^{max}: This parameter is adjusted according to the rules used for adjusting the pickup current, as described in Section 6.2.
- V_t: To adjust the voltage pickup, i.e., V_t, the positive-sequence voltage at the relay location for the two-phase fault at the remote-end of the downstream main relay would be calculated. The voltage pickup should be considered equal to or greater than the calculated value.
- TMS, λ, ξ: To calculate these parameters, the following objective function is defined [8]:

$$t = \min \sum_{s \in S} \left(\sum_i t_i + \sum_j t_j \right) \tag{6.50}$$

where t_i is the time delay of the main relay for the three-phase fault occurring at its near-end and t_j is the operation time of its associated backup relay for the fault occurring at the remote-end of the main relay. Meanwhile, S represents a set of various topologies of the system. The coordination constraint between the main and backup relays could be expressed as

$$t_j - t_i \geq \text{CTI} \tag{6.51}$$

Moreover, the constraints of variables λ and ξ are given by

$$\lambda_{min} \leq \lambda \leq \lambda_{max} \tag{6.52}$$

$$\xi_{min} \leq \xi \leq \xi_{max} \tag{6.53}$$

where ξ_{min}, ξ_{max}, λ_{min}, and λ_{max} are positive constant parameters.

Using VS-OC functions could significantly improve the relay sensitivity, especially when the fault current decreases in adjacent to IBDGs. However, the main issue of using such functions is their complexities for coordination and setting adjustment. Furthermore, the coordination between such a function with fuses and reclosers is a challenge for distribution systems.

6.2.2.3 Fault-Induced Delayed Voltage Recovery

As explained in Chapter 4, the OCR of a healthy feeder may operate incorrectly due to the fault-induced delayed voltage recovery (FIDVR) phenomenon. A current-based technique is presented in this section to detect FIDVR and block the relay to avoid its maloperation [13]. This algorithm includes two major parts. The first one detects the possibility of FIDVR occurrence and discriminates between the FIDVR and other disturbances. In addition, the second part blocks

OCRs for a duration during the FIDVR phenomenon. In the following, these parts are expressed.

6.2.2.3.1 First Part – FIDVR Detection An IM decelerates during short-circuit faults and its speed decreases. After the fault clearing instant, it begins to reaccelerate and thus its current magnitude increases. Thus, the FIDVR phenomenon may occur in the system after clearing a short-circuit fault, particularly in the presence of high penetration of IMs and a long fault duration. The FIDVR occurrence signal (FOS) is adopted in this section to discriminate this phenomenon from the other disturbances. Figure 6.11 illustrates the algorithm for determining the FOS status [13].

FIDVR usually occurs after a system short-circuit fault. If the fault occurs on the downstream system, the current passing through the relay will increase. For such a condition, an overcurrent function 50 (block (1) in Fig. 6.11) is employed for fault detection whose threshold is I_{th}^{50}. Moreover, when the fault takes place

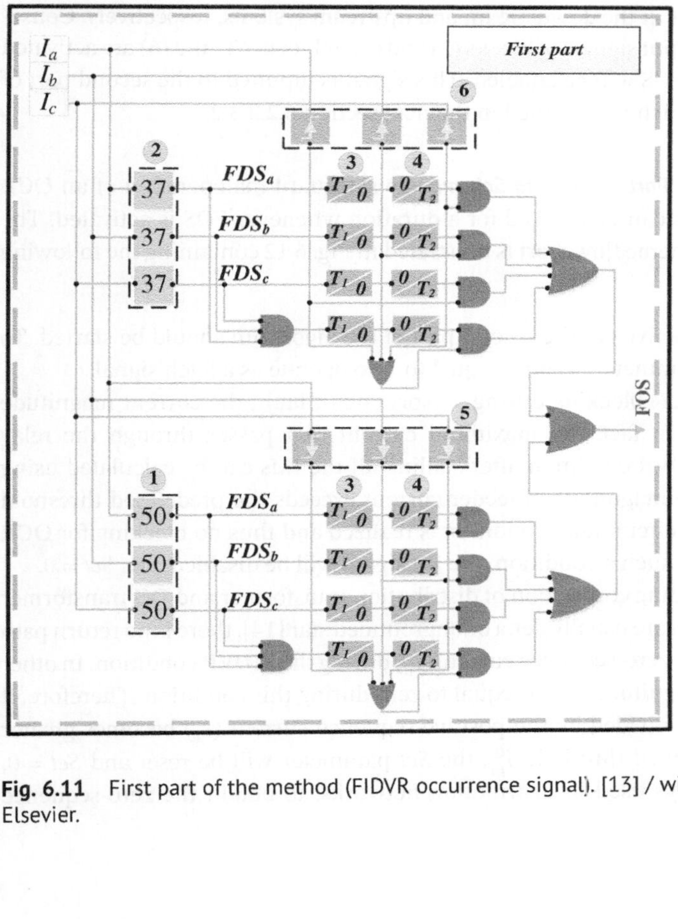

Fig. 6.11 First part of the method (FIDVR occurrence signal). [13] / with permission of Elsevier.

at the upstream system or other feeders connected to the same subtransmission substation, the bus voltage drops and thus the load current passing through the relay will decrease, especially before motor stalling. In this condition, an undercurrent function 37, i.e., block (2), is used to detect the voltage drop, whose threshold is I_{th}^{37}.

As shown in Fig. 6.11, if $I > I_{th}^{50}$ or $I < I_{th}^{37}$ conditions are fulfilled, a fault event would be detected and the associated fault detection signal (FDS) is set to 1. The output signals of blocks (1) and (2), i.e., FDSs, are the inputs of block (3). If FDS = 1 for a predefined duration time T_1, the output signal of block (3) will be enabled representing a prolonged fault event, which may result in the FIDVR condition. Block (4) is defined as a reset module which is utilized to reset FDS signal after T_2 s. Distribution feeder short-circuit faults are usually cleared between 0.1 and 1 s; thus, T_2 is considered as 1 s.

FOS should be issued after the fault clearance. To realize this instant, the change of current magnitude is utilized. Therefore, a falling edge detector, block (5), and a rising edge detector, block (6), are employed to identify the fault clearing instant after short circuits at the downstream and upstream systems, respectively. Consequently, if the output signals of blocks (3) and (5) or blocks (3) and (6) are activated simultaneously, FOS will be enabled. This signal is inputted to the second part of the algorithm, which is explained in the next section 6.2.2.3.2.

6.2.2.3.2 Second Part – Blocking Scheme To safeguard maloperation of an OCR against FIDVR, it can be blocked for a duration whenever FOS is activated. The OCR blocking scheme flowchart is illustrated in Fig. 6.12 containing the following steps [13]:

1) Once FOS is activated, the second part of the algorithm should be started. To do so, a *Set* parameter becomes equal to 1, to operate as a latch signal.
2) To prevent OCR blocking during a short-circuit fault, the current magnitude is employed. In fact, the maximum current that passes through the relay during an FIDVR condition after stalling all the IMs can be calculated using (5). When the magnitude of feeder current exceeds the predefined threshold (I_{MR}), a short-circuit fault condition is realized and thus no blocking for OCR is required. In such a condition, the *Set* signal will be disabled, i.e., *Set* = 0.
3) Since the winding connection of distribution transformer and DG transformer at the HV sides are usually delta or ungrounded star [14], there is no return path for passing the zero-sequence current (I_0) during the FIDVR condition. In other words, the magnitude of I_0 is equal to zero during this condition. Therefore, if the ratio of I_0 divided by the positive-sequence current (I_1) becomes greater than a predefined threshold I_{th}^{0}, the *Set* parameter will be reset and *Set* = 0. In digital relays used in distribution networks, to obtain the zero-sequence

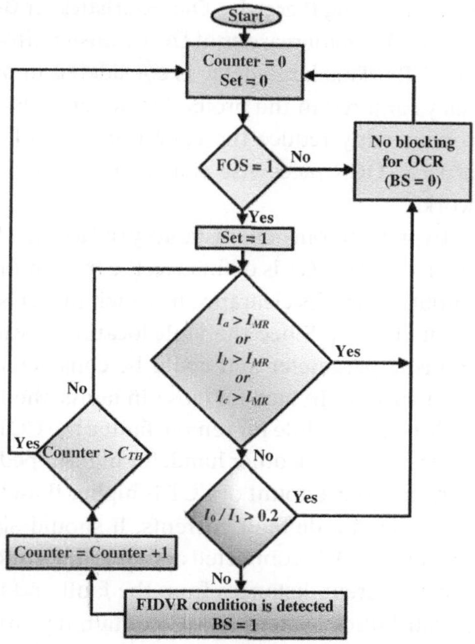

Fig. 6.12 Second part of the method (blocking scheme flowchart). [13] / with permission of Elsevier.

current, sometimes the outputs of three-phase current transformers (CT) are summed up and employed, due to the lack of core-balance or neutral CTs. This results in measurement errors of zero-sequence current. To cope with this issue, the appropriate value for I_{th}^{0} threshold is suggested to be 0.2 pu.

4) The FIDVR condition usually lasts in the system for 3–30 s [15, 16]. Thus, the *Set* signal will be disabled after an appropriate time interval (T_3). To do so, a reset counter is employed, which its threshold (C_{th}) can be calculated as

$$C_{th} = f_c \times T_3 \tag{6.54}$$

where f_c is the sampling frequency.

6.3 DG Protection Solutions to Improve System Stability

6.3.1 Transient Stability

As mentioned in Chapter 4, transient stability of DGs is a serious concern for distribution systems and MGs, especially during the fault occurrence and after the fault clearance. In this section, some techniques are presented to enhance the transient stability of DGs and prevent their instability.

6.3.1.1 Using Piecewise Characteristics for Overcurrent Relays

One of the main reasons of DG's transient instability is the large intentional delays of OCRs. This large delay is considered to prevent miscoordination between the relays and retain the protection system selectivity. To avoid miscoordination and concurrently reduce the operation time of the relays and enhance the stability of DGs, OCR piecewise characteristics have been proposed by various research works.

To prevent transient instability of DGs, the fault should be cleared at most within the DG CCT. CCT is defined as the maximum time between the short-circuit fault initiation and its clearance in which the DGs do not lose its stability and synchronism [17, 18]. Since the fault location is variable, the CCT value is not a deterministic parameter and could be considered as a function of fault location and fault current. In fact, for close-in faults, the DG becomes unstable quickly, and its CCT is short, while for remote faults, the CCT is larger and the DG stability margin increases. On the other hand, for high-impedance faults with low-magnitude fault currents, the amount of CCT is higher than that of the low-impedance faults with large-magnitude fault currents. It should also be noted that the amount of CCT for various DGs connected at different locations of the system is not similar due to their different distances from the fault and their various specifications. Thus, for a distribution system or MG containing various DGs, the CCT constraint should also be considered for each DG in the OCRs optimal coordination problems.

For each DG located in the system, the following constraints should be considered in the relay coordination problem [2, 3]:

$$t_R|_{@fi} + t_{margin} \leq \mathrm{CCT}_{DGn}|_{@fi} \tag{6.55}$$

$$t_R|_{@fj} + t_{margin} \leq \mathrm{CCT}_{DGn}|_{@fj} \tag{6.56}$$

where fi and fj indicate close-in and remote-end faults, respectively. CCT_{DGn} is the CCT of DGn, which is the nearest DG to the fault location and t_{margin} is a time margin to cover unintentional delays such as the delays of CBs and auxiliary relays. As an example, a typical distribution system containing two DGs is depicted in Fig. 6.13. As shown, for close-in faults, the CCT curves are located below the OCR curves. Thus, for these faults, the operation delay of the OCR is larger than the CCT, and thus, the DG may become unstable. To resolve this issue, it is necessary to consider the stability criteria for this system in adjusting the OCRs. These stability criteria are given as [2, 3]

$$\begin{aligned}
&t_{R2}|_{@f1} \leq \mathrm{CCT}_{DG1}|_{@f1} - t_{margin}\\
&t_{R2}|_{@f2} \leq \mathrm{CCT}_{DG1}|_{@f2} - t_{margin}\\
&t_{R3}|_{@f3} \leq \mathrm{CCT}_{DG2}|_{@f3} - t_{margin}\\
&t_{R4}|_{@f4} \leq \mathrm{CCT}_{DG2}|_{@f4} - t_{margin}
\end{aligned} \tag{6.57}$$

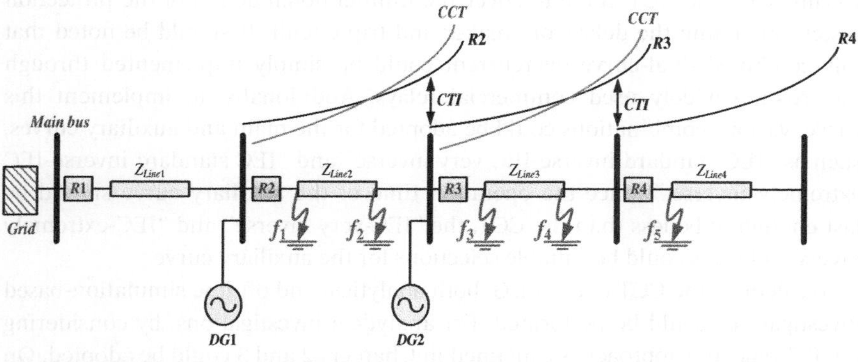

Fig. 6.13 Typical distribution system containing two DGs [2, 3].

Furthermore, the coordination criteria are given as

$$\begin{aligned} t_{R2}|_{@f3} - t_{R3}|_{@f3} &\geq \text{CTI} \\ t_{R2}|_{@f4} - t_{R3}|_{@f4} &\geq \text{CTI} \end{aligned} \tag{6.58}$$

However, it may be impossible to fulfill both the coordination and stability criteria using a single-curve inverse overcurrent. Thus, to meet both criteria, a dual-curve inverse overcurrent is adopted, which is composed of two curves, i.e., the main and auxiliary curves, as depicted in Fig. 6.14. The main curve is used to provide proper backup for the downstream relays and also clear the low-magnitude current faults at the far-end of the main protection region, where the CCT is relatively larger than the main relay operation delay. Furthermore, for close-in faults, where the fault current is relatively high and the CCT value is smaller than the main curve operation time, an auxiliary curve is exploited to clear the faults. The auxiliary curve should have a proper margin, for example,

Fig. 6.14 Overcurrent relay main and auxiliary curves and CCT curve [2, 3].

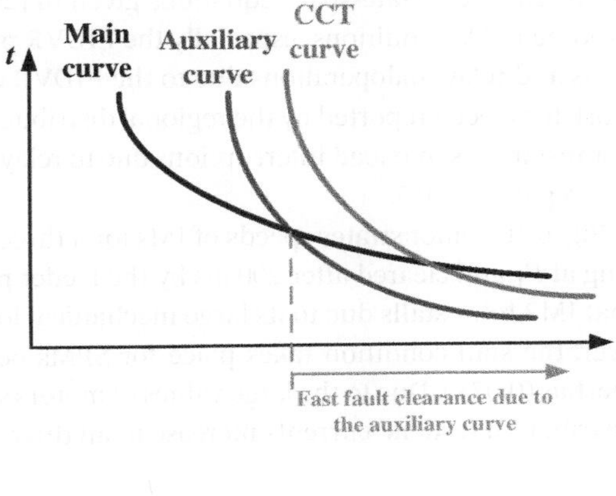

100 ms, with the CCT curve to cover the unintentional delays of the protection system including the delays of breaker and trip circuit. It should be noted that this combined dual-curve overcurrent could be simply implemented through the existing widely used commercial relays. Additionally, to implement this curve, various combinations could be adopted for the main and auxiliary curves, such as "IEC standard inverse-IEC very inverse" and "IEC standard inverse-IEC extremely inverse." Since the operation time of the auxiliary curve should be fast enough to be less than the CCT, the "IEC-very inverse" and "IEC-extremely inverse" choices could be suitable selections for the auxiliary curve.

To calculate the CCT of each DG, both analytical and off-line simulation-based investigations could be performed. For analytical investigations, by considering the DG type, the approaches explained in Chapters 2 and 3 could be adopted. On the other hand, for simulation-based studies, the transient model of the system is required to determine the amount of CCT through time-domain simulation of the system for various short-circuit events. Although simulation-based studies may be more accurate, they could be time-consuming and also costly hardware may be required, especially for large distribution systems.

6.3.1.2 Adaptive Load-Shedding

6.3.1.2.1 Impact of Load-Shedding on Transient Stability In Chapter 2, the impact of dynamic loads including IMs on the DG's transient stability has been explained. In this section, this impact and the influence of load-shedding for enhancing the DG's stability are briefly reviewed again. The under-study MG, shown in Fig. 6.15, contains two synchronous-type DGs, two aggregated three-phase IMs (IM1 and IM2), an aggregated single-phase IM (SPM), aggregated static loads, and remote-control switches (RCSs). In this system, the aggregated IM data and feeder information are extracted from real measured data. The detailed information about this real distribution system is given in [19]. To model the single-phase IM, a new dynamic component is developed and implemented in the PSCAD/EMTDC software based on the transient state-space equations given in [20–22]. This real system is highly exposed to LV conditions, especially the FIDVR phenomenon. Various load rejections and relay maloperations due to the FIDVR condition in this part of the network have been reported by the regional distribution company in the recent years. For instance, some load interruptions due to relay maloperations were reported in this region in 2015.

Fig. 6.16 demonstrates speeds of IMs for a three-phase short-circuit fault occurring at F_1 and cleared after 200 ms by the feeder protection relay [19]. This shows that IM2 rotor stalls due to its large mechanical load torque ($T_L = 0.95$ pu). Moreover, the stall condition takes place for SPMs because of their associated small inertias (0.07 s). Due to the large values of motors slips at the fault-clearing instant, distribution feeders' currents increase in all three phases after the fault clearance,

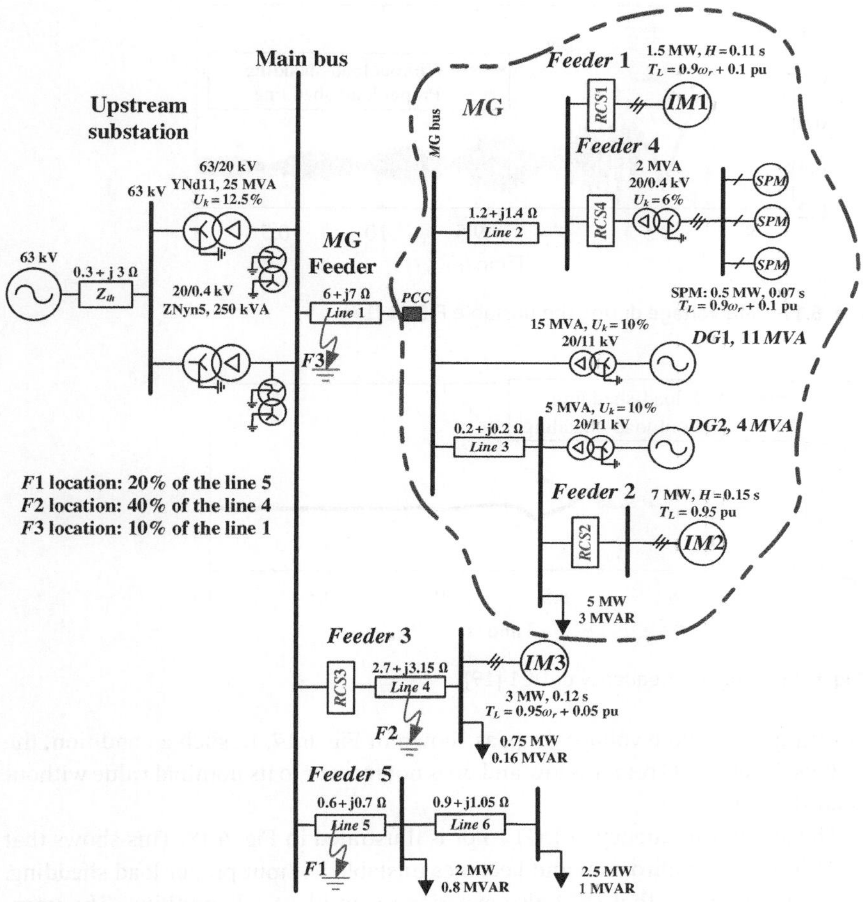

Fig. 6.15 Single-line diagram of a real distribution network. [19] / with permission of IEEE.

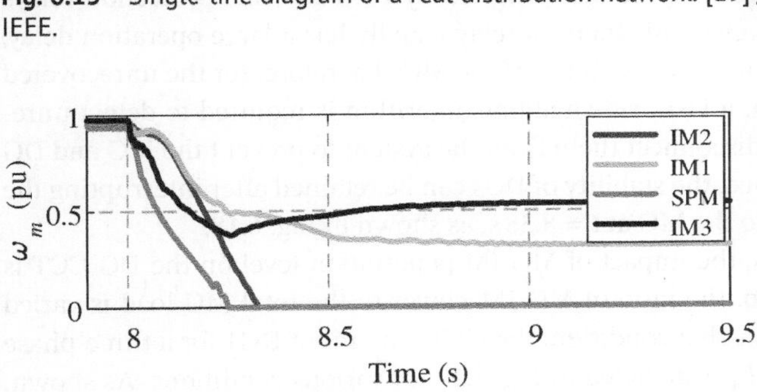

Fig. 6.16 IM rotors speeds during the unstable FIDVR [19].

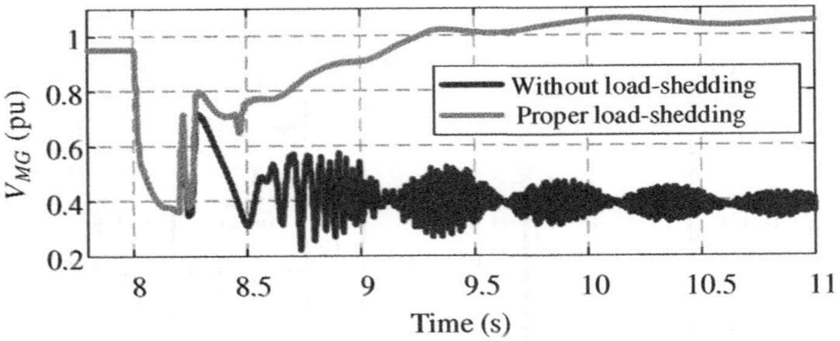

Fig. 6.17 MG voltage during the unstable FIDVR [19].

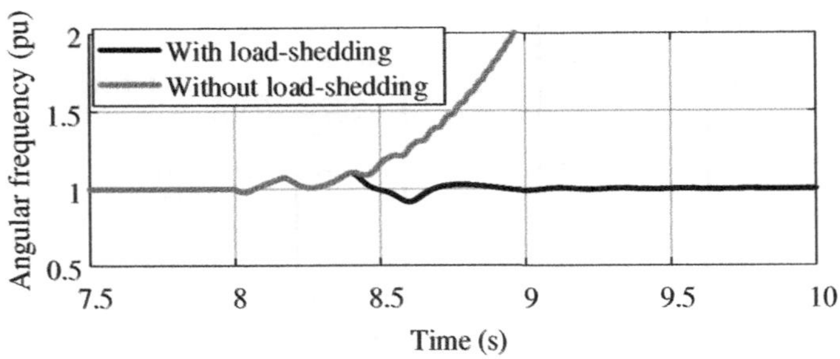

Fig. 6.18 Angular frequency of DG1 [19].

resulting in a severe voltage drop, as shown in Fig. 6.17. In such a condition, the voltage level of MG remains low and does not recover to its nominal value without load shedding.

The angular frequency of DG1 rotor is illustrated in Fig. 6.18. This shows that DG1 loses its synchronism and becomes unstable without proper load shedding. It should be noted that DG2 also experiences an identical condition. The unrecovered IMs may eventually be disconnected by their thermal protection relays (ANSI code 49). However, this thermal relay usually has a large operation delay, which may lead to transient instability of the MG. Therefore, for the unrecovered FIDVR phenomenon, a fast load-shedding algorithm is required to detect unrecovered feeders and disconnect them from the system to prevent the MG and DG instability. For instance, the stability of DGs can be retained after interrupting the IM loads connected to the MG at $t = 8.38$ s, as shown in Fig. 6.18.

For further studies, the impact of MG IM penetration level on the DG CCT is investigated. To do so, the ratio of MG IM power to the total MG load is varied from 0% to 100%. In such a condition, the CCT amount of DG1 for a three-phase short-circuit fault at F_1 is depicted in Fig. 6.19 for various conditions. As shown,

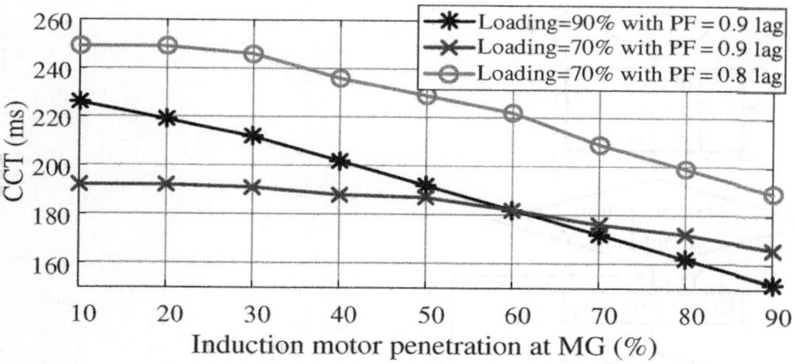

Fig. 6.19 CCT values of DG1 for various amounts of IM penetration in MG [19].

the amount of CCT significantly reduces with increasing the IM penetration. Moreover, the CCT is influenced by various real-time factors, including FCT, fault location, and DG loading conditions. Therefore, the load-shedding strategy to tackle the MG transient instability problem should be designed, considering such parameters. This reveals that an adaptive solution is required to shed proper amount of IM loads. In this regard, considering the investigation of IM load dynamic behavior discussed in Chapter 2, a simple and effective adaptive load-shedding scheme is presented in the next subsection 6.3.1.2.2, to disconnect proper loads and retain the transient stability of DGs.

6.3.1.2.2 Adaptive Load-Shedding Scheme In Chapter 2, an index has been introduced for identifying the IM reacceleration condition, named reacceleration detection index (*RDI*). Considering the *RDI*, an adaptive load-shedding scheme is presented for MGs whose flowchart is illustrated in Fig. 6.20, including the following parts:

- Part (1): This step is employed to initialize the algorithm.
- Part (2): If a short-circuit fault persists in the system for some duration, the FIDVR phenomenon may occur after the fault clearance due to the IM behavior. In such a condition, the system admittance Y seen by the load-shedding relay increases. In fact, part (2) is utilized to detect the possibility of FIDVR occurrence after the clearance of a short-circuit fault. To do so, the $Y > Y_{th}$ condition should be fulfilled after the fault clearance. Parameter Y_{th} is a constant threshold. If the DG load angle becomes greater than a threshold φ_{th} at the fault-clearing instant, i.e., $\varphi_c > \varphi_{th}$, DG will lose its stability as the load shedding cannot help retain the DG stability. In such a condition, DG should be disconnected. This threshold could be determined using off-line simulation studies or through analytical investigations.

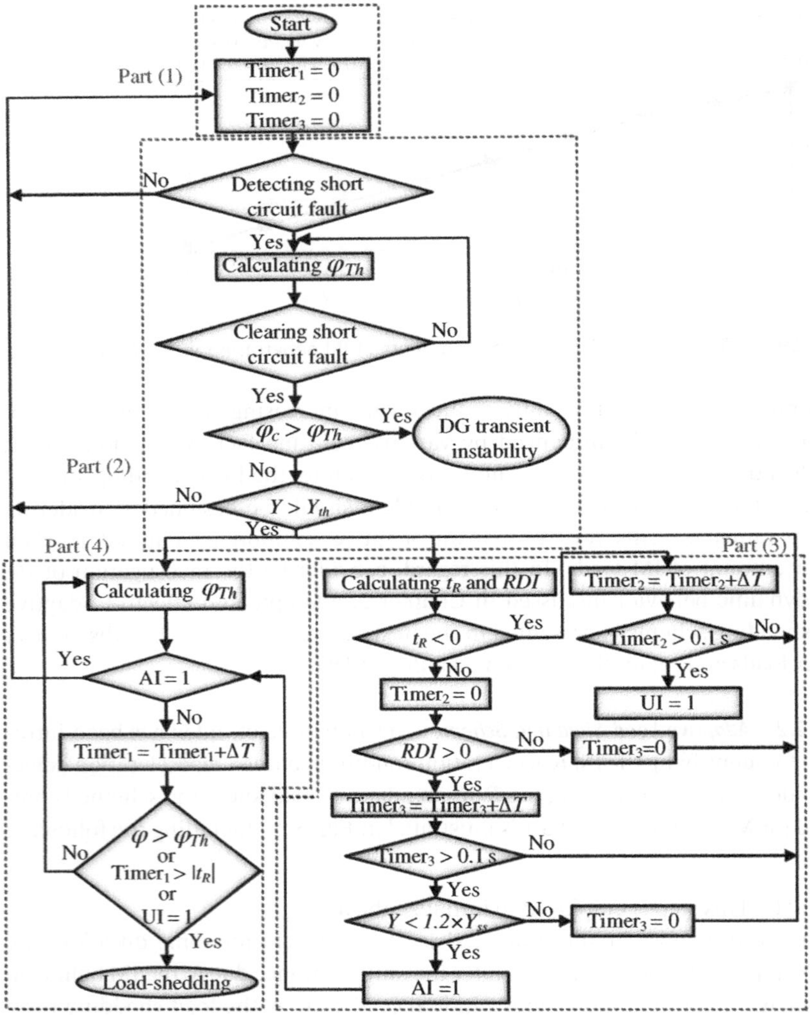

Fig. 6.20 Adaptive load-shedding scheme flowchart. [19] / with permission of IEEE.

- Part (3): This part is employed to distinguish the recovered and unrecovered IMs based on the RDI and the investigations performed in Chapter 2. If the reacceleration time of the aggregated IM, t_R – equation (6.30), Chapter 2 – is negative for 100 ms, the unrecoverable IM condition is detected and the unrecoverable index would be enabled (UI = 1). As such, the location of load shedding would be the unrecovered IM feeders. In addition, if $RDI > 0$ for 100 ms and $Y < 1.2 \times Y_{ss}$, a motor reacceleration condition is identified, and thus, the acceleration index AI would be set to 1. It should be noted that the parameter Y_{ss} denotes steady-state admittance of the feeder.

- Part (4): This part is employed to check the DG stability criterion. If the measured load angle of a DG is greater than its own threshold φ_{th}, the unrecovered IMs should be disconnected, even before elapsing t_R or activating UI. In other words, the algorithm does not wait for t_R seconds in such a condition, and the IMs that do not recover to their nominal speeds are disconnected to prevent the DG transient instability. If motor reacceleration is detected, i.e., for AI = 1, the algorithm would be reset. In addition, If the IM speed does not recover after t_R seconds or UI = 1, the load-shedding command will be issued. Therefore, similar to the previous step, the location of load shedding would be the unrecovered IM feeders.

6.3.2 Loss of Field Protection

The challenges associated with the loss of field (LOF) protection in SSSGs have been discussed in Chapter 4. Before presenting solutions to address these challenges, the key issues related to this protection scheme are briefly revisited.

6.3.2.1 SSSG Behavior During LOF Events

A LOF event for an SSSG can occur either totally or partially. Total LOF occurs when the field winding excitation voltage is completely short- or open-circuited. However, in the case of partial LOF, the flux produced by the field winding would be decreased, e.g., due to the excitation voltage drop. After occurrence of total or partial LOF events, automatic voltage regulator (AVR) of the SSSG tries to boost the excitation level. For partial LOF event, the MG instability could be prevented by fast operation, i.e., fast boosting of AVR. In other words, if some portion of the rotor winding is short-circuited, increasing the DC field current can compensate the reduction of the field winding effective turns. Since the SSSG output voltage can be maintained near the predefined value and the generator synchronism with the system would be kept, this case is referred as a recoverable LOF (RLOF). As the SSSG can operate for some duration under such a condition, its LOF function action is not necessary [23]. However, if MG stability cannot be maintained, even with rapid excitation boosting, as in the case of a total LOF event, the situation is classified as an unrecoverable LOF (ULOF) event. As discussed in Chapter 4, the main issues of this protection function could be summarized as:

- Delay in detection of ULOF event
- Maloperation during islanding event
- Maloperation during RLOF.

In this subsection 6.3.2.1, an algorithm is presented to resolve these issues in MGs and provide the following benefits:

- Fast operation during ULOF events
- High security against non-LOF events
- Continual secure operation of SSSG during RLOF events
- Preserving SSSG in the grid to improve load supply reliability
- Providing an adaptive trip time to prevent motor stalling and SSSG operation under severe conditions
- Straightforward implantation for SSSG units.

In this algorithm, the output signal (E_c) of the SSSG's AVR (shown in Fig. 6.21) is exploited. The impact of AVR signal output on the continuation of SSSG operation during LOF events is described as follows [23]:

- For the total LOF event, AVR boost command cannot increase the air gap flux. In this case, referred as ULOF, the SSSG should be immediately disconnected. In such a condition, terminal voltage V_t gradually decreases where its decrement rate depends on the strength of the upstream system.
- For the partial LOF event, the field current and internal voltage are usually increased by applying the AVR boost command. For this case, referred as RLOF, the SSSG voltage may be restored to its nominal value, and thus, the SSSG remains stable. However, if the system voltage or SSSG angular frequency has largely deviated from their permissible values, SSSG should be disconnected. This condition is also considered a ULOF event.

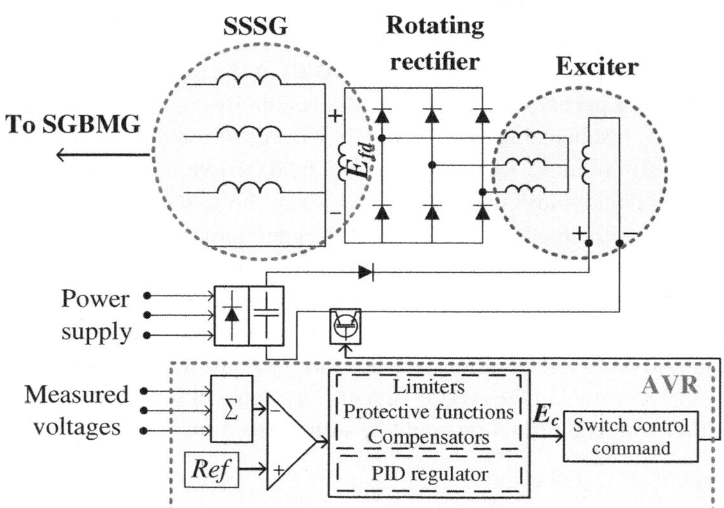

Fig. 6.21 Equivalent model of SSSG indirect excitation system. [23] / with permission of IEEE.

To develop an appropriate approach for discriminating ULOF events from other system disturbances, especially RLOF and islanding events, the real distribution system shown in Fig. 6.22 is investigated, whose information has been presented in Chapter 4. The SSSG1's AVR output is depicted in Fig. 6.23 for the following scenarios:

1) MG islanding: The PCC breaker is opened at $t = 20$ s, and SSSG1 operation mode is changed to voltage/frequency control mode after the islanding event.
2) ULOF: E_{fd} decreases to 20% of its prefault value at $t = 20$ s.
3) RLOF: E_{fd} decreases to 40% of its prefault value at $t = 20$ s.

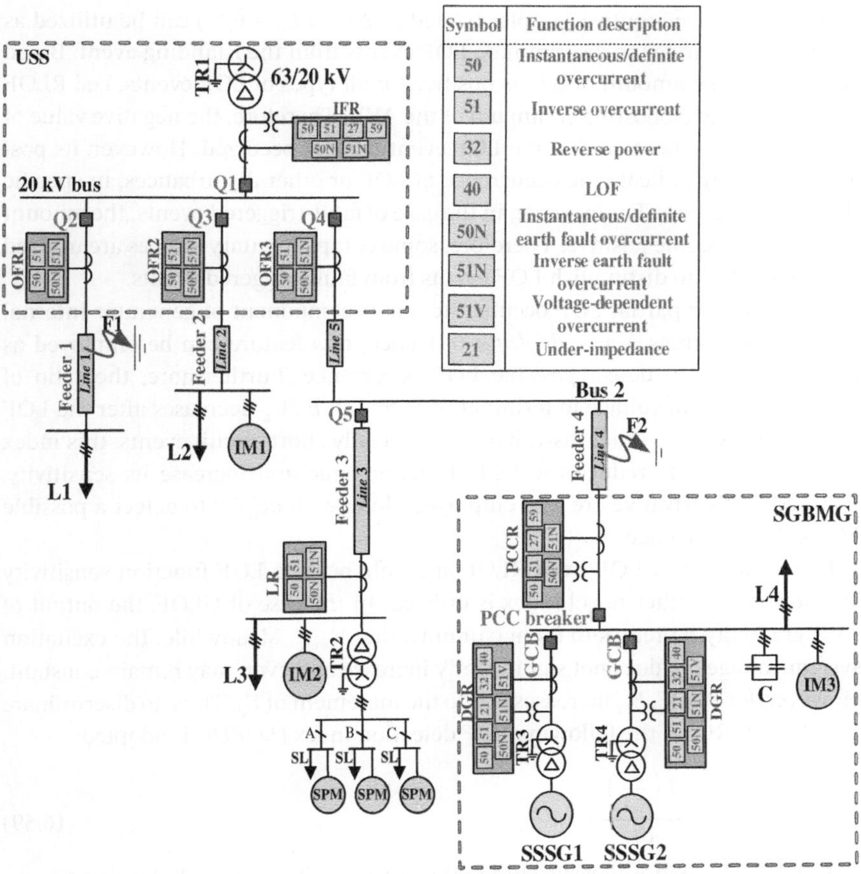

Fig. 6.22 Sample test system.

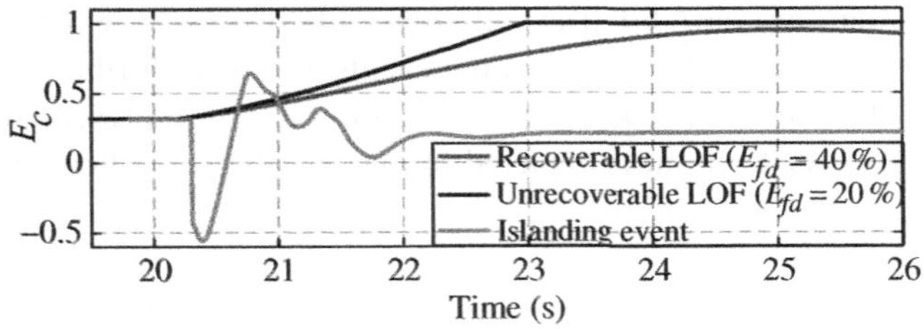

Fig. 6.23 AVR output (E_c) for different LOF and islanding events. [23] / with permission of IEEE.

As shown in Fig. 6.23, the E_c value increases during the both LOF events to compensate the SSSG voltage drop after the LOF events. Thus, the difference of E_c values for two consecutive samples (stated as $\Delta E_c = E_c^i - E_c^{i-1}$) can be utilized as an appropriate index to discriminate LOF events from the islanding event. Based on Fig. 6.23, the amount of ΔE_c is positive for all types of LOF events, i.e., RLOF and ULOF due to boosting command of the AVR. Therefore, the negative value of ΔE_c is just an indication that the LOF event has not occurred. However, its positive value may indicate the occurrence of LOF or other disturbances, in specific short-circuit events. For instance, in the case of fault-triggered events, the amount of ΔE_c may become positive. Therefore, some complementary indices are utilized along with ΔE_c to distinguish LOF events from fault-triggered events.

When total or partial LOF occurs, the q-axis component of generator internal voltage (e_q') decreases, i.e., $de_q'/dt < 0$. Hence, this feature can be employed as a simple index to detect possible LOF occurrence. Furthermore, the ratio of generator internal voltage to terminal voltage, i.e., E_i/V_t, decreases after the LOF event. However, for other disturbances, especially short-circuit events, this index increases. Thus, to reduce the LOF detection time and increase its sensitivity, E_i/V_t and its derivative are also employed along with de_q'/dt to detect a possible LOF event occurrence.

To discriminate ULOF from RLOF and enhance the LOF function sensitivity and security, another novel index is utilized. In the case of ULOF, the output of AVR gradually increases to its maximum value V_{max}. Meanwhile, the excitation system voltage E_{fd} does not significantly increase and even may remain constant. However, for RLOF, E_{fd} increases due to the increment of E_c. Thus, to discriminate ULOF from RLOF, the following LOF detection index ($LOFDI$) is adopted:

$$LOFDI = \frac{d\left(\frac{E_c}{E_{fd}}\right)}{dt} \tag{6.59}$$

Commonly E_{fd} is not measured directly and its estimation includes some complexities. Instead of that, the q-axis component of generator internal voltage (e_q'),

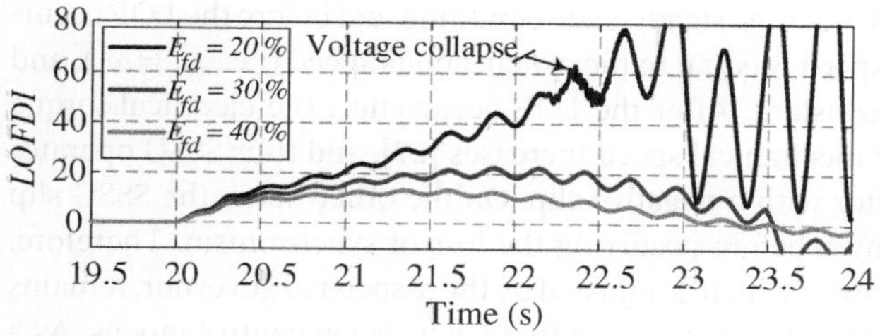

Fig. 6.24 *LOFDI* index for RLOF and ULOF events. [23] / with permission of IEEE.

which is directly proportional to E_{fd}, can be easily calculated [24, 25] and used. Thus, *LOFDI* can be rewritten as

$$LOFDI = \frac{d\left(\frac{E_c}{e'_q}\right)}{dt} \tag{6.60}$$

The amount of *LOFDI* is depicted in Fig. 6.24 for various events. This reveals that in the case of RLOF condition (E_{fd} = 30% and E_{fd} = 40%), the amount of *LOFDI* becomes negative before the SSSG loses its synchronism or a voltage collapse occurs in the MG. To predict the SSSG loss of synchronism and severe voltage drop after the LOF occurrence, an adaptive time delay (T_{th}) is adopted. To do so, after detecting a possible LOF occurrence using the aforementioned indices, a timer would be activated. If the amount of *LOFDI* becomes negative before the adaptive time delay has elapsed, i.e., the time count reaches T_{th}, the RLOF condition would be realized. However, the amount of *LOFDI* gradually increases for ULOF condition (E_{fd} = 20%) and its value does not become negative before elapsing T_{th}. In such a condition, the amount of *LOFDI* remains positive before the voltage collapse. Thus, if this value remains positive for T_{th}, an ULOF condition is detected. The threshold T_{th} is adaptively determined based on two constraints: (i) preserving SSSG synchronism with the upstream system and (ii) preventing motor stalling and severe voltage drop. The proper procedure for adaptively determining the T_{th} parameter based on the mentioned criteria is developed in the following subsections 6.3.2.2 and 6.3.2.3.

6.3.2.2 Preserving SSSG Synchronism During LOF Event

For the grid-connected SSSGs, the rotor position (θ) is defined as "the angle by which rotor direct axis is ahead of the magnetic axis of phase *a*," which can be presented as [26]

$$\theta = \omega_r t + \delta + \pi/2 \tag{6.61}$$

where ω_r is the rotor mechanical speed in pu, *t* denotes time in seconds, and rotor angle (δ) is the displacement of the rotor quadrature axis from the synchronously rotating reference axis. Under the steady-state condition and before the LOF occurrence, the mechanical speed is equal to the synchronous speed (i.e., $\omega_r = \omega_s$), and rotor angle δ remains constant. After the LOF occurrence, the electrical torque decreases and the rotor mechanical speed increases [27], and thus SSSG operates as an induction generator with a negative slip. On the other hand, the SSSG slip during LOF remains small before occurring the loss of synchronism. Therefore, the SSSG mechanical power, which is adjusted by the associated governor, remains constant in both constant active power and frequency droop control modes. As a result, the SSSG output active power (P_t) remains constant after the LOF occurrence and before occurring loss of synchronism and pole slip [28]. In other words, since the P_m input power is constant and, during the abovementioned time period, the generator is operating synchronously, the P_t output power should remain constant as well. In such a condition, to investigate the SSSG behavior during the LOF event, the common swing equations can be employed as [24]

$$\frac{d\delta}{dt} = \omega_0 \times \Delta\omega = \omega_0 \times (\omega_r - \omega_s) \tag{6.62}$$

$$\frac{d\Delta\omega}{dt} = P_m - P_e \tag{6.63}$$

Since P_e is almost equal to P_m before the loss of synchronism, the amount of ${d\Delta\omega}/{dt}$ will be almost equal to zero based on (6.63), and thus the amount of $\Delta\omega$ remains almost constant. Moreover, due to the small slip of the generator in this period, the mechanical speed is no longer equal to the synchronous speed ($\omega_r > \omega_s$). In other words, although $\Delta\omega$ remains almost constant during the LOF event, its value is not equal to zero; and thus, based on (6.62), the amount of ${d\delta}/{dt}$ remains constant (i.e., ${d\delta}/{dt} = k$) and positive after the LOF occurrence and before the loss of synchronism. In this condition, the amounts of SSSG active (P_t) and reactive (Q_t) powers can be calculated by

$$P_t = \frac{V_t E_i \sin\delta_{LOF}}{X'_d}$$
$$Q_t = \frac{V_t(E_i \cos\delta_{LOF} - V_t)}{X'_{qq}} \tag{6.64}$$

where X'_{dd} and X'_{qq} are the subtransient d- and q-axis reactances of SSSG. Meanwhile, δ_{LOF} is the SSSG rotor angle during the LOF event, which can be determined as follows based on (6.64):

$$\delta_{LOF} = tg^{-1}(u) \tag{6.65}$$

where $u = (X'_d P_t)/\left(X'_q Q_t + V_t^2\right)$ and the increment rate of rotor angle after LOF occurrence k can be obtained using

$$k = \frac{dtg^{-1}(u)}{dt} = \frac{1}{1+u^2} \times \frac{du}{dt} \tag{6.66}$$

In these equations, P_t is almost constant and the increment of δ_{LOF} is mainly related to the decrement of V_t, E_i, and Q_t. To preserve SSSG synchronism with the upstream system, the required control action, i.e., AVR boosting should be performed before δ_{LOF} reaches to its associated maximum critical value (δ_{max}). If $LOFDI > 0$ and the rotor angle is not returned back to its permissible operation range ($\delta_{LOF} < \pi/2$) in T_{ps}, the SSSG would lose its synchronism and ULOF condition would be detected. According to (6.62), the parameter T_{ps} can be adaptively determined by

$$T_{ps} = \frac{(\delta_{\max} - \delta_0)}{k} \tag{6.67}$$

6.3.2.3 Preventing Motor Stalling During LOF Event

After the LOF occurrence, both the SSSG reactive power and terminal voltage decrease due to the reduction of its flux linkage. Thus, the speed of adjacent IMs decreases and they may stall during the LOF event. To scrutinize such an event, all IMs can be considered an aggregated IM seen from the SSSG terminal [29]. The amount of speed reduction of the aggregated motor depends on the severity of LOF event. According to Equation (2.24), T_{Stall}, which is equal to deceleration time from s_{nom} to $s_{critical}$, can be determined by

$$T_{Stall} = [\ln(1 + \Delta s/(F(V_t) - s_{critical})]/M(V_t) \tag{6.68}$$

where

$$F(V_t) = \frac{A + B\omega_s^{IM}}{k_p V_t^2 + B\omega_s^{IM}}, M(V_t) = \frac{k_p V_t^2 + B\omega_s^{IM}}{J\omega_s^{IM}}, k_p = \frac{3}{R'_r \omega_s^{IM}} \tag{6.69}$$

and $\Delta s = s_{critical} - s_{nom}$. Therefore, if proper control actions are not performed, the aggregated IM slip will become greater than $s_{critical}$ after T_{Stall}. For this condition, SSSG should be disconnected to prevent severe voltage drop in the MG. It should be noted that T_{Stall} is adaptively determined based on the SSSG terminal voltage. In addition, the parameters of the aggregated IM can be determined through off-line calculations and field measured data using the procedure explained in [30, 31].

6.3.2.4 Algorithm Flowchart

Fig. 6.25 presents the flowchart of the LOF detection algorithm, which includes the following parts [23]:

- Part (1): This is employed to initialize the algorithm ($Timer_1 = 0$).
- Part (2): This part is utilized to discriminate LOF condition from other system disturbances, especially islanding events and short circuits. As described in Section 6.3.2.1, ΔE_c is not negative for LOF conditions; however, during short circuits, ΔE_c may become positive. Thus, to distinguish LOF event from other disturbances, the following conditions shall be fulfilled simultaneously:

1) The amount of e'_q decreases and its derivative, i.e., de'_q/dt, becomes negative $\left(de'_q/dt < 0\right)$. To increase the algorithm security, especially for the practical applications, and prevent the nuisance triggering of the main algorithm due to measurement errors, signal noises, and also some non-LOF events (especially stable power swing conditions), a predefined negative threshold ($Th1$) is employed. Therefore, the aforementioned condition $\left(de'_q/dt < 0\right)$ can be rewritten as $de'_q/dt < Th1 < 0$.
2) The amount of E_i/V_t becomes less than its normal steady-state value, which is represented by a threshold, namely, K_{th}, or its derivative, i.e., $d(E_i/V_t)/dt$ becomes negative. Similar to the previous case and to provide more security, a predefined negative threshold ($Th2$) is considered, and this condition can be rewritten as $d(E_i\ /Vt)/dt < Th2 < 0$. In addition, the amount of E_i/V_t during the steady-state operation can be easily calculated using the KVL rule. Since the amount of E_i/V_t is commonly greater than 1 for the grid-connected SSSGs, the setting for K_{th} is assumed to be equal to 1.

Parameters $Th1$ and $Th2$ can be determined by investigating the amount of $de'_q/dt < 0$ and $d(E_i/V_t)/dt$ during various LOF events using off-line simulation studies. Based on the performed simulation-based investigations for various combinations of SSSG-AVR-excitation systems with different parameters and various types of LOF events, parameters $Th1$ and $Th2$ can be properly considered equal to -0.15 pu/s and -0.25%/s, respectively. Since the values of parameters for various combinations of SSSG-AVR-excitation systems commonly vary in a specified range, these setting values can be appropriately applied to different systems without requiring new simulations.

After detecting the LOF occurrence, Timer$_1$ will be started. If Timer$_1 > T_{th}$, an ULOF is detected and the trip command should be issued. Parameter T_{th} is calculated by

$$T_{th} = 0.9 \times \min\ \{T_{Stall}, T_{ps}\} \tag{6.70}$$

- Part (3): If *LOFDI* becomes less than zero for 200 ms, SSSG remains stable and an RLOF is identified. In such a condition, a proper alarm can be also issued to warn the SSSG operators that a partial LOF has occurred.

Based on the given flowchart, the logic diagram of this method is presented in Fig. 6.25(b), which can be employed for practical implementation. This logic diagram contains three parts including input module, preprocessing module, and logic module. The voltage and current signals of the SSSG and its AVR command (E_c) would be injected into the LOF function through the input module. Then, the SSSG current and voltage phasors are extracted by the preprocessing module using discrete Fourier transform algorithm. Afterward, the required parameters including e'_q, E_i, and k are calculated using these phasors. The measured phasors,

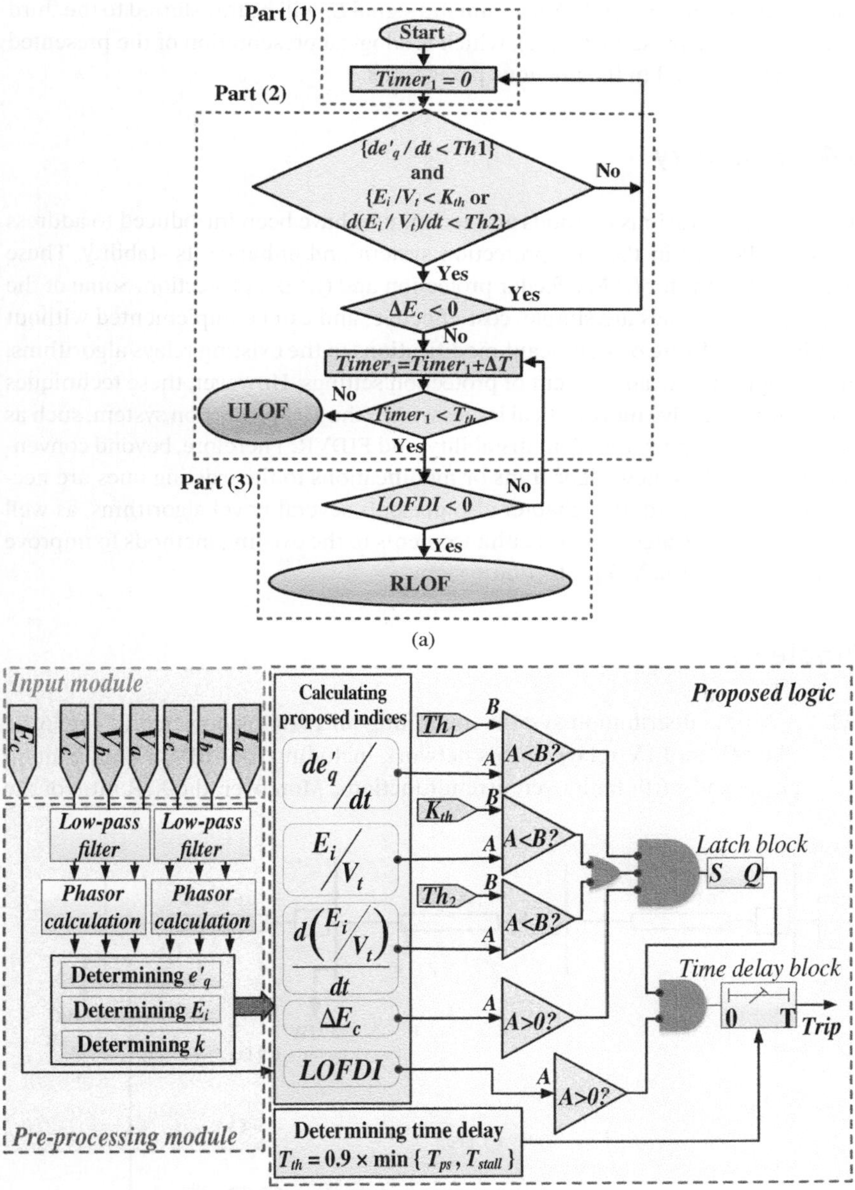

Fig. 6.25 LOF algorithm: (a) flowchart and (b) logic diagram. [23] / with permission of IEEE.

calculated parameters, and AVR command signal E_c will be transferred to the third module (i.e., the presented logic), which is a logical presentation of the presented flowchart described in the previous paragraphs.

6.4 Summary

In this chapter, various methods and techniques have been introduced to address major challenges in the MG protection system and enhance its stability. These include solutions for (i) MV feeder protection and (ii) DG protection. Some of the proposed approaches are simple, cost-effective, and can be implemented without additional hardware or significant modifications to the existing relays algorithms, such as appropriate adjustment of protection settings. However, these techniques alone cannot resolve more critical issues within the MG protection system, such as sympathetic tripping, transient instability, and FIDVR. Therefore, beyond conventional approaches, new algorithms or modifications to the existing ones are necessary. In this regard, this chapter also presents several novel algorithms, as well as practical yet straightforward enhancements to the existing methods to improve the reliability of the MG protection system.

Problems

6.1 A simple distribution system containing an IBDG is depicted in Fig. 6.26. The MV and LV relays of this network including the IBDG relays contain phase and earth fault overcurrent functions. Moreover, the X/R ratio of the

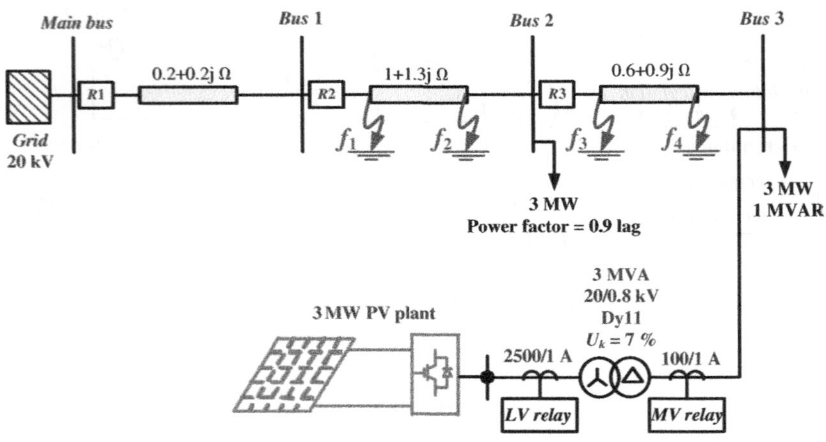

Fig. 6.26 Distribution system of Problem 6.1.

grid is considered 5, and the short-circuit level is assumed to be 20 kA. In addition, single-phase short-circuit level is assumed to be limited to 1.5 kA at the *Main bus*. The CT ratio for *R*1, *R*2, and *R*3 is 250/1 A. The maximum load of the system is also shown in Fig. 6.26. Answer the following questions:

a) Discuss whether earth fault protection is required at the LV side of the IBDG transformer or not. If not, explain how the earth fault event is detected on the LV side.
b) Calculate proper setting values for the phase fault and earth fault (if needed) overcurrent functions of the MV and LV relays for the whole system.
c) Discuss whether directional OCRs are required or if using nondirectional elements would suffice.
d) Calculate the operation time of the protection system for solid three-phase faults at f_1, f_2, f_3, and f_4.
e) Investigate if the IBDG is disconnected, whether it would be necessary to change the settings of relays *R*1, *R*2, and *R*3 or not.

6.2 The single-line diagram of an SSSG including its unit transformer is depicted in Fig. 6.27. Suppose this SSSG gets connected to Bus 2 of the system shown in Fig. 6.26.

a) Discuss whether earth fault protection is required at the LV side of the SSSG transformer or not. If not, explain how the earth fault event is detected at the LV side.
b) Calculate the settings of the LV and MV relays of the SSSG transformer.
c) Revise the setting values calculated in Problem 6.1.
d) Discuss whether directional OCRs are required or if using nondirectional elements would suffice.

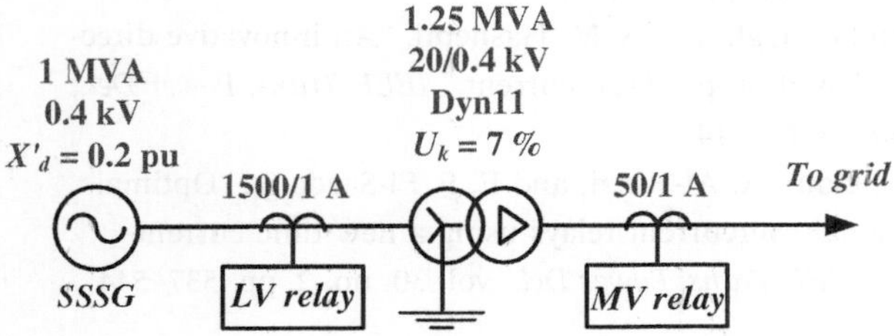

Fig. 6.27 Single-line diagram of the SSSG and its unit transformer for Problem 6.2.

6.3 The CCT of the SSSG for three-phase faults at f_3 and f_4 are 210 and 550 ms, respectively. Discuss whether the SSSG remains stable or not using the calculated settings in Problem 6.2. If not, modify the settings to retain the SSSG stability.

References

1 Tavanir, "Technical guidelines for protection of distribution networks," Iran, 2020. Available: https://wamp.tavanir.org.ir/rule/index.php.

2 T. S. Aghdam, H. K. Karegar, and H. H. Zeineldin, "Transient stability constrained protection coordination for distribution systems with DG," *IEEE Trans. Smart Grid*, vol. 9, no. 6, pp. 5733–5741, 2017.

3 T. S. Aghdam, H. K. Karegar, and H. H. Zeineldin, "Optimal coordination of double-inverse overcurrent relays for stable operation of DGs," *IEEE Trans. Ind. Inf.*, vol. 15, no. 1, pp. 183–192, 2018.

4 M. Y. Shih, C. A. C. Salazar, and A. C. Enr'ıquez, "Adaptive directional overcurrent relay coordination using ant colony optimization," *IET Gener. Transm. Distrib.*, vol. 9, no. 14, pp. 2040–2049, 2015.

5 M. Y. Shih, A. Conde, Z. Leonowicz, and L. Martirano, "An adaptive over-current coordination scheme to improve relay sensitivity and overcome drawbacks due to distributed generation in smart grids," *IEEE Trans. Ind. Appl.*, vol. 53, no. 6, pp. 5217–5228, 2017.

6 K. T. Martin, A. C. Marchesan, O. C. B. de Araújo, G. Cardoso, and M. F. da Silva, "Mixed integer linear programming applied to adaptive directional overcurrent protection considering N-1 contingency," *IEEE Trans. Ind. Appl.*, vol. 59, no. 3, pp. 2807–2821, 2023.

7 M. Singh, T. Vishnuvardhan, and S. Srivani, "Adaptive protection coordination scheme for power networks under penetration of distributed energy resources," *IET Gener. Transm. Distrib.*, vol. 10, no. 15, pp. 3919–3929, 2016.

8 A. K. Soni, A. Mohapatra, and S. N. Singh, "Protection coordination in ac microgrid via novel voltage-supervised directional over-current relays," *IEEE Trans. Power Del.*, vol. 39, no. 3, pp. 1549–1562, 2024.

9 A. Jalilian, A. M. Tarafdar Hagh, and S. M. Hashemi, "An innovative directional relaying scheme based on postfault current," *IEEE Trans. Power Del.*, vol. 29, no. 6, pp. 2640–2647, 2014.

10 K. A. Saleh, H. H. Zeineldin, A. Al-Hinai, and E. F. El-Saadany, "Optimal coordination of directional overcurrent relays using a new time current-voltage characteristic," *IEEE Trans. Power Del.*, vol. 30, no. 2, pp. 537–544, Apr. 2015.

11 A. K. Soni, A. Kumar, R. K. Panda, A. Mohapatra, and S. N. Singh, "Adaptive coordination of relays in AC microgrid considering operational and topological changes," *IEEE Syst. J.*, vol. 17, no. 2, pp. 3071–3082, 2023.

12 M. Ojaghi and V. Mohammadi, "Use of clustering to reduce the number of different setting groups for adaptive coordination of overcurrent relays," *IEEE Trans. Power Del.*, vol. 33, no. 3, pp. 1204–1212, 2017.

13 R. Bekhradian, M. Davarpanah, and M. Sanaye-Pasand, "Current-based blocking scheme to stabilize distribution network relays against FIDVR," *Int. J. Electr. Power Energy Syst.*, vol. 132, 2021.

14 T. H. Chen, M. S. Chen, T. Inoue, P. Kotas, and E. A. Chebli, "Three-phase cogenerator and transformer models for distribution system analysis," *IEEE Trans. Power Del.*, vol. 6, no. 4, pp. 1671–1681, Oct. 1991.

15 "Fault-induced delayed voltage recovery," Technical Reference Paper, NERC, 2009.

16 E. Hajipour *et al.*, "An improved aggregated model of residential air conditioners for FIDVR studies," *IEEE Trans. Power Syst.*, vol. 35, no. 2, pp. 909–919, 2019.

17 R. Razzaghi, M. Davarpanah, and M. Sanaye-Pasand, "A novel protective scheme to protect small-scale synchronous generators against transient instability," *IEEE Trans. Ind. Electron.*, vol. 60, no. 4, pp. 1659–1667, 2012.

18 P. Kundur *et al.*, "Definition and classification of power system stability, IEEE/CIGRE joint task force on stability terms and definitions," *IEEE Trans. Power Syst.*, vol. 19, no. 3, pp. 1387–1401, 2004.

19 R. Bekhradian, M. Sanaye-Pasand, and M. Davarpanah, "Innovative load shedding scheme to restore synchronous generator-based microgrids during FIDVR," *IEEE Trans. Smart Grid*, vol. 14, no. 1, pp. 388–399, 2023.

20 P. Krause, O. Wasynczuk, and S. D. Sudhoff, *Analysis of Electric Machinery and Drive Systems*, vol. 75. Wiley, 2013.

21 N. Lu, Y. Xie, and Z. Huang, *Air Conditioner Compressor Performance Model.* Washington DC, USA: Pacific Northwest National Laboratory, 2008.

22 Y. Liu, V. Vittal, J. Undrill, and J. H. Eto, "Transient model of air-conditioner compressor single phase induction motor," *IEEE Trans. Power Syst.*, vol. 28, no. 4, pp. 4528–4536, 2013.

23 R. Bekhradian, M. Sanaye-Pasand, M. Davarpanah, and M. Abedini, "Innovative loss of field protection algorithm for small-scale synchronous generators," *IEEE Trans. Power Del.*, vol. 38, no. 4, pp. 2661–2672, 2023.

24 M. Abedini, M. Sanaye-Pasand, and M. Davarpanah, "An analytical approach to detect generator loss of excitation based on internal voltage calculation," *IEEE Trans. Power Del.*, vol. 32, no. 5, pp. 2329–2338, 2016.

25 M. Abedini, M. Sanaye-Pasand, M. Davarpanah, and R. Iravani, "A loss-of-field detection relay based on rotor signals estimation," *IEEE Trans. Power Del.*, vol. 33, no. 2, pp. 779–788, 2017.

26 H. Saadat, *Power System Analysis*, 3rd ed, 1999.

27 A. Das, S. Dhar, S. Royburman, and A. Sanyal, "The efficacy of generator protection under sudden loss of excitation using offset-type MHO-relay," *J. Inst. Eng.*, vol. 98, no. 1, pp. 115–120, 2017.

28 D. Reimert, *Protective Relaying for Power Generation Systems*, 3rd ed. Taylor & Francis, 2006.

29 J. V. Milanovic, M. T. Aung, and S. C. Vegunta, "The influence of induction motors on voltage sag propagation—Part I: Accounting for the change in sag characteristics," *IEEE Trans. Power Del.*, vol. 23, no. 2, pp. 1063–1071, 2008.

30 H. Bai, P. Zhang, and V. Ajjarapu, "A novel parameter identification approach via hybrid learning for aggregate load modeling," *IEEE Trans. Power Syst.*, vol. 24, no. 3, pp. 1145–1154, Aug. 2009.

31 H. Bai and V. Ajjarapu, "A novel load-shedding strategy for mitigating fault-induced delayed voltage recovery," *IEEE Trans. Power Syst.*, vol. 26, no. 1, pp. 294–304, 2011.

Index

Printed and bound by CPI Group (UK) Ltd, Croydon, CR0 4YY

06/08/2026

14937661-0001